VOLUME ONE HUNDRED AND EIGHT

CURRENT TOPICS IN DEVELOPMENTAL BIOLOGY

Mechanisms of Regeneration

CURRENT TOPICS IN DEVELOPMENTAL BIOLOGY

"A meeting-ground for critical review and discussion of developmental processes"
A.A. Moscona and Alberto Monroy (Volume 1, 1966)

VOLUME ONE HUNDRED AND EIGHT

Current Topics in DEVELOPMENTAL BIOLOGY

Mechanisms of Regeneration

Edited by

BRIGITTE GALLIOT

Department of Genetics and Evolution, Faculty of Sciences
University of Geneva
Geneva, Switzerland

ELSEVIER

AMSTERDAM • BOSTON • HEIDELBERG • LONDON
NEW YORK • OXFORD • PARIS • SAN DIEGO
SAN FRANCISCO • SINGAPORE • SYDNEY • TOKYO
Academic Press is an imprint of Elsevier

Academic Press is an imprint of Elsevier
525 B Street, Suite 1800, San Diego, CA 92101-4495, USA
225 Wyman Street, Waltham, MA 02451, USA
The Boulevard, Langford Lane, Kidlington, Oxford, OX5 1GB, UK
32 Jamestown Road, London NW1 7BY, UK
Radarweg 29, PO Box 211, 1000 AE Amsterdam, The Netherlands

First edition 2014

ISBN: 978-0-12-391498-9
ISSN: 0070-2153

Printed and bound by CPI Group (UK) Ltd, Croydon, CR0 4YY

CONTENTS

CONTRIBUTORS

Alejandro Sánchez Alvarado
Howard Hughes Medical Institute & Stowers Institute for Medical Research, Kansas City, Missouri, USA

Joelle A. Baddour
Department of Chemical and Materials Engineering and Center for Tissue Regeneration and Engineering, University of Dayton, Dayton, Ohio, USA

Andrea Barbero
Department of Biomedicine, University of Basel, and Department of Surgery, University Hospital Basel, Tissue engineering group, Hebelstrasse Basel, Switzerland

Lutz Brusch
Center for Information Services and High Performance Computing (ZIH), Technische Universität Dresden, Dresden, Germany

Matteo Centola
Department of Biomedicine, University of Basel, and Department of Surgery, University Hospital Basel, Tissue engineering group, Hebelstrasse Basel, Switzerland

Osvaldo Chara
Center for Information Services and High Performance Computing (ZIH), Technische Universität Dresden, Dresden, Germany

Brigitte Galliot
Department of Genetics and Evolution, University of Geneva, Switzerland

José E. García-Arrarás
Department of Biology, University of Puerto Rico, Rio Piedras Campus, Rio Piedras, Puerto Rico, USA

Miranda D. Grounds
School of Anatomy, Physiology and Human Biology, The University of Western Australia, Perth, Australia

Hai Huang
National Laboratory of Plant Molecular Genetics, Shanghai Institute of Plant Physiology and Ecology, Shanghai Institutes for Biological Sciences, Chinese Academy of Sciences, Shanghai, China

Kenneth D. Irvine
Howard Hughes Medical Institute, Waksman Institute and Department of Molecular Biology and Biochemistry, Rutgers The State University of New Jersey, Piscataway, New Jersey, USA

Won-Jae Lee
School of Biological Science, Seoul National University and National Creative Research Initiative Center for Symbiosystem, Seoul, South Korea

Ivan Martin
Department of Biomedicine, University of Basel, and Department of Surgery, University Hospital Basel, Tissue engineering group, Hebelstrasse Basel, Switzerland

Vladimir S. Mashanov
Department of Biology, University of Puerto Rico, Rio Piedras Campus, Rio Piedras, Puerto Rico, USA

Masayuki Miura
Department of Genetics, Graduate School of Pharmaceutical Sciences, The University of Tokyo, and Core Research for Evolutional Science and Technology, Japan Science and Technology Agency, Tokyo, Japan

Upendra Nongthomba
Department of Molecular Reproduction, Development and Genetics, Indian Institute of Science, Bangalore, India

Mamta Rai
Department of Molecular Reproduction, Development and Genetics, Indian Institute of Science, Bangalore, India

Silke Reiter
Department of Genetics and Evolution, University of Geneva, Switzerland

Sofia M.C. Robb
Department of Plant Pathology & Microbiology; Institute for Integrative Genome Biology, and Department of Botany & Plant Sciences, University of California-Riverside, California, USA

Konstantinos Sousounis
Department of Biology and Center for Tissue Regeneration and Engineering, University of Dayton, Dayton, Ohio, USA

Ying Hua Su
State Key Laboratory of Crop Biology, College of Life Sciences, Shandong Agricultural University, Taian, Shandong, China

Gongping Sun
Howard Hughes Medical Institute, Waksman Institute and Department of Molecular Biology and Biochemistry, Rutgers The State University of New Jersey, Piscataway, New Jersey, USA

Elly M. Tanaka
Center for Regenerative Therapies Dresden (CRTD), Dresden, Germany

Beatrice Tonnarelli
Department of Biomedicine, University of Basel, and Department of Surgery, University Hospital Basel, Tissue engineering group, Hebelstrasse Basel, Switzerland

Panagiotis A. Tsonis
Department of Biology and Center for Tissue Regeneration and Engineering, University of Dayton, Dayton, Ohio, USA

Sophie Vriz
Collège de France, Center for Interdisciplinary Research in Biology (CIRB), and University Paris-Diderot, Paris, France

Lin Xu
National Laboratory of Plant Molecular Genetics, Shanghai Institute of Plant Physiology and Ecology, Shanghai Institutes for Biological Sciences, Chinese Academy of Sciences, Shanghai, China

Rolf Zeller
Department of Biomedicine, Developmental Genetics, University of Basel, Mattenstrasse, Basel, Switzerland

Xian Sheng Zhang
State Key Laboratory of Crop Biology, College of Life Sciences, Shandong Agricultural University, Taian, Shandong, China

Olga Zueva
Department of Biology, University of Puerto Rico, Rio Piedras Campus, Rio Piedras, Puerto Rico, USA

PREFACE

The initial concept of this special issue was to invite a duo of experts for each chapter to discuss specific aspects of the mechanisms underlying regenerative processes. In some cases, this invitation came at the right time and two colleagues mutually agreed to join their efforts and deliver their views on a particular topic, within this largely multidisciplinary field that regeneration covers. In other cases, this dialectic pairing could not materialize and close collaborators joined to help write some chapters. In all cases, however, I would like to express my gratitude to every contributor of this special issue. Indeed, they all went far beyond their favorite model to discuss some of the most recent findings and strategies in their field, with a transversal view. As a result, this special issue offers both novel concepts and stimulating ideas on the various mechanisms at work during regeneration.

In Chapter 1, Lin Xu and Hai Huang (Shanghai, China) provide an overview of the genetic and epigenetic control of the different regenerative processes in plants, not only in *Arabidopsis thaliana* but also in rice and moss. With clear words and excellent illustrations, the authors have made three regenerative contexts where regeneration operates in plants, that is, tissue repair, *de novo* organogenesis, and somatic embryogenesis, accessible to any biologists. Each of these processes corresponds to a cell fate transition, from cells that retain partial pluripotency when the root tip repairs, through the formation of an intermediate blastema-like structure named callus during *de novo* organogenesis, or through typical dedifferentiation of somatic cells during embryogenesis. Cell fate transition arises from a complex network of interactions involving chromatin, transcription factors, and hormones. In the case of somatic embryogenesis, these processes start to be well known and show some similarity with the mechanisms involved in the dedifferentiation of induced pluripotent stem cells from differentiated animal cells. In Chapter 2, Ying Hua Su and Xian Sheng Zhang (Shandong, China) continue along the same line with a specific focus on how hormones that behave as morphogens in plants control *de novo* organogenesis and somatic embryogenesis in *Arabidopsis*. The authors review recent advances on three important aspects, the biosynthetic regulation of hormonal functions, the cross talk between the auxin and cytokinin hormone signaling pathways, and the interactions between developmental and environmental cues. The authors of these two chapters leave us with a series of questions: What kinds of wound signals are released in the various contexts of plant regeneration? Which are the target genes of the plant hormones? How is the action of transcription and epigenetic factors coordinated? Which cells undergo callus initiation in natural conditions?

The potential role of chromatin modifications in tissue repair and during animal regeneration remains obscure. In Chapter 3, Sofia Robb and Alejandro Sanchez Alvardo (Kansas City, USA) discuss recent findings concerning the role of histone modifications in the planarian flatworm *Schmidtea mediterranea*, which can regenerate any missing part due to its abundant and dynamic population of adult stem cells (ASCs) referred to as neoblasts. After an inventory of the histones and histone-modifying enzymes present in *S. mediterranea*, the authors show that a number of these enzymes are distributed in a way reminiscent of either stem cells or nerve cells. Functional studies involving one such histone deacetylase reveal that the regulation of the level of histone acetylation is essential to maintain the stem cell function of neoblasts, that is, self-renewal and differentiation of cell progeny. Therefore, planarians may provide a promising model for deciphering the epigenetic mechanisms that either underlie or accompany the regenerative potential in animals.

In Chapter 4, Gongping Sun and Kenneth Irvine (Piscataway, USA) discuss the control of growth, both at the initiation and the termination phases of regeneration. *Drosophila* offers a versatile model to address this central question in regeneration, as wound healing can be tested not only in the embryo in the absence of any regeneration but also in the larvae either at the cuticle level or in the imaginal discs, where tissue ablation triggers regeneration. Finally, it can also be assessed in the adult cuticle or intestine, which displays an efficient tissue repair response. In each of these biological contexts, like in mammals regenerating their liver, the JNK pathway plays a central role in the injury response triggering the activation of various downstream signaling pathways, which may vary between organs. Recent quantitative genetic analyses performed in *Drosophila* show that injured tissues are more sensitive to a decrease in key signaling components as JAK or Yki, when compared to developing tissues or organs. Consequently, if the signaling pathways involved in the development and the regeneration of a given structure are the same, their interplay and their signaling strengths may importantly differ. This novel concept will surely generate further experiments in the future.

In Chapter 5, Sophie Vriz (Paris, France), Silke Reiter, and Brigitte Galliot (Geneva, Switzerland) consider cell death as a possible motor for regeneration. After describing the various animal contexts where cell death positively influences regeneration, these authors discuss recent studies in *Drosophila*, zebrafish, and *Hydra*, all pointing to the production of ROS upon injury and its potential to activate the cell death program through the JNK and/or MAPK pathways. Cell death, however, does not always occur at the time and in the area of injury and may also be observed in more distant tissues with some delay. An important issue is to understand the different roles of these dying cells. The signaling molecules released by apoptotic cells are diverse and can instruct survivor cells. Dying cells can also provide a strong *sentiment de vide* (feeling of emptiness) to the neighboring tissues, which can rapidly develop a compensatory response involving enigmatic signaling components. Injury-induced cell death might thus be quite versatile, activating and/or deactivating a variety of mechanisms to trigger a regenerative program.

In Chapter 6, Won-Jae Lee (Seoul, South Korea) and Masayuki Miura (Tokyo, Japan) discuss the impact of systemic wound response (SWR), a phenomenon they recently discovered whereby a local wound in the *Drosophila* adult skin (integument) can trigger long-distance responses. So far, these responses were identified in two distinct locations in the adult fly, in neuronal and intestinal cells by the Lee and Miura laboratories, respectively. Interestingly, both SWRs involve ROS signaling and are required for the survival of the adult animal. Consequently, we may no longer consider wound healing as a local process only, but also realize that injuries produce toxic systemic substances that engage reactions within the whole organism. What is the nature of these systemic factors? Which types of signaling do they elicit in each organ? Understanding how such toxic systemic factors circulating after a local damage affect the entire organism will help characterize the parameters that positively or negatively modulate the adult regenerative potential.

The evolution of regenerative processes is poorly understood. In Chapter 7, Vladimir Mashanov, Olga Zueva and José García-Arrarás (San Juan, Puerto Rico) provide a comparative analysis of intestine regeneration across eumetazoans. Indeed, while the gut is the target of repeated environmental stresses, it nevertheless maintains day after day its essential capacity to absorb nutrients. Adult mammals and *Drosophila* efficiently repair their intestine epithelium due to a large stock of active stem cells. In addition, progenitors and some differentiated cells can revert to a stem cell stage in mammals. While these organisms are capable of repairing their gut, they are unable to regenerate any portion of it. Why? The authors use the sea cucumber *Holothuria glaberrima*, which spontaneously regenerates its whole gut after evisceration, to address this question. *H. glaberrima* regenerates its gut by recruiting both resident enterocytes coming from the luminal epithelium left in the stumps, and external cells evaginating from the surrounding connective tissue, thus providing new enterocytes after transdifferentiation. This dual contribution of resident and external cells is not restricted to echinoderms and is also observed in insects and vertebrates. Therefore, two distinct strategies, different from tissue self-renewal, and each relying on differentiated and/or stem cells, likely emerged in the course of animal evolution. Future studies on the interactions between the gut epithelium and the connective tissue should help understand the underlying mechanisms. They may also contribute to the development of tissue bioengineering in humans.

Aging processes are an important factor that downregulates the strength of regeneration in mammals. In Chapter 8, Konstantinos Sousounis, Joelle Baddour, and Panagiotis Tsonis (Dayton, USA) synthesize the results of studies assessing the impact of aging in five distinct types of adult mammalian stem cells (endothelial, hematopoietic, skeletal muscle, neural, mesenchymal). They also discuss the age-dependent changes of the regenerative potential of those organs capable of regeneration or repair, such as the lungs, the skeletal muscle, the liver, the nervous system, or the cochlear hair cell. They conclude that both the self-renewal of stem cells and regeneration are dramatically altered upon aging in mammals. This negative impact of aging factors may be better understood in the future, either through parabiosis approaches or by looking at nonmammalian model displaying less sensitivity to aging. For example, newts keep regenerating their lens with a similar efficiency over decades and hence comparative studies between aging and "nonaging" organisms may help elucidate which are the factors negatively impacting tissue repair and regeneration in the aged organisms.

In Chapter 9, Mamta Rai, Upendra Nongthomba (Bangalore, India), and Miranda Grounds (Perth, Australia) compare the factors that regulate the maintenance, degeneration, and regeneration of skeletal muscles, focusing on limb muscles in mice and indirect flight muscles in *Drosophila*. Mice make use of ASCs, the quiescent satellite cells, to regenerate their muscles after injury, but probably do not use them to maintain the muscle function in case of moderate usage. Similarly, the maintenance of muscular function in *Drosophila* does not rely on myogenesis, since adult muscles lack ASCs. In fact, mice and *Drosophila* skeletal muscles share numerous characteristics including the genetic networks used for their development, a similar cellular organization and the metabolic pathways necessary to maintain their activity. Both species exhibit neuromuscular diseases and sarcopenia, the age-dependent loss of muscles. As a consequence, flies may offer an amenable model to investigate the respective contributions of intracellular oxidative stress, mitochondrial defects, accumulation of protein aggregates, and the modifications of the extracellular matrix on these processes. Flies can be trained to exercise and, beside local benefits, this muscular training provides a systemic positive effect.

In Chapter 10, Osvaldo Chara, Elly Tanaka, and Lutz Brusch (Dresden, Germany) look at regeneration from a different prospective, reviewing existing mathematical models and identifying a series of parameters to progress from injury signaling to tissue restoration. To model the initiation phase of regeneration, they integrate various models applied to four distinct cellular events: cell proliferation, cell differentiation, apoptosis, and cell migration. Regarding the patterning phase of regeneration, the authors take us through a journey from the source-sink model to different versions of the reaction-diffusion model. They also consider cell-based models that better integrate cellular properties such as cell polarity, cell shape and cell mechanics. Finally, they propose a multicultural model to terminate the regenerative response, which integrates the feedback of mechanical forces, a regulation of growth based on reaction-diffusion as well as some slope sensor of concentration gradient. The authors raise a series of open problems, since current models do not fully address the complexity of adult regeneration. Compared to developing organisms, regenerating tissues or organs are much larger and quite heterogenous in terms of cell types, a feature which will have to be integrated in future models.

Finally, in Chapter 11, Beatrice Tonnarelli, Matteo Centola, Andrea Barbero, Rolf Zeller, and Ivan Martin (Basel, Switzerland) show how engineers and developmental biologists can elaborate strategies to implement developmental mechanisms into artificial materials, which will instruct tissues or organs to regenerate in adult bodies. By taking regeneration of long bones as model, they go through three levels of complexity using first progenitor cells cultured in bioreactor to produce bone tissue through membranal ossification. This rather simple form of ossification, however, does not recapitulate the genuine endochondral mode of long bone formation. Therefore, in a second step, they implement endochondral ossification by applying to endogenous skeletal progenitors the signaling involved in the development and growth of long bones, using morphogenetic gradients. The third step requires to adapt this engineered development to the adult parameters, that is, the environmental cues, the aging of progenitors, the innate and adaptative immune responses involved in tissue repair, etc. Ultimately, the authors plan to reengineer a developmental process able to adapt to adult specificities. We are still far from this but the proposed strategy, integrating developmental, biotechnological, and biophysical information, definitely opens the way to restaure stable functional structures, noninvasive and prone to adapt to their new environment.

These complex engineering approaches are admittedly far from clinical applications. They nevertheless represent an ambitious and necessary task for the years to come. Their future successes may significantly improve the lives of people affected by pathologies destroying their tissues or organs. This includes disabled children recovering after major diseases or traumatisms occurring during their infancy, adult fighting chronic diseases, or aging people. However, significant breakthrough in this field will likely result from integrative strategies involving a deep knowledge of the basic cellular, developmental, and regenerative mechanisms. These essential steps will require the combined use of multiple competences as well as diverse mammalian and nonmammalian animal models. We hope this special issue illustrates the importance of these various approaches and will raise some more interest for this fascinating field of research.

Brigitte Galliot

CHAPTER ONE

Genetic and Epigenetic Controls of Plant Regeneration

Lin Xu[1], Hai Huang
National Laboratory of Plant Molecular Genetics, Shanghai Institute of Plant Physiology and Ecology, Shanghai Institutes for Biological Sciences, Chinese Academy of Sciences, Shanghai, China
[1]Corresponding author: e-mail address: xulin01@sibs.ac.cn

Contents

Abstract

Plants have evolved powerful regeneration abilities to recover from damage. Studies on plant regeneration are of high significance as the underlying mechanisms of plant regeneration are not only linking to the fundamental researches in many fields but also to the development of widely used plant biotechnology. Higher plants show three main types of regeneration: tissue regeneration, *de novo* organogenesis, and somatic embryogenesis. In this review, we summarize recent research on plant regeneration, mainly focusing on *Arabidopsis thaliana* and moss. New data suggest that plant hormones trigger regeneration and that several key transcription factors respond to hormone signals to determine cell-fate transition. Cell-fate transition requires genome-wide changes in gene expression, which are regulated via epigenetic pathways. Certain epigenetic factors may be recruited by transcription factors to relocate

Current Topics in Developmental Biology, Volume 108
ISSN 0070-2153
http://dx.doi.org/10.1016/B978-0-12-391498-9.00009-7

to new loci and regulate gene expression. Cross talk among hormone signaling, transcription factors, and epigenetic factors is involved in different types of plant regeneration, suggesting that elegant and complex regulatory mechanisms control which type of regeneration is triggered in plants under different circumstances. Since regeneration is initiated by wounding, identification of the wound signal is an important objective for future research.

ABBREVIATIONS

2,4-D 2,4-dichlorophenoxyacetic acid
2-IP 6-(γ,γ-dimethylallylamino) purine
ABA abscisic acid
AGL15 AGAMOUS-LIKE15
ALF4 ABERRANT LATERAL ROOT FORMATION 4
AP2 APETALA2
BBM BABY BOOM
CDKA cyclin-dependent kinase A
CIM callus-inducing medium
CLF CURLY LEAF
CUC2 CUP-SHAPED COTYLEDON2
EMF2 EMBRYONIC FLOWER2
EMK EMBRYOMAKER
ES cell embryonic stem cell
ESR1 ENHANCER OF SHOOT REGENERATION1
EZ elongation zone
FUS3 FUSCA3
GA gibberellin
H2Aub histone H2A ubiquitination
H3K27me3 histone H3 lysine 27 trimethylation
iPS cell induced pluripotent stem cell
JA jasmonic acid
KYP KRYPTONITE
LBD LATERAL ORGAN BOUNDARIES DOMAIN
LEC LEAFY COTYLEDON
LHP1 LIKE HETEROCHROMATIN PROTEIN1
MET1 METHYLTRANSFERASE1
NAC NAM, ATAF1,2, and CUC2
NPA naphthylphthalamic acid
OC organizing center
PcG Polycomb group
PIN PIN-FORMED
PKL PICKLE
PKR2 PICKLE-RELATED2
PLT PLETHORA
PRC Polycomb repressive complex
QC quiescent center

RAM root apical meristem
RGD3 ROOT GROWTH DEFECTIVE3
RID3 ROOT INITIATION DEFECTIVE3
RIM root-inducing medium
RLE Repressive LEC2 Element
ROS reactive oxygen species
SAM shoot apical meristem
SCR SCARECROW
SERK SOMATIC EMBRYOGENESIS RECEPTOR-LIKE KINASE
SHR SHORTROOT
SIM shoot-inducing medium
STM SHOOTMERISTEMLESS
SWN SWINGER
VAL1 VP1/ABI3-LIKE1
WIND1 WOUND INDUCED DEDIFFERENTIATION 1
WOX5 WUSCHEL-RELATED HOMEOBOX 5
WUS WUSCHEL

1. INTRODUCTION

Plants have evolved various kinds of survival strategies to grow under severe natural conditions. Recovery from damage is an important aspect of survival. In animals, regeneration is mainly required to repair organs or tissues after injury. In contrast, regeneration in plants is not only related to tissue and organ repair, but also to the formation of new plants (Birnbaum & Sanchez Alvarado, 2008; Sugimoto, Gordon, & Meyerowitz, 2011). For example, if branches are cut from a tree trunk, new buds can regenerate (Fig. 1.1A). Many species in the Crassulaceae and Cactaceae families are able to regenerate roots from their detached leaves on the soil (Fig. 1.1B), and subsequently regenerate shoots to form a new plant.

The regenerative abilities of plants have been widely exploited in modern agriculture (Sussex, 2008). Tissue culture, in which plant tissues are cultured on medium supplemented with various plant hormones (Fig. 1.1C and D), exploits the pluripotency and totipotency of plant cells to enable rapid propagation of plant populations. Plant tissue culture has a long history (Gautheret, 1983; Thorpe, 2006, 2007) and this technology is now widely used in agriculture. However, there is still much to learn about the cellular and molecular mechanisms underlying regeneration. In the past few years, molecular genetic approaches have been used to explore plant regeneration, especially that of

Figure 1.1 Plant regeneration in nature and in tissue culture. (A) Adventitious shoots regenerating from a damaged branch of a tree (*Prunus* × *Yedoensis*). (B) Leaf of a jade plant (*Crassula ovata*) regenerating roots on the soil when detached. (C) Callus regenerated from *Arabidopsis* leaf explants on CIM containing 2.2 μM 2,4-dichlorophenoxyacetic acid (2,4-D) and 0.2 μM kinetin. (D) Shoot regeneration from *Arabidopsis* root explants on SIM containing 12.5 μM 6-(γ,γ-dimethylallylamino) purine (2-IP) and 0.57 μM indole-3-acetic acid (IAA). (See color plate.)

model plants, leading to significant progress in the field. In this review, we summarize the recent advances in research on plant regeneration and discuss several new questions related to the plant regeneration process.

2. TYPES OF REGENERATION IN HIGHER PLANTS

Based on differences in cell-fate transition, regeneration of higher plants can be classified into three main categories: tissue regeneration, *de novo* organogenesis, and somatic embryogenesis.

Tissue regeneration (also called tissue repair) refers to the ability to repair wounded sites on the plant body from which tissues have been removed (Fig. 1.2A). This usually occurs after damage to young plant tissues, such as the root or leaf tips. The major function of tissue regeneration is to restore the damaged or removed part; this is comparable to processes in some animals, for example, limb regeneration of amphibians or head regeneration of hydrae and planarians (Sena & Birnbaum, 2010; Sugimoto et al., 2011). In these animals, however, regeneration requires movement of stem cells, whereas plant cells are unable to move. Thus, the mechanisms of regeneration at least partly differ between animals and plants.

Different from tissue regeneration, *de novo* organogenesis results in the development of new organs either from detached organs or from the original plant (Duclercq, Sangwan-Norreel, Catterou, & Sangwan, 2011). "Cuttage" technology, in which new plants are propagated from cuttings, is an example of *de novo* organogenesis for rapid production of new plants in agriculture (Fig. 1.2B). To propagate a new plant from a cutting, a piece of stem is cut and inserted into the soil. Adventitious roots form from the bottom wounds and adventitious shoots form from top or side wounds. These processes represent *de novo* root or shoot organogenesis (also known as *de novo* root or shoot regeneration). In tissue culture, callus formation is also a type of *de novo* organogenesis (Fig. 1.2B). Using *Arabidopsis thaliana* as an example, callus can be induced to form from various detached explants on callus-inducing medium (CIM), which contains a high level of auxin and a low level of cytokinin (Fig. 1.1C and 1.2B). Adventitious roots or shoots can be induced when callus is placed on root-inducing medium (RIM) or shoot-inducing medium (SIM) (Fig. 1.1D and 1.2B), respectively. RIM contains a low level of auxin without cytokinin, while SIM usually contains a high level of cytokinin and a low level of auxin. Since organ formation from callus can be strictly controlled by adding plant hormones to the medium, calli are used for *de novo* organogenesis for different purposes.

Somatic embryogenesis refers to the process in which a single somatic cell dedifferentiates into an embryo cell when cultured under appropriate conditions. The new embryo is able to further develop to form a whole plant (Yang & Zhang, 2010; Zimmerman, 1993) (Fig. 1.2C). The regeneration of a whole plant via somatic embryogenesis reflects the proposed totipotency of plant cells (Verdeil, Alemanno, Niemenak, & Tranbarger, 2007). This process causes a change in cell fate from a somatic cell back into an embryo stem cell and appears to resemble the induction of pluripotent stem (iPS) or embryonic stem (ES) cells from somatic cells in animals (Okita, Ichisaka, & Yamanaka, 2007; Takahashi & Yamanaka, 2006).

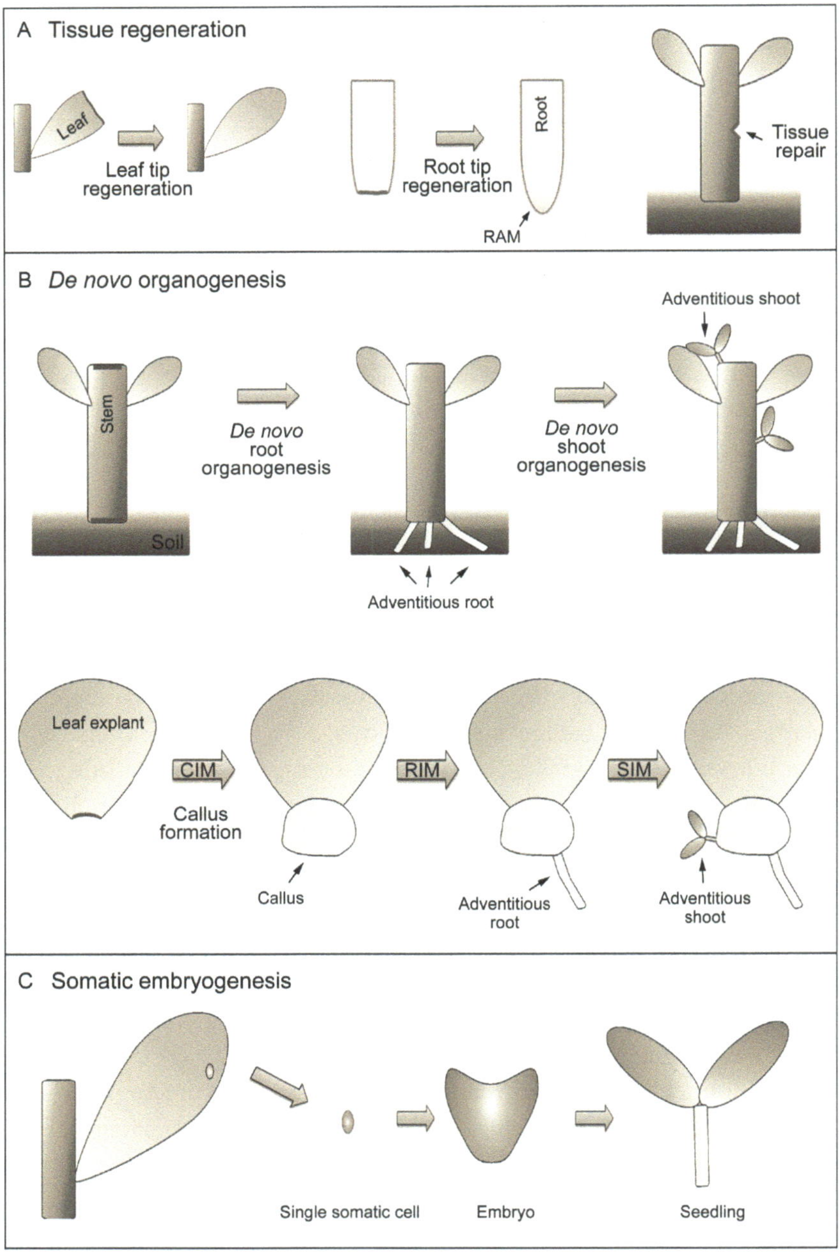

Figure 1.2 Different types of plant regeneration. (A) Tissue regeneration. Excised tips of very young leaves (left) and roots (middle) can re-grow via tissue regeneration. Stem tissues can also be repaired after damage (right). (B) *De novo* organogenesis. Upper panels show *de novo* root and shoot organogeneses during the propagation of plant materials via cuttings. Lower panels show a plant regenerated from leaf explants induced to form callus, adventitious roots, and adventitious shoots on CIM, RIM, and SIM, respectively, in tissue culture. (C) Somatic embryogenesis. A single somatic cell can be induced by auxin to form an embryo, which develops into a whole plant, reflecting the totipotency of plant cells. Red lines in (A) and (B) indicate wounding sites. (See color plate.)

3. TISSUE REGENERATION

Damaged tissues usually undergo tissue regeneration. In *Arabidopsis*, studies on tissue regeneration have mainly focused on regeneration of the root meristem and the cellular repair of stem tissues. In both systems, the plant hormone auxin plays a central role in the regeneration processes through regulating key transcription factors.

3.1. Root tip regeneration

The root stem cell niche is at the tip of the root apical meristem (RAM), which comprises a quiescent center (QC) and surrounding stem cells (Aichinger, Kornet, Friedrich, & Laux, 2012; Scheres, 2007). The QC usually has two to four cells that divide very slowly and is marked and controlled by the transcription factor gene *WUSCHEL-RELATED HOMEOBOX 5* (*WOX5*) (Sarkar et al., 2007). An early study on maize showed that, after excision of the root meristem zone, the root quickly regenerated a new QC, leading to subsequent root tip regeneration. These results suggested that the root tip has competence for tissue repair (Feldman, 1976).

Recent studies on root tips of *Arabidopsis* revealed the molecular basis of root tip regeneration (Sena, Wang, Liu, Hofhuis, & Birnbaum, 2009; Xu et al., 2006). After laser-induced ablation of the QC cells, there were dynamic changes in auxin polar transport and in the levels of key transcription factors during the regeneration of the new QC (Xu et al., 2006) (Fig. 1.3A). Before ablation of its cells, the QC was the site of maximum auxin accumulation, as PIN-FORMED (PIN)-mediated polar transport moved auxin from the proximal region of the root to the QC cells. When the QC cells were ablated, auxin instead accumulated in the cells proximally adjacent to the original QC position. The high auxin level led to upregulation of the *PLETHORA* (*PLT*) genes, which promoted the SHORTROOT (SHR) proteins to localize into the nucleus to activate the *SCARECROW* (*SCR*) gene. *PLT*s and *SCR* then acted together to switch the cell fates of the QC-adjacent cells into new QC cells through reorganizing PIN proteins and reactivating *WOX5* expression in the new QC (Xu et al., 2006). Therefore, it appears that the whole process of QC formation is mainly controlled by the auxin level, and a certain high concentration of auxin is the key factor

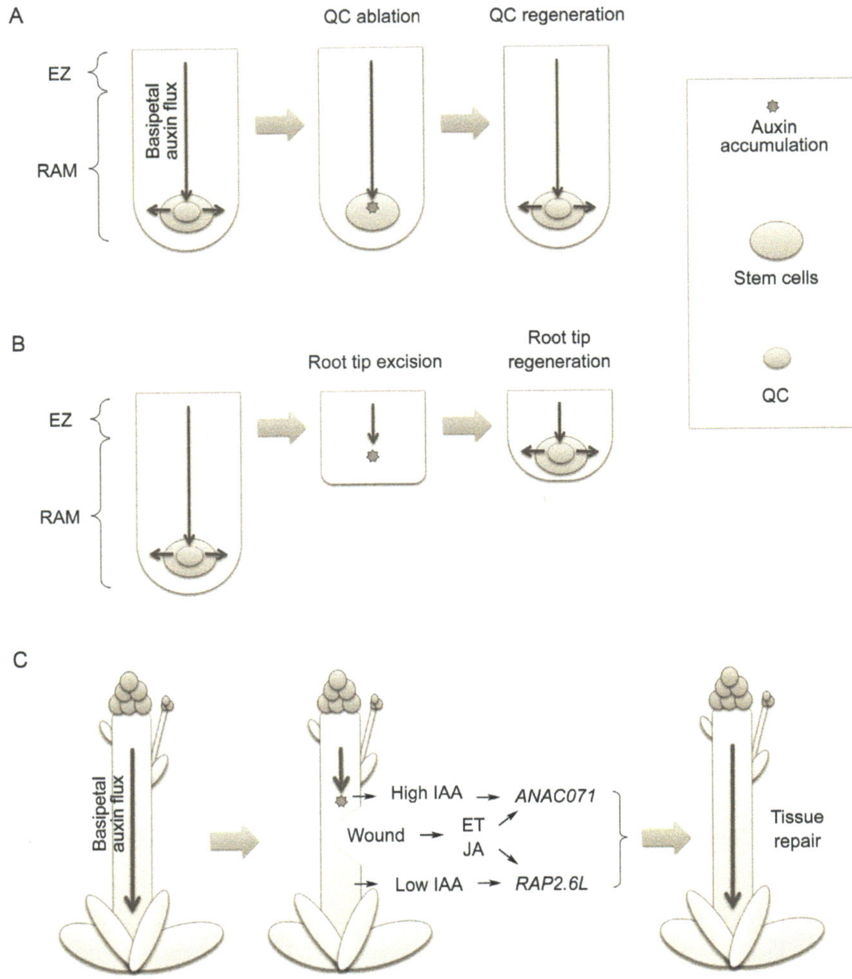

Figure 1.3 Tissue regeneration in *Arabidopsis*. (A) Root tip regeneration after ablation of QC cells (Xu et al., 2006). (B) Root tip regeneration after excision of the root tip (Sena et al., 2009). (C) Tissue repair in the stem (Asahina et al., 2011; Reid & Ross, 2011). Note that the basipetal auxin flux is disrupted by wounding. This results in different auxin levels between upper and lower regions of the wound, triggering the regeneration process to repair the wound. EZ, elongation zone. (For color version of this figure, the reader is referred to the online version of this chapter.)

determining the QC cell fate. However, it is not yet known whether *PLTs* are the direct target of auxin signaling.

Instead of laser ablation, Sena et al. used a RAM-excision approach to analyze the root regeneration process (Sena et al., 2009) (Fig. 1.3B). The difference between QC ablation and RAM excision is that only the QC is destroyed by ablation (Xu et al., 2006) whereas the entire stem cell niche is removed by excision (Sena et al., 2009). Therefore, the excision experiment could determine whether a stem cell niche was required to regenerate a new QC. Roots with the RAM removed were able to generate a new QC. Similar to the QC regeneration in the ablation experiment, the newly regenerated QC was controlled by accumulation of high levels of auxin. Treatment with the auxin polar transport-inhibitor naphthylphthalamic acid (NPA) blocked QC regeneration, suggesting that auxin accumulation is pivotal for this process. Since all stem cells were removed in the excision experiment, the newly regenerated QC cells were likely derived from partially differentiated cells. In addition, several stem cell niche-deficient mutants retained the ability of regeneration, indicating that QC regeneration may not require a functional stem cell niche (Sena et al., 2009). However, Sena et al. also found that the QC regeneration capacity was limited to the very distal region of the root tip (the terminal 130 μm region), and the regeneration ability was extremely reduced when a 270-μm portion of the tip was removed (Sena et al., 2009). These findings indicated that fully differentiated cells may not readily change back into QC cells in response to high auxin concentrations, and that cells within the 130 μm of the root tip might retain partial pluripotency.

3.2. Tissue repair in the stem

Recent studies using incision experiments demonstrated how plant tissues repair themselves after wounding (Asahina et al., 2011; Reid & Ross, 2011) (Fig. 1.3C). When the stem is wounded, the basipetal auxin stream is blocked at the wounded position. This results in accumulation of a high level of auxin above the wound and a relatively low auxin level below it. The high auxin level induces expression of the NAC (NAM, ATAF1,2, and CUC2) family transcription factor gene *ANAC071*, while the low auxin level promotes the APETALA2 (AP2)/ERF family transcription factor gene *RAP2.6L*. Both genes are essential for wound healing; however, the molecular roles of the two transcription factors are not yet clear. In addition to the key function of auxin, wound-induced jasmonic acid (JA) and ethylene are also required to regulate *RAP2.6L* (Asahina et al., 2011).

Studies on regeneration in cucumber and tomato revealed that another plant hormone, gibberellin (GA), is involved in wound healing of hypocotyls (Asahina et al., 2002, 2007). Interestingly, when cotyledons of these plants were removed, the damaged hypocotyls were unable to heal. This phenomenon was caused by the loss of GA derived from the cotyledons, as treatment with GA could rescue the healing defects. These results suggested that GA is also an important hormone for tissue repair. However, the *Arabidopsis* GA-deficient mutant did not show any defects in tissue repair in the stem (Asahina et al., 2011), suggesting that different tissues may require different mechanisms for repair.

4. *De novo* ORGANOGENESIS

De novo organogenesis is a strategy commonly used by plants to survive damage. In nature, *de novo* organogenesis usually occurs to form roots and/or shoots, allowing detached tissues, organs, or wounded plant bodies to develop into new organs or plants. In tissue culture, detached plant tissues or organs on CIM usually first form a pluripotent cell mass, a "callus," from which roots and shoots may grow. It is possible that *de novo* organogenesis in nature also begins with the formation of a tiny pluripotent cell mass, and that the main difference of callus induced in tissue culture is that it is an extremely pronounced form of the "natural" pluripotent cell mass.

4.1. Callus formation

In the 1950s, Skoog and Miller formulated a medium that was suitable for both callus induction and *de novo* organogenesis, and proposed that shoot or root initiation in tissue culture mainly relies on different ratios of auxin and cytokinin in the medium (Skoog & Miller, 1957). After this discovery, callus induction became a widely used method for plant regeneration. Although techniques for callus formation have been used for more than 50 years, the cellular and molecular basis of callus formation was largely unknown until recently. Many plant biology textbooks define callus as a group of undifferentiated totipotent cells, because of its ability to regenerate different types of plant organs. Thus, callus formation from explants was thought to be a dedifferentiation process. However, recent studies in *Arabidopsis* have demonstrated that callus is a mass of root meristem tip cells and that callus induction resembles lateral root formation (Atta et al., 2009; Che, Lall, & Howell, 2007; He, Chen, Huang, & Xu, 2012; Sugimoto, Jiao, & Meyerowitz, 2010) (Fig. 1.4).

The first indication was from phenotypic studies using root and hypocotyl explants to generate callus. Callus formation from these root and hypocotyl explants initiated with divisions of xylem-pole pericycle cells (Atta et al., 2009; Che et al., 2007). Ablation of xylem-pole pericycle cells completely blocked callus formation (Che et al., 2007). The xylem-pole pericycle is the site from which lateral roots form (Benkova & Bielach, 2010; Peret et al., 2009). Therefore, these pericycle cells may serve as pluripotent stem cells that are required for both lateral root initiation and callus formation. The results also suggested that the two biological processes may share similar genetic pathways.

Further important evidence supporting this possibility was reported by Sugimoto et al. (2010). First, a genome-wide gene expression analysis showed that the transcription profile of callus resembled that of the

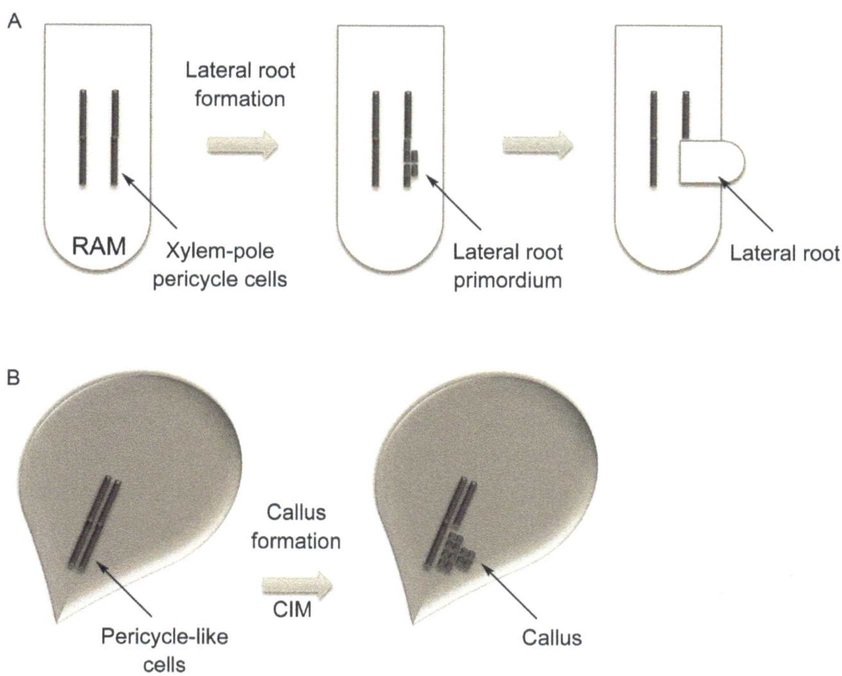

Figure 1.4 Cell-fate transition in callus formation. (A) Model of lateral root formation initiated from xylem-pole pericycle cells. (B) Model of callus formation from a leaf explant. Callus is a cluster of cells resembling root meristem tip cells, which are initiated from pericycle-like cells. (For color version of this figure, the reader is referred to the online version of this chapter.)

RAM, rather than the shoot apical meristem (SAM) or the embryo, suggesting that callus is a group of root meristem tip cells. Second, expressions of root meristem genes were induced in callus even when it was derived from aerial organs such as petals and cotyledons. Third, callus was shown to be derived from certain cells surrounding the vasculature of aerial organs. These cells express a root xylem-pole pericycle marker J0121 (Laplaze et al., 2005), and therefore, are known as pericycle-like cells. Finally, the lateral-root-initiation deficient mutant *aberrant lateral root formation 4-1* (*alf4-1*) was unable to form callus, suggesting that regulation of callus and lateral root formation at least partially share the same underlying mechanism. All of the results mentioned earlier show that callus formation from aerial organs is not via a dedifferentiation process as previously thought, but forms via a transdifferentiation process. Consequently, callus is no longer considered to be a population of totipotent cells, but rather a pluripotent cell mass (Sugimoto et al., 2010, 2011).

To investigate genetic regulation during callus formation, the gene expression profiles of leaf explants and callus were analyzed (He et al., 2012). The results showed that the leaf-to-callus process comprised at least three stages, each of which was characterized by altered gene expressions. In stage 1, auxin response genes were upregulated; in stage 2, leaf genes were downregulated; and in stage 3, root genes were upregulated (He et al., 2012). These changes in gene expression profiles are consistent with the current hypothesis that callus is a group of root meristem tip cells. Since epigenetic regulation is usually involved in genome-wide regulation of gene expression, He et al. analyzed several mutants corresponding to different histone methylation pathways during callus formation from leaf and cotyledon explants (He et al., 2012). The plant materials in those experiments included the Polycomb group (PcG) mutants *curly leaf-50 swinger-1* (*clf-50 swn-1*) and *embryonic flower2* (*emf2*), both of which are defective in callus formation from leaf explants. CLF, SWN, and EMF2 are components of the Polycomb repressive complex 2 (PRC2) (Chanvivattana et al., 2004; Goodrich et al., 1997; Yang, Chen, & Sung, 1995; Yoshida et al., 2001), which represses gene expression via trimethylation of histone H3 at lysine 27 (H3K27me3) at the target loci (Schubert et al., 2006). In the *clf-50 swn-1* and *emf2* mutants, the levels of H3K27me3 were reduced at the loci of many auxin-response genes and root genes, whereas those at many leaf gene loci were elevated. The proposed role of PcG in the leaf-to-callus transition is to repress leaf genes by depositing H3K27me3 on their loci, because many leaf genes cannot be silenced during callus formation from leaf explants in the PcG mutants (He et al., 2012). However, it is not clear whether demethylation of H3K27me3 also plays a role in derepressing the auxin-response genes and root genes. It is possible that in addition to the PcG pathway, many other epigenetic pathways are also involved in gene regulation during callus formation. Furthermore, in future research, it will be important to determine how epigenetic pathways specifically regulate their targets.

An important finding about gene regulation during callus formation was that several *LATERAL ORGAN BOUNDARIES DOMAIN* (*LBD*) family genes play roles in the process, providing a molecular link between auxin signaling and the establishment of root-cell fate (Fan, Xu, Xu, & Hu, 2012). *LBD16*, *17*, *18*, and *29* are key genes in lateral root formation (Berckmans et al., 2011; Feng, Sun, Wang, Liu, & Zhu, 2012; Feng, Zhu, Du, & Cui, 2012; Lee, Kim, Kim, Lee, & Kim, 2013; Okushima, Fukaki, Onoda, Theologis, & Tasaka, 2007), and recent research showed that they are also critical in callus formation (Fan et al., 2012). Overexpression of *LBD* genes resulted in spontaneous callus formation without an increase in exogenous auxin levels, suggesting that *LBD*s act downstream of auxin signaling (Fan et al., 2012). Previous studies showed that these *LBD* genes are the direct targets of auxin signaling, and that LBD proteins control cell proliferation and cell wall loosening during lateral root initiation (Berckmans et al., 2011; Lee et al., 2013; Okushima et al., 2007). Therefore, it is possible that during callus formation, the *LBD* genes in explant tissues respond to high auxin levels in the CIM and initiate cell division to form callus (Fan et al., 2012).

Another interesting discovery was that in rice, genome-wide DNA methylation patterns always change during callus formation (Stroud et al., 2013). These changes are inherited across generations after the callus regenerates into whole plants. Therefore, it seems that the plants regenerated from callus may harbor epigenetic information different from that of the original (preculture) plant (Stroud et al., 2013). It is still unknown whether other epigenetic regulation mechanisms such as histone modifications and chromatin remodeling coordinate with DNA methylation during rice callus formation.

The process of callus formation from explants is correlated with cell-fate transition, accompanied by large-scale changes in gene expression (Che, Lall, Nettleton, & Howell, 2006; He et al., 2012; Pischke, Huttlin, Hegeman, & Sussman, 2006; Sugimoto et al., 2010; Xu et al., 2012). Although it has been established that the cell-fate transition during callus formation resembles that during lateral root formation and several key factors in this process have been identified, the mechanisms by which these processes are precisely and gradually programmed still need to be investigated in the future.

4.2. *De novo* root organogenesis

De novo root organogenesis occurs commonly in many plant species when detached organs meet suitable conditions, for example, when organs fall on wet and nutrient-rich soil. Some plants, such as those in the Crassulaceae and Cactaceae families, use *de novo* root organogenesis to proliferate their populations (Fig. 1.1B). However, the molecular mechanism directing this rooting process is unclear. In recent studies on *Arabidopsis*, various systems have been established to study *de novo* root organogenesis from excised leaves and hypocotyls (Correa Lda, Troleis, Mastroberti, Mariath, & Fett-Neto, 2012). Interestingly, adventitious roots originate from the vascular procambium or cambium in *Arabidopsis*, suggesting that the procambium or cambium cells may serve as potential pluripotent stem cells for regeneration (Ahkami et al., 2009; Correa Lda et al., 2012; Greenwood, Cui, & Xu, 2001). In addition, inhibition of auxin polar transport completely blocked the rooting process of leaf explants, suggesting that auxin biology must be involved in *de novo* root organogenesis (our unpublished data). It is possible that *de novo* root organogenesis is similar to lateral or adventitious root formation from hypocotyls. However, *de novo* root organogenesis from detached organs is triggered by the wound signal, which is absent during lateral or adventitious root formation from hypocotyls. The molecular nature of the wound signal has not yet been defined. Also, it is still unknown how the wounding signal controls auxin biology and cell-fate transition in the procambium or cambium. Understanding these processes is crucial to unraveling the molecular mechanisms of *de novo* root organogenesis.

4.3. *De novo* shoot organogenesis

In nature, *de novo* shoot organogenesis usually occurs on wounded plant stems (see example in Fig. 1.1A). In tissue culture, callus can also be induced to undergo shoot organogenesis (see example in Fig. 1.1D). Histological studies using different explants showed that two types of cells give rise to regenerated shoots. In *Arabidopsis* root explants, adventitious shoots emerged from the xylem-pole pericycle cells on SIM (Atta et al., 2009; Che et al., 2007). In peach palm shoot apex explants, adventitious shoots were initiated from the preprocambium (de Almeida, de Almeida,

Mendes Graner, Ebling Brondani, & Fiori de Abreu-Tarazi, 2012). These results suggest that xylem-pole pericycle cells in roots and preprocambium cells in aerial organs contain pluripotent stem cells for *de novo* shoot regeneration. Their origin in the xylem-pole pericycle and the preprocambium might be important for the newly formed shoots to capture water and nutrients from the explants via the vascular system.

In fact, xylem-pole pericycle, preprocambium, procambium, and cambium cells all have stem cell features. Procambium and cambium are pluripotent vascular meristem in primary development and secondary development of vascular tissues, respectively (Elo, Immanen, Nieminen, & Helariutta, 2009). Preprocambium is the progenitor of procambium (Elo et al., 2009). As well, in terms of cell lineage, the cambium in roots develops partly from xylem-pole pericycle cells (Rost, Barbour, Stocking, & Murphy, 1997). It is possible that only stem cells are competent to regenerate roots or shoots in *de novo* organogenesis. Since *de novo* shoot organogenesis is initiated from the preprocambium or xylem-pole pericycle cells, it seems that this regeneration process shares some similarities with that of lateral root formation, rather than lateral shoot formation. Lateral shoots originate from external cells on the adaxial side of leaf axils, while adventitious shoots formed via *de novo* shoot organogenesis initiate from internal xylem-pole pericycle or preprocambium cells.

In tissue culture, callus can be easily induced to undergo *de novo* shoot organogenesis on SIM containing a high level of cytokinin, suggesting that cytokinin is the key hormone for shoot induction, just like auxin is the key hormone for root induction (Duclercq, Sangwan-Norreel, et al., 2011; Skoog & Miller, 1957). This is consistent with several features of SAM and RAM development in *Arabidopsis*. In the RAM, the highest level of auxin is in the QC region, suggesting a role of auxin in controlling or maintaining the QC (Sabatini et al., 1999). Similarly, in the SAM, cytokinin has an essential role to control the organizing center (OC) (Bartrina, Otto, Strnad, Werner, & Schmulling, 2011; Gordon, Chickarmane, Ohno, & Meyerowitz, 2009; Leibfried et al., 2005; Rupp, Frank, Werner, Strnad, & Schmulling, 1999; Sablowski, 2009; Yanai et al., 2005; Zhao et al., 2010). On the other hand, auxin could be an assistant factor in shoot regeneration, just like cytokinins can be an assistant factor in rooting (Cheng et al., 2013; Skoog & Miller, 1957).

Several *Arabidopsis* mutants that are defective in *de novo* shoot organogenesis are also defective in their cytokinin or auxin pathways (Che, Gingerich, Lall, & Howell, 2002; Che, Lall, & Howell, 2008; Che et al., 2006;

Hwang & Sheen, 2001; Inoue et al., 2001; Pernisova et al., 2009; Sakai et al., 2001). However, the mechanism by which hormone signaling controls the shooting process is unclear. One possibility is that these hormones upregulate key SAM-related genes, such as *CUP-SHAPED COTYLEDON2* (*CUC2*), *WUSCHEL* (*WUS*), and *SHOOTMERISTEMLESS* (*STM*). A detailed analysis of gene expression using real-time fluorescence techniques revealed expression patterns of several key genes during shooting from callus on SIM (Gordon et al., 2007). Correct distributions of *WUS* and *CUC2* appeared to be essential for the shoot progenitor cells, because *CUC2* was expressed in the shoot progenitor that was surrounded by cells expressing *WUS*. The proliferation of the shoot progenitor led to the radial patterning of the newly formed shoot promeristem, accompanied by *STM* expression in the region surrounding it. At this stage, PIN1 was also distributed in the surface of the shoot promeristem, suggesting that polar auxin transport is also required for the organization of the SAM (Gordon et al., 2007).

Upregulations of *WUS* and *STM* are important for shooting on SIM (Chatfield et al., 2013; Cheng, Zhu, Gao, & Zhang, 2010; Gordon et al., 2007; Tamaki et al., 2009). Interestingly, upregulation of *WUS* requires epigenetic regulation, as mutations in the DNA methyltransferase gene *METHYLTRANSFERASE1* (*MET1*), the histone H3K9 methyltransferase gene *KRYPTONITE* (*KYP*), the H3K4 demethylase JMJ14, and the histone acetyltransferase *HAC1* all resulted in altered *WUS* expression and affected the shooting rate (Li et al., 2011). It was proposed that DNA methylation and H3K9me2 function to repress *WUS* expression in callus. Thus, removal of these two epigenetic modifications resulted in *WUS* derepression, which facilitated shooting (Li et al., 2011).

Several other genes are also involved in regulating *de novo* shoot organogenesis in tissue culture. *ENHANCER OF SHOOT REGENERATION1* (*ESR1*) and *ESR2* both encode AP2/ERF domain transcription factors that control shooting (Banno, Ikeda, Niu, & Chua, 2001; Matsuo, Makino, & Banno, 2011). *ESR2* promotes shooting via upregulation of *CUC1* expression (Ikeda, Banno, Niu, Howell, & Chua, 2006). The *hoc* mutant, which is defective in the class III HD-ZIP family gene *ATHB15/CORONA/INCURVATA4*, overproduces cytokinin, facilitating shoot regeneration from root explants (Catterou et al., 2002; Duclercq, Assoumou Ndong, Guerineau, Sangwan, & Catterou, 2011). In addition, the temperature-sensitive mutants *srds*, *root initiation defective3* (*rid3*) and *root growth defective3* (*rgd3*) were shown to be defective in *de novo* shoot organogenesis (Ohtani, Demura, & Sugiyama, 2008; Ohtani & Sugiyama, 2005; Ozawa, Yasutani, Fukuda, Komamine, & Sugiyama, 1998; Tamaki et al., 2009; Yasutani, Ozawa, Nishida, Sugiyama, & Komamine, 1994). Although genes that regulate *de novo* shoot organogenesis have been identified, the network from hormones to key genes, including epigenetic regulations, has not yet been established. Evidence for direct molecular interactions is also lacking.

5. SOMATIC EMBRYOGENESIS

In theory, somatic embryogenesis is a typical dedifferentiation process that returns a differentiated somatic cell back to a totipotent embryo stem cell state. The early experiments in which carrot cells were induced to undergo somatic embryogenesis (Steward, Mapbs, & Mears, 1958) provided strong evidence for the plant totipotency theory in early plant cell biology studies. Recent studies suggest that somatic embryogenesis is a complex process involving hormone actions, transcription factors, and epigenetic regulations (Yang & Zhang, 2010).

5.1. Altered hormone balance in somatic embryogenesis

The balance between two plant hormones, GA and abscisic acid (ABA), is pivotal in controlling cells in the embryo or the postembryo state (de Castro & Hilhorst, 2006; Hays, Mandel, & Pharis, 2001; Hu et al., 2008; Ogawa et al., 2003; Phillips et al., 1997; Vahdati, Bayat, Ebrahimzadeh, Jariteh, & Mirmasoumi, 2008; White, Proebsting, Hedden, & Rivin, 2000; Yamaguchi, Kamiya, & Nambara, 2007) (Fig. 1.5). Embryo cells usually have a low ratio of GA to ABA, while this ratio is higher in somatic cells (Braybrook & Harada, 2008). These findings suggested that the cell-fate transition from somatic cells to embryo cells requires an increased ABA level and a decreased GA level.

Auxin, especially the synthetic auxin 2,4-D, is commonly used in culture medium to trigger somatic embryogenesis (Bai, Su, Yuan, & Zhang, 2013; Elhiti et al., 2013; Gaj, 2004; Ikeda-Iwai, Satoh, & Kamada, 2002; Jimenez, 2005; Luo & Koop, 1997; Michalczuk, Ribnicky, Cooke, & Cohen, 1992; Raghavan, 2004; Su & Zhang, 2009; Su et al., 2009). It was proposed that auxin might induce and initiate somatic embryogenesis, whereas changes in the GA/ABA ratio could provide a suitable environment for cells to become competent to form a somatic embryo (Braybrook & Harada, 2008). In addition, ethylene was also shown to

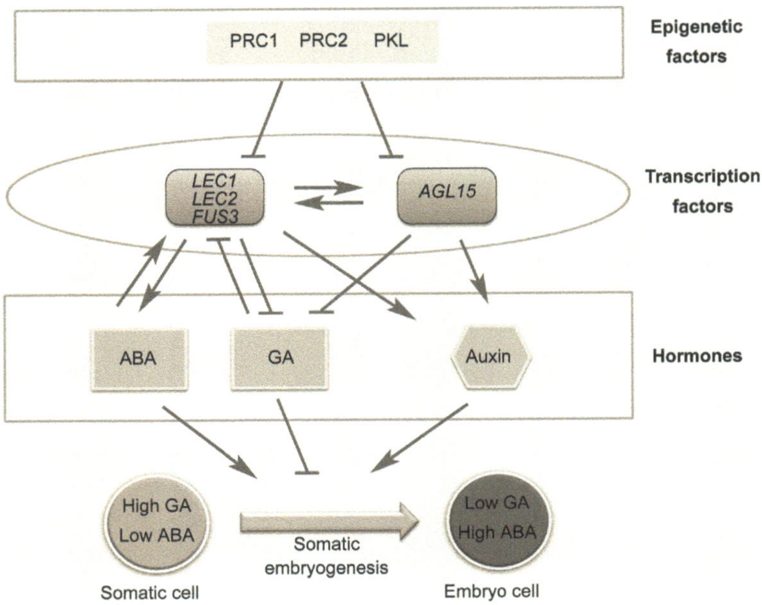

Figure 1.5 Regulation of somatic embryogenesis. Epigenetic and transcription factors and plant hormones are involved in somatic embryogenesis. Embryo and somatic cells are characterized by different GA/ABA ratios, and auxin triggers the induction of somatic embryogenesis. Plant hormones and transcription factors form a network with cross talk, and epigenetic pathways involving PcG and PKL repress expression of embryo genes to prevent somatic cells from bearing embryo traits. (For color version of this figure, the reader is referred to the online version of this chapter.)

be involved in the somatic cell-to-embryonic cell transition (Bai et al., 2013; Piyatrakul et al., 2012; Zheng, Zheng, & Perry, 2013).

5.2. Key transcription factors in somatic embryogenesis

Two types of transcription factors play essential roles in somatic embryo formation in *Arabidopsis*: *LEAFY COTYLEDON* (*LEC*) genes and the *AGAMOUS-LIKE15* (*AGL15*) gene (Braybrook & Harada, 2008; Harding, Tang, Nichols, Fernandez, & Perry, 2003) (Fig. 1.5). The *LEC* genes include *LEC1*, which encodes the HAP3 subunit of the CCAAT-binding transcription factor (Kwong et al., 2003; Lotan et al., 1998), and *LEC2* and *FUSCA3* (*FUS3*), which encode B3-domain proteins (Luerssen, Kirik, Herrmann, & Misera, 1998; Stone et al., 2001). *AGL15* encodes a MADS-box transcription factor (Harding et al., 2003; Heck, Perry, Nichols, & Fernandez, 1995; Rounsley, Ditta, & Yanofsky, 1995).

These genes are expressed exclusively in embryos. Ectopic overexpression of each of these genes resulted in embryonic traits in somatic tissues or the formation of somatic embryos (Harding et al., 2003; Lotan et al., 1998; Stone et al., 2001). These properties are reminiscent of those of Yamanaka factors, which induce adult cells to become iPS cells or ES cells in animals (Okita et al., 2007; Takahashi & Yamanaka, 2006).

LEC and *AGL15* genes can promote expression of each other and determine embryo cell fate by cross talk among hormone pathways (Braybrook & Harada, 2008; Braybrook et al., 2006; Zheng, Ren, Wang, Stromberg, & Perry, 2009). Molecular analyses suggested that *AGL15* inhibits the GA pathway via promoting *GA2ox6* expression and repressing *GA3ox2* expression (Wang, Caruso, Downie, & Perry, 2004; Zheng et al., 2009), and upregulates the auxin signaling gene *IAA30* (Zheng et al., 2009). *FUS3* inhibits the GA pathway via repressing the GA biosynthesis genes *GA3ox1* and *GA3ox2* (Curaba et al., 2004; Gazzarrini, Tsuchiya, Lumba, Okamoto, & McCourt, 2004). In addition, the *FUS3* pathway and the ABA action pathway positively regulate each other, although the molecular mechanism remains unclear (Gazzarrini et al., 2004; Kagaya, Okuda, et al., 2005; Kagaya, Toyoshima, et al., 2005). *LEC2* promotes the auxin pathway by upregulating the auxin biosynthesis genes *YUC2* and *YUC4* and the *IAA30* gene (Braybrook et al., 2006; Stone et al., 2008). All of the transcription factors described earlier make up a genetic network that promotes the ABA and auxin pathways but inhibits the GA pathway. On the other hand, their expressions also respond to changes in the levels of these hormones. Therefore, somatic embryogenesis requires this highly complex network.

Several other transcription factors also play roles in regulating somatic embryogenesis. *PGA37/MYB118* and *MYB115* positively regulate *LEC1* expression (Wang et al., 2009). Overexpression of *BABY BOOM* (*BBM*) and *EMBRYOMAKER* (*EMK*), both of which encode AP2-domain transcription factors, caused somatic embryogenesis (Boutilier et al., 2002; Tsuwamoto, Yokoi, & Takahata, 2010). *ABI3*, which encodes a B3 domain transcription factor, was positively regulated by *LEC*s and *AGL15* (Giraudat et al., 1992; Kagaya, Toyoshima, et al., 2005; Zheng et al., 2009). A double mutation of the B3 domain genes *VP1/ABI3-LIKE1* (*VAL1*) and *VAL2* resulted in somatic embryogenesis with ectopic expressions of *LEC1*, *ABI3*, and *FUS3*, suggesting that *VAL1* and *VAL2* are negative regulators of somatic embryogenesis (Suzuki, Wang, & McCarty, 2007). *WUS* and other SAM-related genes were also shown to play roles in promoting somatic embryogenesis (Bouchabke-Coussa et al., 2013; Elhiti, Tahir, Gulden, Khamiss, & Stasolla, 2010; Mordhorst, Hartog, El Tamer, Laux, & de Vries, 2002; Su et al., 2009; Zuo, Niu, Frugis, & Chua, 2002). In addition, SOMATIC EMBRYOGENESIS RECEPTOR-LIKE KINASEs (SERKs) might be involved in a signaling pathway in somatic embryogenesis via forming a protein complex with AGL15 (Hecht et al., 2001; Karlova et al., 2006).

5.3. Epigenetic regulation of somatic embryogenesis

Epigenetic pathways are critical to retain somatic features during post-embryonic development of *Arabidopsis* via repressing embryo-specific genes. There are two major epigenetic pathways that prevent ectopic expression of embryo genes in somatic cells: the PcG and PICKLE (PKL) pathways (Fig. 1.5). Mutations in these pathways can result in somatic embryogenesis.

The PcG pathway is important to retain somatic cell identity. Two PcG complexes, PRC1 and PRC2, have been identified from *Arabidopsis*. Genome-wide reprogramming of PRC2-mediated H3K27me3 is not only involved in leaf-to-callus formation as mentioned earlier (He et al., 2012), but also in repressing embryonic genes during the embryo-to-seedling phase transition (Bouyer et al., 2011). Mutations in PRC2 components resulted in an incomplete transition from embryo to seedling, and the aberrant mutant seedling exhibited disorganized cell divisions and callus-like tissues with embryo traits (Bouyer et al., 2011; Chanvivattana et al., 2004). *LEC1*, *LEC2*, and *FUS3* were ectopically expressed in the PRC2 component double mutant *clf swn* (Makarevich et al., 2006). Genome-wide analysis of H3K27me3 distribution suggested that these embryo-specific genes are highly trimethylated in somatic cells (Zhang, Clarenz, et al., 2007). Recently, a *cis* element, named *Repressive LEC2 Element* (*RLE*), was identified in the *LEC2* promoter; the function of *RLE* is to recruit PRC2 for trimethylation at the *LEC2* locus (Berger, Dubreucq, Roudier, Dubos, & Lepiniec, 2011).

PRC1 is currently thought to contain four major proteins in *Arabidopsis*: LIKE HETEROCHROMATIN PROTEIN1 (LHP1, also called TERMINAL FLOWER2) (Gaudin et al., 2001; Kotake, Takada, Nakahigashi, Ohto, & Goto, 2003), AtRING1a/b (Xu & Shen, 2008), AtBMI1a/b (Bratzel, Lopez-Torrejon, Koch, Del Pozo, & Calonje, 2010; Chen,

Molitor, Liu, & Shen, 2010), and EMF1 (Aubert et al., 2001; Bratzel et al., 2010). LHP1 serves as a reader to recognize H3K27me3 (Turck et al., 2007; Zhang, Germann, et al., 2007), and the RING-domain proteins AtRING1a/b and AtBMI1a/b are responsible for H2A ubiquitination (H2Aub) after PRC1 binds to its targets (Bratzel et al., 2010). Mutations in AtRING1a/b and AtBMI1a/b resulted in strongly ectopic embryo traits in somatic tissues and ectopic expression of *LECs*, *FUS3*, *AGL15*, and *ABI3* (Bratzel et al., 2010; Chen et al., 2010). A recent study showed that VAL proteins might recruit PRC1-mediated H2Aub to initiate repression of the embryo genes, and the repression state is then maintained by PRC2-mediated H3K27me3 (Yang et al., 2013). These results suggested that PRC1 and PRC2 may function together to control the same embryonic genes.

PKL is another epigenetic factor that plays a critical role in preventing somatic cells from bearing embryonic traits. *PKL* encodes a CHD-type ATP-dependent chromatin remodeling factor, and loss of function in *PKL* was first identified as a GA-deficient mutant (Ogas, Cheng, Sung, & Somerville, 1997; Ogas, Kaufmann, Henderson, & Somerville, 1999). Chromatin remodeling factors function to regulate gene expression via rearranging nucleosome positions (Clapier & Cairns, 2009). Mutations in *PKL* resulted in ectopic expression of *LEC1*, *LEC2*, and *FUS3* in the root tip, and therefore, caused the formation of a pickle-like root phenotype (Dean Rider et al., 2003). A recent study showed that *PICKLE-RELATED2* (*PKR2*), a *PKL* homolog in *Arabidopsis*, plays redundant roles with *PKL* in repressing embryonic traits in the root. A mutation in *PKR2* in the *pkl* mutant background enhanced the pickle-like root phenotype (Aichinger et al., 2009). However, the molecular mechanism by which PKL represses *LEC* genes is still arguable. It was proposed that PKL might repress these embryonic genes via promoting H3K27me3 (Zhang, Bishop, Ringenberg, Muir, & Ogas, 2012; Zhang et al., 2008), whereas other data suggested that PKL promotes expression of PRC2 genes to indirectly repress *LEC1* and *FUS3* (Aichinger et al., 2009). In future research, it will be important to clarify the molecular role of PKL in somatic embryogenesis. A genome-wide ChIP-seq analysis to determine the localization of PKL may be a useful strategy toward this goal. On the other hand, PKL also negatively regulates cytokinin responses in callus formation (Furuta et al., 2011), suggesting its multiple roles in plant regeneration.

6. WOUND SIGNAL IN PLANT REGENERATION

All three types of plant regeneration are triggered by wounding, indicating that this is the original signal(s). This wound signal causes changes in hormone biology, leading to changes in gene expression. However, the molecular nature of the wound signal remains unclear. It is possible that it is a complex mixture of different chemicals and that the signal varies among different tissues/organs or under different wounding conditions. This is because wounding can induce different types of regeneration when different kinds or portions of tissues/organs are damaged or detached. Studies on wounding have suggested several possible candidates, such as plasma transmembrane potential, Ca^{2+}, reactive oxygen species (ROS), plant hormones, and changes in various metabolic processes (Leon, Rojo, & Sanchez-Serrano, 2001; Maffei, Mithofer, & Boland, 2007). The wound signal triggers not only regeneration but also the defense response; therefore, its signal transduction pathway could be very complex.

Recent studies have focused on the *Arabidopsis* wound-responsive gene, *WOUND INDUCED DEDIFFERENTIATION 1* (*WIND1*) (Iwase et al., 2011; Iwase, Ohme-Takagi, & Sugimoto, 2011). *WIND1* encodes an AP2/ERF domain transcription factor, and expression of *WIND1* is induced at the wounded region. Ectopic expression of *WIND1* promotes cell proliferation and callus formation from somatic cells. *WIND1* appeared to function in regulating plant regeneration via the cytokinin pathway (Iwase, Mitsuda, et al., 2011; Iwase, Ohme-Takagi, & Sugimoto, 2011). However, it is still not clear how wounding induces *WIND1*. Detailed analysis of the *WIND1* promoter may reveal aspects of its upstream signaling pathway.

7. REGENERATION IN MOSS

The moss *Physcomitrella patens* is an ideal model plant for studying plant regeneration because many different types of tissues or cells are easily triggered to undergo cell-fate transition and regeneration. For example, a single protoplast from moss can regenerate into the protonema apical cell. This phenomenon was proposed to be the result of a reprogramming process from differentiated cells into stem cells with expressional changes of thousands of genes (Xiao, Zhang, Yang, Zhu, & He, 2012).

Another example of moss regeneration is the *de novo* regeneration of chloronema apical cells from excised gametophyte leaf cells (Ishikawa et al., 2011). Regeneration occurred rapidly without addition of any exogenous hormones; thus, the cell-fate transition probably depends on its own endogenous hormones. Ishikawa et al. used different cell proliferation inhibitors to test the relationship between cell division and cell-fate transition during moss regeneration. An interesting discovery was that expression of the *cyclin-dependent kinase A* (*CDKA*) gene was induced prior to cell-fate transition and cell division. Furthermore, addition of the DNA synthesis inhibitor aphidicolin to stop the cell cycle neither prevented the cell-fate transition from leaf cell to chloronema apical cell, nor blocked the induction of protonema-specific genes. This suggested that cell-fate transition occurs before, and is independent of, cell division during moss regeneration. However, addition of the CDKA-inhibitor roscovitine to the medium blocked both cell-fate transition and cell division, suggesting that CDKA has dual roles in controlling cell-fate transition and cell division during regeneration (Ishikawa et al., 2011). It is not known whether these separate cell-fate transition and cell division processes that occur during moss regeneration commonly occur in other types of plant regeneration. It is possible that the initiation of cell-fate transition is first triggered by plant hormones and key transcription factor genes, and then cell division proceeds to form new tissues and organs.

8. CONCLUDING REMARKS AND PERSPECTIVES

The underlying theme of regeneration is cell-fate transition (cell-fate reprogramming), and different types of regeneration involve different kinds of cell-fate transition. For example, root QC regeneration involves cell-fate transition from the root meristem or partially differentiated cell into the QC cell; callus regeneration involves cell-fate transition from pericycle-like cells into root meristem cells; and somatic embryogenesis involves cell-fate transition from somatic cells into embryo cells. Not all of these fate transitions represent a dedifferentiation process; in fact, callus formation from aerial organs is a transdifferentiation process. Cell-fate transitions are usually accompanied by the actions of plant hormones, which trigger transcription factors and epigenetic factors to reprogram the genome-wide gene expression profile. Transcription factors and epigenetic factors also regulate hormone actions via effects on their biosyntheses and signaling, forming feedback networks. Therefore, plant regeneration could be a good platform to study molecular mechanisms of cell-fate transition in plants. It is interesting that some regeneration processes share similar regulatory mechanisms with those of developmental processes. For example, callus formation resembles lateral root formation. Therefore, studies on regeneration may also provide clues about plant developmental biology.

Although recent studies on plant regeneration have greatly increased our understanding of the plasticity of plant cells, many questions remain unanswered. What is the wound signal that triggers the regeneration process? Which gene(s) is the direct target of plant hormones for cell-fate transition? How do transcription and epigenetic factors corporately act to regulate their downstream events? In addition, Sugimoto et al. provided important findings about callus formation from aerial organs, but the callus initiators, the so-called pericycle-like cells, have not been identified. All these points should be explored in future research.

ACKNOWLEDGMENTS

The authors apologize for references not cited due to space limitations. This work was supported by grants from the National Basic Research Program of China (973 Program, 2012CB910503).

REFERENCES

Ahkami, A. H., Lischewski, S., Haensch, K. T., Porfirova, S., Hofmann, J., Rolletschek, H., et al. (2009). Molecular physiology of adventitious root formation in *Petunia hybrida* cuttings: Involvement of wound response and primary metabolism. *The New Phytologist, 181*, 613–625.

Aichinger, E., Kornet, N., Friedrich, T., & Laux, T. (2012). Plant stem cell niches. *Annual Review of Plant Biology, 63*, 615–636.

Aichinger, E., Villar, C. B., Farrona, S., Reyes, J. C., Hennig, L., & Kohler, C. (2009). CHD3 proteins and polycomb group proteins antagonistically determine cell identity in Arabidopsis. *PLoS Genetics, 5*, e1000605.

Asahina, M., Azuma, K., Pitaksaringkarn, W., Yamazaki, T., Mitsuda, N., Ohme-Takagi, M., et al. (2011). Spatially selective hormonal control of RAP2.6L and ANAC071 transcription factors involved in tissue reunion in *Arabidopsis*. *Proceedings of the National Academy of Sciences of the United States of America, 108*, 16128–16132.

Asahina, M., Iwai, H., Kikuchi, A., Yamaguchi, S., Kamiya, Y., Kamada, H., et al. (2002). Gibberellin produced in the cotyledon is required for cell division during tissue reunion in the cortex of cut cucumber and tomato hypocotyls. *Plant Physiology, 129*, 201–210.

Asahina, M., Yamauchi, Y., Hanada, A., Kamiya, Y., Kamada, H., Satoh, S., et al. (2007). Effects of the removal of cotyledons on endogenous gibberellin levels in hypocotyls of young cucumber and tomato seedlings. *Plant Biotechnol, 24*, 99–106.

Atta, R., Laurens, L., Boucheron-Dubuisson, E., Guivarc'h, A., Carnero, E., Giraudat-Pautot, V., et al. (2009). Pluripotency of Arabidopsis xylem pericycle underlies shoot regeneration from root and hypocotyl explants grown *in vitro*. *The Plant Journal, 57*, 626–644.

Aubert, D., Chen, L., Moon, Y. H., Martin, D., Castle, L. A., Yang, C. H., et al. (2001). EMF1, a novel protein involved in the control of shoot architecture and flowering in Arabidopsis. *Plant Cell, 13*, 1865–1875.
Bai, B., Su, Y. H., Yuan, J., & Zhang, X. S. (2013). Induction of somatic embryos in *Arabidopsis* requires local *YUCCA* expression mediated by the down-regulation of ethylene biosynthesis. *Molecular Plant, 6*, 1247–1260.
Banno, H., Ikeda, Y., Niu, Q. W., & Chua, N. H. (2001). Overexpression of Arabidopsis *ESR1* induces initiation of shoot regeneration. *Plant Cell, 13*, 2609–2618.
Bartrina, I., Otto, E., Strnad, M., Werner, T., & Schmulling, T. (2011). Cytokinin regulates the activity of reproductive meristems, flower organ size, ovule formation, and thus seed yield in *Arabidopsis thaliana*. *Plant Cell, 23*, 69–80.
Benkova, E., & Bielach, A. (2010). Lateral root organogenesis—From cell to organ. *Current Opinion in Plant Biology, 13*, 677–683.
Berckmans, B., Vassileva, V., Schmid, S. P., Maes, S., Parizot, B., Naramoto, S., et al. (2011). Auxin-dependent cell cycle reactivation through transcriptional regulation of *Arabidopsis E2Fa* by lateral organ boundary proteins. *Plant Cell, 23*, 3671–3683.
Berger, N., Dubreucq, B., Roudier, F., Dubos, C., & Lepiniec, L. (2011). Transcriptional regulation of *Arabidopsis LEAFY COTYLEDON2* involves *RLE*, a *cis*-element that regulates trimethylation of histone H3 at lysine-27. *Plant Cell, 23*, 4065–4078.
Birnbaum, K. D., & Sanchez Alvarado, A. (2008). Slicing across kingdoms: Regeneration in plants and animals. *Cell, 132*, 697–710.
Bouchabke-Coussa, O., Obellianne, M., Linderme, D., Montes, E., Maia-Grondard, A., Vilaine, F., et al. (2013). Wuschel overexpression promotes somatic embryogenesis and induces organogenesis in cotton (*Gossypium hirsutum* L.) tissues cultured in vitro. *Plant Cell Reports, 32*, 675–686.
Boutilier, K., Offringa, R., Sharma, V. K., Kieft, H., Ouellet, T., Zhang, L., et al. (2002). Ectopic expression of BABY BOOM triggers a conversion from vegetative to embryonic growth. *Plant Cell, 14*, 1737–1749.
Bouyer, D., Roudier, F., Heese, M., Andersen, E. D., Gey, D., Nowack, M. K., et al. (2011). Polycomb repressive complex 2 controls the embryo-to-seedling phase transition. *PLoS Genetics*, 7, e1002014.
Bratzel, F., Lopez-Torrejon, G., Koch, M., Del Pozo, J. C., & Calonje, M. (2010). Keeping cell identity in *arabidopsis* requires PRC1 RING-finger homologs that catalyze H2A monoubiquitination. *Current Biology, 20*, 1853–1859.
Braybrook, S. A., & Harada, J. J. (2008). LECs go crazy in embryo development. *Trends in Plant Science, 13*, 624–630.
Braybrook, S. A., Stone, S. L., Park, S., Bui, A. Q., Le, B. H., Fischer, R. L., et al. (2006). Genes directly regulated by LEAFY COTYLEDON2 provide insight into the control of embryo maturation and somatic embryogenesis. *Proceedings of the National Academy of Sciences of the United States of America, 103*, 3468–3473.
Catterou, M., Dubois, F., Smets, R., Vaniet, S., Kichey, T., Van Onckelen, H., et al. (2002). *hoc*: An *Arabidopsis* mutant overproducing cytokinins and expressing high *in vitro* organogenic capacity. *The Plant Journal, 30*, 273–287.
Chanvivattana, Y., Bishopp, A., Schubert, D., Stock, C., Moon, Y. H., Sung, Z. R., et al. (2004). Interaction of Polycomb-group proteins controlling flowering in *Arabidopsis*. *Development, 131*, 5263–5276.
Chatfield, S. P., Capron, R., Severino, A., Penttila, P. A., Alfred, S., Nahal, H., et al. (2013). Incipient stem cell niche conversion in tissue culture: Using a systems approach to probe early events in *WUSCHEL*-dependent conversion of lateral root primordia into shoot meristems. *The Plant Journal, 73*, 798–813.
Che, P., Gingerich, D. J., Lall, S., & Howell, S. H. (2002). Global and hormone-induced gene expression changes during shoot development in *Arabidopsis*. *Plant Cell, 14*, 2771–2785.

Che, P., Lall, S., & Howell, S. H. (2007). Developmental steps in acquiring competence for shoot development in *Arabidopsis* tissue culture. *Planta, 226*, 1183–1194.

Che, P., Lall, S., & Howell, S. H. (2008). Acquiring competence for shoot development in *Arabidopsis*: ARR2 directly targets A-type *ARR* genes that are differentially activated by CIM preincubation. *Plant Signaling & Behavior, 3*, 99–101.

Che, P., Lall, S., Nettleton, D., & Howell, S. H. (2006). Gene expression programs during shoot, root, and callus development in Arabidopsis tissue culture. *Plant Physiology, 141*, 620–637.

Chen, D., Molitor, A., Liu, C., & Shen, W. H. (2010). The *Arabidopsis* PRC1-like ring-finger proteins are necessary for repression of embryonic traits during vegetative growth. *Cell Research, 20*, 1332–1344.

Cheng, Z. J., Wang, L., Sun, W., Zhang, Y., Zhou, C., Su, Y. H., et al. (2013). Pattern of auxin and cytokinin responses for shoot meristem induction results from the regulation of cytokinin biosynthesis by AUXIN RESPONSE FACTOR3. *Plant Physiology, 161*, 240–251.

Cheng, Z. J., Zhu, S. S., Gao, X. Q., & Zhang, X. S. (2010). Cytokinin and auxin regulates *WUS* induction and inflorescence regeneration in vitro in *Arabidopsis*. *Plant Cell Reports, 29*, 927–933.

Clapier, C. R., & Cairns, B. R. (2009). The biology of chromatin remodeling complexes. *Annual Review of Biochemistry, 78*, 273–304.

Correa Lda, R., Troleis, J., Mastroberti, A. A., Mariath, J. E., & Fett-Neto, A. G. (2012). Distinct modes of adventitious rooting in *Arabidopsis thaliana*. *Plant Biology (Stuttgart, Germany), 14*, 100–109.

Curaba, J., Moritz, T., Blervaque, R., Parcy, F., Raz, V., Herzog, M., et al. (2004). *AtGA3ox2*, a key gene responsible for bioactive gibberellin biosynthesis, is regulated during embryogenesis by *LEAFY COTYLEDON2* and *FUSCA3* in Arabidopsis. *Plant Physiology, 136*, 3660–3669.

de Almeida, M., de Almeida, C. V., Mendes Graner, E., Ebling Brondani, G., & Fiori de Abreu-Tarazi, M. (2012). Pre-procambial cells are niches for pluripotent and totipotent stem-like cells for organogenesis and somatic embryogenesis in the peach palm: A histological study. *Plant Cell Reports, 31*, 1495–1515.

de Castro, R. D., & Hilhorst, H. W. M. (2006). Hormonal control of seed development in GA- and ABA-deficient tomato (*Lycopersicon esculentum* Mill. cv. Moneymaker) mutants. *Plant Science, 170*, 462–470.

Dean Rider, S., Jr., Henderson, J. T., Jerome, R. E., Edenberg, H. J., Romero-Severson, J., & Ogas, J. (2003). Coordinate repression of regulators of embryonic identity by *PICKLE* during germination in *Arabidopsis*. *The Plant Journal, 35*, 33–43.

Duclercq, J., Assoumou Ndong, Y. P., Guerineau, F., Sangwan, R. S., & Catterou, M. (2011). *Arabidopsis* shoot organogenesis is enhanced by an amino acid change in the ATHB15 transcription factor. *Plant Biology (Stuttgart, Germany), 13*, 317–324.

Duclercq, J., Sangwan-Norreel, B., Catterou, M., & Sangwan, R. S. (2011). *De novo* shoot organogenesis: From art to science. *Trends in Plant Science, 16*, 597–606.

Elhiti, M., Hebelstrup, K. H., Wang, A., Li, C., Cui, Y., Hill, R. D., et al. (2013). Function of type-2 Arabidopsis hemoglobin in the auxin-mediated formation of embryogenic cells during morphogenesis. *The Plant Journal, 74*, 946–958.

Elhiti, M., Tahir, M., Gulden, R. H., Khamiss, K., & Stasolla, C. (2010). Modulation of embryo-forming capacity in culture through the expression of Brassica genes involved in the regulation of the shoot apical meristem. *Journal of Experimental Botany, 61*, 4069–4085.

Elo, A., Immanen, J., Nieminen, K., & Helariutta, Y. (2009). Stem cell function during plant vascular development. *Seminars in Cell & Developmental Biology, 20*, 1097–1106.

Fan, M., Xu, C., Xu, K., & Hu, Y. (2012). LATERAL ORGAN BOUNDARIES DOMAIN transcription factors direct callus formation in *Arabidopsis* regeneration. *Cell Research, 22*, 1169–1180.
Feldman, L. J. (1976). The *de novo* origin of the quiescent center in regenerating root apices of *Zea mays. Planta, 128*, 207–212.
Feng, Z., Sun, X., Wang, G., Liu, H., & Zhu, J. (2012). LBD29 regulates the cell cycle progression in response to auxin during lateral root formation in *Arabidopsis thaliana. Annals of Botany, 110*, 1–10.
Feng, Z., Zhu, J., Du, X., & Cui, X. (2012). Effects of three auxin-inducible LBD members on lateral root formation in *Arabidopsis thaliana. Planta, 236*, 1227–1237.
Furuta, K., Kubo, M., Sano, K., Demura, T., Fukuda, H., Liu, Y. G., et al. (2011). The CKH2/PKL chromatin remodeling factor negatively regulates cytokinin responses in Arabidopsis calli. *Plant & Cell Physiology, 52*, 618–628.
Gaj, M. D. (2004). Factors influencing somatic embryogenesis induction and plant regeneration with particular reference to *Arabidopsis thaliana* (L.) Heynh. *Plant Growth Regulation, 43*, 27–47.
Gaudin, V., Libault, M., Pouteau, S., Juul, T., Zhao, G., Lefebvre, D., et al. (2001). Mutations in *LIKE HETEROCHROMATIN PROTEIN 1* affect flowering time and plant architecture in *Arabidopsis. Development, 128*, 4847–4858.
Gautheret, R. J. (1983). Plant tissue culture: A history. *Bot. Mat. Tokyo, 96*, 393–410.
Gazzarrini, S., Tsuchiya, Y., Lumba, S., Okamoto, M., & McCourt, P. (2004). The transcription factor FUSCA3 controls developmental timing in *Arabidopsis* through the hormones gibberellin and abscisic acid. *Developmental Cell, 7*, 373–385.
Giraudat, J., Hauge, B. M., Valon, C., Smalle, J., Parcy, F., & Goodman, H. M. (1992). Isolation of the *Arabidopsis ABI3* gene by positional cloning. *Plant Cell, 4*, 1251–1261.
Goodrich, J., Puangsomlee, P., Martin, M., Long, D., Meyerowitz, E. M., & Coupland, G. (1997). A polycomb-group gene regulates homeotic gene expression in *Arabidopsis. Nature, 386*, 44–51.
Gordon, S. P., Chickarmane, V. S., Ohno, C., & Meyerowitz, E. M. (2009). Multiple feedback loops through cytokinin signaling control stem cell number within the *Arabidopsis* shoot meristem. *Proceedings of the National Academy of Sciences of the United States of America, 106*, 16529–16534.
Gordon, S. P., Heisler, M. G., Reddy, G. V., Ohno, C., Das, P., & Meyerowitz, E. M. (2007). Pattern formation during de novo assembly of the *Arabidopsis* shoot meristem. *Development, 134*, 3539–3548.
Greenwood, M. S., Cui, X., Xu, F. (2011). Response to auxin changes during maturation-related loss of adventitious rooting competence in loblolly pine (*Pinus taeda*) stem cuttings. *Physiologia Plantarum, 111*, 373–380.
Harding, E. W., Tang, W., Nichols, K. W., Fernandez, D. E., & Perry, S. E. (2003). Expression and maintenance of embryogenic potential is enhanced through constitutive expression of *AGAMOUS-Like 15. Plant Physiology, 133*, 653–663.
Hays, D., Mandel, R., & Pharis, R. (2001). Hormones in zygotic and microspore embryos of *Brassica napus. Plant Growth Regulation, 35*, 47–58.
He, C., Chen, X., Huang, H., & Xu, L. (2012). Reprogramming of H3K27me3 is critical for acquisition of pluripotency from cultured *Arabidopsis* tissues. *PLoS Genetics, 8*, e1002911.
Hecht, V., Vielle-Calzada, J. P., Hartog, M. V., Schmidt, E. D., Boutilier, K., Grossniklaus, U., et al. (2001). The Arabidopsis *SOMATIC EMBRYOGENESIS RECEPTOR KINASE 1* gene is expressed in developing ovules and embryos and enhances embryogenic competence in culture. *Plant Physiology, 127*, 803–816.
Heck, G. R., Perry, S. E., Nichols, K. W., & Fernandez, D. E. (1995). AGL15, a MADS domain protein expressed in developing embryos. *Plant Cell, 7*, 1271–1282.

Hu, J., Mitchum, M. G., Barnaby, N., Ayele, B. T., Ogawa, M., Nam, E., et al. (2008). Potential sites of bioactive gibberellin production during reproductive growth in *Arabidopsis*. *Plant Cell*, *20*, 320–336.

Hwang, I., & Sheen, J. (2001). Two-component circuitry in *Arabidopsis* cytokinin signal transduction. *Nature*, *413*, 383–389.

Ikeda, Y., Banno, H., Niu, Q. W., Howell, S. H., & Chua, N. H. (2006). The *ENHANCER OF SHOOT REGENERATION 2* gene in *Arabidopsis* regulates *CUP-SHAPED COTYLEDON 1* at the transcriptional level and controls cotyledon development. *Plant & Cell Physiology*, *47*, 1443–1456.

Ikeda-Iwai, M., Satoh, S., & Kamada, H. (2002). Establishment of a reproducible tissue culture system for the induction of *Arabidopsis* somatic embryos. *Journal of Experimental Botany*, *53*, 1575–1580.

Inoue, T., Higuchi, M., Hashimoto, Y., Seki, M., Kobayashi, M., Kato, T., et al. (2001). Identification of CRE1 as a cytokinin receptor from *Arabidopsis*. *Nature*, *409*, 1060–1063.

Ishikawa, M., Murata, T., Sato, Y., Nishiyama, T., Hiwatashi, Y., Imai, A., et al. (2011). *Physcomitrella* cyclin-dependent kinase A links cell cycle reactivation to other cellular changes during reprogramming of leaf cells. *Plant Cell*, *23*, 2924–2938.

Iwase, A., Mitsuda, N., Koyama, T., Hiratsu, K., Kojima, M., Arai, T., et al. (2011). The AP2/ERF transcription factor WIND1 controls cell dedifferentiation in *Arabidopsis*. *Current Biology*, *21*, 508–514.

Iwase, A., Ohme-Takagi, M., & Sugimoto, K. (2011). WIND1: A key molecular switch for plant cell dedifferentiation. *Plant Signaling & Behavior*, *6*, 1943–1945.

Jimenez, V. M. (2005). Involvement of plant hormones and plant growth regulators on *in vitro* somatic embryogenesis. *Plant Growth Regulation*, *47*, 91–110.

Kagaya, Y., Okuda, R., Ban, A., Toyoshima, R., Tsutsumida, K., Usui, H., et al. (2005). Indirect ABA-dependent regulation of seed storage protein genes by FUSCA3 transcription factor in *Arabidopsis*. *Plant & Cell Physiology*, *46*, 300–311.

Kagaya, Y., Toyoshima, R., Okuda, R., Usui, H., Yamamoto, A., & Hattori, T. (2005). *LEAFY COTYLEDON1* controls seed storage protein genes through its regulation of *FUSCA3* and *ABSCISIC ACID INSENSITIVE3*. *Plant & Cell Physiology*, *46*, 399–406.

Karlova, R., Boeren, S., Russinova, E., Aker, J., Vervoort, J., & de Vries, S. (2006). The Arabidopsis SOMATIC EMBRYOGENESIS RECEPTOR-LIKE KINASE1 protein complex includes BRASSINOSTEROID-INSENSITIVE1. *Plant Cell*, *18*, 626–638.

Kotake, T., Takada, S., Nakahigashi, K., Ohto, M., & Goto, K. (2003). *Arabidopsis TERMINAL FLOWER 2* gene encodes a heterochromatin protein 1 homolog and represses both *FLOWERING LOCUS T* to regulate flowering time and several floral homeotic genes. *Plant & Cell Physiology*, *44*, 555–564.

Kwong, R. W., Bui, A. Q., Lee, H., Kwong, L. W., Fischer, R. L., Goldberg, R. B., et al. (2003). *LEAFY COTYLEDON1-LIKE* defines a class of regulators essential for embryo development. *Plant Cell*, *15*, 5–18.

Laplaze, L., Parizot, B., Baker, A., Ricaud, L., Martiniere, A., Auguy, F., et al. (2005). GAL4-GFP enhancer trap lines for genetic manipulation of lateral root development in *Arabidopsis thaliana*. *Journal of Experimental Botany*, *56*, 2433–2442.

Lee, H. W., Kim, M. J., Kim, N. Y., Lee, S. H., & Kim, J. (2013). LBD18 acts as a transcriptional activator that directly binds to the *EXPANSIN14* promoter in promoting lateral root emergence of Arabidopsis. *The Plant Journal*, *73*, 212–224.

Leibfried, A., To, J. P., Busch, W., Stehling, S., Kehle, A., Demar, M., et al. (2005). WUSCHEL controls meristem function by direct regulation of cytokinin-inducible response regulators. *Nature*, *438*, 1172–1175.

Leon, J., Rojo, E., & Sanchez-Serrano, J. J. (2001). Wound signalling in plants. *Journal of Experimental Botany, 52*, 1–9.

Li, W., Liu, H., Cheng, Z. J., Su, Y. H., Han, H. N., Zhang, Y., et al. (2011). DNA methylation and histone modifications regulate *De Novo* shoot regeneration in *Arabidopsis* by modulating *WUSCHEL* expression and auxin signaling. *PLoS Genetics*, 7, e1002243.

Lotan, T., Ohto, M., Yee, K. M., West, M. A., Lo, R., Kwong, R. W., et al. (1998). *Arabidopsis LEAFY COTYLEDON1* is sufficient to induce embryo development in vegetative cells. *Cell, 93*, 1195–1205.

Luerssen, H., Kirik, V., Herrmann, P., & Misera, S. (1998). *FUSCA3* encodes a protein with a conserved VP1/AB13-like B3 domain which is of functional importance for the regulation of seed maturation in *Arabidopsis thaliana. The Plant Journal, 15*, 755–764.

Luo, Y., & Koop, H. U. (1997). Somatic embryogenesis in cultured immature zygotic embryos and leaf protoplasts of *Arabidopsis thaliana* ecotypes. *Planta, 202*, 387–396.

Maffei, M. E., Mithofer, A., & Boland, W. (2007). Before gene expression: Early events in plant-insect interaction. *Trends in Plant Science, 12*, 310–316.

Makarevich, G., Leroy, O., Akinci, U., Schubert, D., Clarenz, O., Goodrich, J., et al. (2006). Different *Polycomb* group complexes regulate common target genes in *Arabidopsis. EMBO Reports, 7*, 947–952.

Matsuo, N., Makino, M., & Banno, H. (2011). *Arabidopsis ENHANCER OF SHOOT REGENERATION (ESR)1* and *ESR2* regulate *in vitro* shoot regeneration and their expressions are differentially regulated. *Plant Science, 181*, 39–46.

Michalczuk, L., Ribnicky, D. M., Cooke, T. J., & Cohen, J. D. (1992). Regulation of indole-3-acetic Acid biosynthetic pathways in carrot cell cultures. *Plant Physiology, 100*, 1346–1353.

Mordhorst, A. P., Hartog, M. V., El Tamer, M. K., Laux, T., & de Vries, S. C. (2002). Somatic embryogenesis from *Arabidopsis* shoot apical meristem mutants. *Planta, 214*, 829–836.

Ogas, J., Cheng, J. C., Sung, Z. R., & Somerville, C. (1997). Cellular differentiation regulated by gibberellin in the *Arabidopsis thaliana pickle* mutant. *Science, 277*, 91–94.

Ogas, J., Kaufmann, S., Henderson, J., & Somerville, C. (1999). PICKLE is a CHD3 chromatin-remodeling factor that regulates the transition from embryonic to vegetative development in *Arabidopsis. Proceedings of the National Academy of Sciences of the United States of America, 96*, 13839–13844.

Ogawa, M., Hanada, A., Yamauchi, Y., Kuwahara, A., Kamiya, Y., & Yamaguchi, S. (2003). Gibberellin biosynthesis and response during *Arabidopsis* seed germination. *Plant Cell, 15*, 1591–1604.

Ohtani, M., Demura, T., & Sugiyama, M. (2008). Differential requirement for the function of SRD2, an snRNA transcription activator, in various stages of plant development. *Plant Molecular Biology, 66*, 303–314.

Ohtani, M., & Sugiyama, M. (2005). Involvement of SRD2-mediated activation of snRNA transcription in the control of cell proliferation competence in *Arabidopsis. The Plant Journal, 43*, 479–490.

Okita, K., Ichisaka, T., & Yamanaka, S. (2007). Generation of germline-competent induced pluripotent stem cells. *Nature, 448*, 313–317.

Okushima, Y., Fukaki, H., Onoda, M., Theologis, A., & Tasaka, M. (2007). ARF7 and ARF19 regulate lateral root formation via direct activation of *LBD/ASL* genes in *Arabidopsis. Plant Cell, 19*, 118–130.

Ozawa, S., Yasutani, I., Fukuda, H., Komamine, A., & Sugiyama, M. (1998). Organogenic responses in tissue culture of *srd* mutants of *Arabidopsis thaliana. Development, 125*, 135–142.

Peret, B., De Rybel, B., Casimiro, I., Benkova, E., Swarup, R., Laplaze, L., et al. (2009). *Arabidopsis* lateral root development: An emerging story. *Trends in Plant Science, 14*, 399–408.

Pernisova, M., Klima, P., Horak, J., Valkova, M., Malbeck, J., Soucek, P., et al. (2009). Cytokinins modulate auxin-induced organogenesis in plants via regulation of the auxin efflux. *Proceedings of the National Academy of Sciences of the United States of America, 106*, 3609–3614.
Phillips, J., Artsaenko, O., Fiedler, U., Horstmann, C., Mock, H. P., Muntz, K., et al. (1997). Seed-specific immunomodulation of abscisic acid activity induces a developmental switch. *The EMBO Journal, 16*, 4489–4496.
Pischke, M. S., Huttlin, E. L., Hegeman, A. D., & Sussman, M. R. (2006). A transcriptome-based characterization of habituation in plant tissue culture. *Plant Physiology, 140*, 1255–1278.
Piyatrakul, P., Putranto, R. A., Martin, F., Rio, M., Dessailly, F., Leclercq, J., et al. (2012). Some ethylene biosynthesis and AP2/ERF genes reveal a specific pattern of expression during somatic embryogenesis in *Hevea brasiliensis*. *BMC Plant Biology, 12*, 244.
Raghavan, V. (2004). Role of 2,4-dichlorophenoxyacetic acid (2,4-D) in somatic embryogenesis on cultured zygotic embryos of *Arabidopsis*: Cell expansion, cell cycling, and morphogenesis during continuous exposure of embryos to 2,4-D. *American Journal of Botany, 91*, 1743–1756.
Reid, J. B., & Ross, J. J. (2011). Regulation of tissue repair in plants. *Proceedings of the National Academy of Sciences of the United States of America, 108*, 17241–17242.
Rost, T. L., Barbour, M. G., Stocking, C. R., & Murphy, T. M. (1997). *Plant biology*. Belmont, California, USA: Wadsworth publishing company.
Rounsley, S. D., Ditta, G. S., & Yanofsky, M. F. (1995). Diverse roles for MADS box genes in Arabidopsis development. *Plant Cell, 7*, 1259–1269.
Rupp, H. M., Frank, M., Werner, T., Strnad, M., & Schmulling, T. (1999). Increased steady state mRNA levels of the *STM* and *KNAT1* homeobox genes in cytokinin overproducing *Arabidopsis thaliana* indicate a role for cytokinins in the shoot apical meristem. *The Plant Journal, 18*, 557–563.
Sabatini, S., Beis, D., Wolkenfelt, H., Murfett, J., Guilfoyle, T., Malamy, J., et al. (1999). An auxin-dependent distal organizer of pattern and polarity in the *Arabidopsis* root. *Cell, 99*, 463–472.
Sablowski, R. (2009). Cytokinin and WUSCHEL tie the knot around plant stem cells. *Proceedings of the National Academy of Sciences of the United States of America, 106*, 16016–16017.
Sakai, H., Honma, T., Aoyama, T., Sato, S., Kato, T., Tabata, S., et al. (2001). ARR1, a transcription factor for genes immediately responsive to cytokinins. *Science, 294*, 1519–1521.
Sarkar, A. K., Luijten, M., Miyashima, S., Lenhard, M., Hashimoto, T., Nakajima, K., et al. (2007). Conserved factors regulate signalling in *Arabidopsis thaliana* shoot and root stem cell organizers. *Nature, 446*, 811–814.
Scheres, B. (2007). Stem-cell niches: Nursery rhymes across kingdoms. *Nature Reviews. Molecular Cell Biology, 8*, 345–354.
Schubert, D., Primavesi, L., Bishopp, A., Roberts, G., Doonan, J., Jenuwein, T., et al. (2006). Silencing by plant polycomb-group genes requires dispersed trimethylation of histone H3 at lysine 27. *The EMBO Journal, 25*, 4638–4649.
Sena, G., & Birnbaum, K. D. (2010). Built to rebuild: In search of organizing principles in plant regeneration. *Current Opinion in Genetics & Development, 20*, 460–465.
Sena, G., Wang, X., Liu, H. Y., Hofhuis, H., & Birnbaum, K. D. (2009). Organ regeneration does not require a functional stem cell niche in plants. *Nature, 457*, 1150–1153.
Skoog, F., & Miller, C. O. (1957). Chemical regulation of growth and organ formation in plant tissues cultured in vitro. *Symposia of the Society for Experimental Biology, 11*, 118–130.

Steward, F. C., Mapbs, M. O., & Mears, K. (1958). Growth and organized development of culture cells. II. Organization in cultures grown from freely suspended cells. *American Journal of Botany*, *45*, 705–708.

Stone, S. L., Braybrook, S. A., Paula, S. L., Kwong, L. W., Meuser, J., Pelletier, J., et al. (2008). *Arabidopsis* LEAFY COTYLEDON2 induces maturation traits and auxin activity: Implications for somatic embryogenesis. *Proceedings of the National Academy of Sciences of the United States of America*, *105*, 3151–3156.

Stone, S. L., Kwong, L. W., Yee, K. M., Pelletier, J., Lepiniec, L., Fischer, R. L., et al. (2001). *LEAFY COTYLEDON2* encodes a B3 domain transcription factor that induces embryo development. *Proceedings of the National Academy of Sciences of the United States of America*, *98*, 11806–11811.

Stroud, H., Ding, B., Simon, S. A., Feng, S., Bellizzi, M., Pellegrini, M., et al. (2013). Plants regenerated from tissue culture contain stable epigenome changes in rice. *Elife*, *2*, e00354.

Su, Y. H., & Zhang, X. S. (2009). Auxin gradients trigger de novo formation of stem cells during somatic embryogenesis. *Plant Signaling & Behavior*, *4*, 574–576.

Su, Y. H., Zhao, X. Y., Liu, Y. B., Zhang, C. L., O'Neill, S. D., & Zhang, X. S. (2009). Auxin-induced *WUS* expression is essential for embryonic stem cell renewal during somatic embryogenesis in Arabidopsis. *The Plant Journal*, *59*, 448–460.

Sugimoto, K., Gordon, S. P., & Meyerowitz, E. M. (2011). Regeneration in plants and animals: Dedifferentiation, transdifferentiation, or just differentiation? *Trends in Cell Biology*, *21*, 212–218.

Sugimoto, K., Jiao, Y., & Meyerowitz, E. M. (2010). *Arabidopsis* regeneration from multiple tissues occurs via a root development pathway. *Developmental Cell*, *18*, 463–471.

Sussex, I. M. (2008). The scientific roots of modern plant biotechnology. *Plant Cell*, *20*, 1189–1198.

Suzuki, M., Wang, H. H., & McCarty, D. R. (2007). Repression of the *LEAFY COTYLEDON 1/B3* regulatory network in plant embryo development by *VP1/ABSCISIC ACID INSENSITIVE 3-LIKE* B3 genes. *Plant Physiology*, *143*, 902–911.

Takahashi, K., & Yamanaka, S. (2006). Induction of pluripotent stem cells from mouse embryonic and adult fibroblast cultures by defined factors. *Cell*, *126*, 663–676.

Tamaki, H., Konishi, M., Daimon, Y., Aida, M., Tasaka, M., & Sugiyama, M. (2009). Identification of novel meristem factors involved in shoot regeneration through the analysis of temperature-sensitive mutants of Arabidopsis. *The Plant Journal*, *57*, 1027–1039.

Thorpe, T. A. (2006). History of plant tissue culture. *Methods in Molecular Biology*, *318*, 9–32.

Thorpe, T. A. (2007). History of plant tissue culture. *Molecular Biotechnology*, *37*, 169–180.

Tsuwamoto, R., Yokoi, S., & Takahata, Y. (2010). *Arabidopsis EMBRYOMAKER* encoding an AP2 domain transcription factor plays a key role in developmental change from vegetative to embryonic phase. *Plant Molecular Biology*, *73*, 481–492.

Turck, F., Roudier, F., Farrona, S., Martin-Magniette, M. L., Guillaume, E., Buisine, N., et al. (2007). Arabidopsis TFL2/LHP1 specifically associates with genes marked by trimethylation of histone H3 lysine 27. *PLoS Genetics*, *3*, e86.

Vahdati, K., Bayat, S., Ebrahimzadeh, H., Jariteh, M., & Mirmasoumi, M. (2008). Effect of exogenous ABA on somatic embryo maturation and germination in Persian walnut (Juglans regia L.). *Plant Cell, Tissue and Organ Culture*, *93*, 163–171.

Verdeil, J. L., Alemanno, L., Niemenak, N., & Tranbarger, T. J. (2007). Pluripotent versus totipotent plant stem cells: Dependence versus autonomy? *Trends in Plant Science*, *12*, 245–252.

Wang, H., Caruso, L. V., Downie, A. B., & Perry, S. E. (2004). The embryo MADS domain protein AGAMOUS-Like 15 directly regulates expression of a gene encoding an enzyme involved in gibberellin metabolism. *Plant Cell*, *16*, 1206–1219.

Wang, X., Niu, Q. W., Teng, C., Li, C., Mu, J., Chua, N. H., et al. (2009). Overexpression of *PGA37/MYB118* and *MYB115* promotes vegetative-to-embryonic transition in *Arabidopsis*. *Cell Research*, *19*, 224–235.

White, C. N., Proebsting, W. M., Hedden, P., & Rivin, C. J. (2000). Gibberellins and seed development in *Maize*. I. Evidence that gibberellin/abscisic acid balance governs germination versus maturation pathways. *Plant Physiology*, *122*, 1081–1088.

Xiao, L., Zhang, L., Yang, G., Zhu, H., & He, Y. (2012). Transcriptome of protoplasts reprogrammed into stem cells in *Physcomitrella patens*. *PLoS One*, *7*, e35961.

Xu, J., Hofhuis, H., Heidstra, R., Sauer, M., Friml, J., & Scheres, B. (2006). A molecular framework for plant regeneration. *Science*, *311*, 385–388.

Xu, K., Liu, J., Fan, M., Xin, W., Hu, Y., & Xu, C. (2012). A genome-wide transcriptome profiling reveals the early molecular events during callus initiation in *Arabidopsis* multiple organs. *Genomics*, *100*, 116–124.

Xu, L., & Shen, W.-H. (2008). Polycomb silencing of *KNOX* genes confines shoot stem cell niches in *Arabidopsis*. *Current Biology*, *18*, 1966–1971.

Yamaguchi, S., Kamiya, Y., & Nambara, E. (2007). Regulation of ABA and GA levels during seed development and germination in *Arabidopsis*. In K. J. Bradford & H. Nonogaki (Eds.), *Seed development, dormancy and germination* (pp. 224–247). Oxford, UK: Blackwell Publishing.

Yanai, O., Shani, E., Dolezal, K., Tarkowski, P., Sablowski, R., Sandberg, G., et al. (2005). *Arabidopsis* KNOXI proteins activate cytokinin biosynthesis. *Current Biology*, *15*, 1566–1571.

Yang, C., Bratzel, F., Hohmann, N., Koch, M., Turck, F., & Calonje, M. (2013). VAL- and AtBMI1-mediated H2Aub initiate the switch from embryonic to postgerminative growth in *Arabidopsis*. *Current Biology*, *23*, 1324–1329.

Yang, C. H., Chen, L. J., & Sung, Z. R. (1995). Genetic regulation of shoot development in *Arabidopsis*: Role of the *EMF* genes. *Developmental Biology*, *169*, 421–435.

Yang, X. Y., & Zhang, X. L. (2010). Regulation of somatic embryogenesis in higher plants. *Critical Reviews in Plant Sciences*, *29*, 36–57.

Yasutani, I., Ozawa, S., Nishida, T., Sugiyama, M., & Komamine, A. (1994). Isolation of temperature-sensitive mutants of *Arabidopsis thaliana* that are defective in the redifferentiation of shoots. *Plant Physiology*, *105*, 815–822.

Yoshida, N., Yanai, Y., Chen, L., Kato, Y., Hiratsuka, J., Miwa, T., et al. (2001). EMBRYONIC FLOWER2, a novel polycomb group protein homolog, mediates shoot development and flowering in Arabidopsis. *Plant Cell*, *13*, 2471–2481.

Zhang, H., Bishop, B., Ringenberg, W., Muir, W. M., & Ogas, J. (2012). The CHD3 remodeler PICKLE associates with genes enriched for trimethylation of histone H3 lysine 27. *Plant Physiology*, *159*, 418–432.

Zhang, X., Clarenz, O., Cokus, S., Bernatavichute, Y. V., Pellegrini, M., Goodrich, J., et al. (2007). Whole-genome analysis of histone H3 lysine 27 trimethylation in *Arabidopsis*. *PLoS Biology*, *5*, e129.

Zhang, X., Germann, S., Blus, B. J., Khorasanizadeh, S., Gaudin, V., & Jacobsen, S. E. (2007). The *Arabidopsis* LHP1 protein colocalizes with histone H3 Lys27 trimethylation. *Nature Structural & Molecular Biology*, *14*, 869–871.

Zhang, H., Rider, S. D., Jr., Henderson, J. T., Fountain, M., Chuang, K., Kandachar, V., et al. (2008). The CHD3 remodeler PICKLE promotes trimethylation of histone H3 lysine 27. *The Journal of Biological Chemistry*, *283*, 22637–22648.

Zhao, Z., Andersen, S. U., Ljung, K., Dolezal, K., Miotk, A., Schultheiss, S. J., et al. (2010). Hormonal control of the shoot stem-cell niche. *Nature*, *465*, 1089–1092.

Zheng, Y., Ren, N., Wang, H., Stromberg, A. J., & Perry, S. E. (2009). Global identification of targets of the *Arabidopsis* MADS domain protein AGAMOUS-Like15. *Plant Cell*, *21*, 2563–2577.

Zheng, Q., Zheng, Y., & Perry, S. E. (2013). AGAMOUS-Like15 promotes somatic embryogenesis in Arabidopsis and soybean in part by the control of ethylene biosynthesis and response. *Plant Physiology*, *161*, 2113–2127.
Zimmerman, J. L. (1993). Somatic embryogenesis: A model for early development in higher plants. *Plant Cell*, *5*, 1411–1423.
Zuo, J., Niu, Q. W., Frugis, G., & Chua, N. H. (2002). The *WUSCHEL* gene promotes vegetative-to-embryonic transition in *Arabidopsis*. *The Plant Journal*, *30*, 349–359.

CHAPTER TWO

The Hormonal Control of Regeneration in Plants

Ying Hua Su, Xian Sheng Zhang[1]
State Key Laboratory of Crop Biology, College of Life Sciences, Shandong Agricultural University, Taian, Shandong, China
[1]Corresponding author: e-mail address: zhangxs@sdau.edu.cn

Contents

Abstract

Plant cells have a profound capacity to regenerate their full array of tissues from already differentiated organs, as best demonstrated in *in vitro* regeneration systems. Although critical breakthroughs in *in vitro* organogenesis have outlined the role of hormones and their interactions in determination of cultured plant cell developmental fates, the underlying molecular mechanisms are still largely unexplored. Investigations have recently been empowered by the identification of key genes that function in regeneration, involved in hormonal biosynthesis, transport, signaling, and hormone interactions. The establishment of differential hormone-responsive patterns in organ regeneration

Current Topics in Developmental Biology, Volume 108
ISSN 0070-2153
http://dx.doi.org/10.1016/B978-0-12-391498-9.00010-3

zones is critical for *de novo* organ initiation. The present review focuses on recent findings providing insights into hormone-regulated plant regeneration at the molecular level and the formation of hormonal-response environments required for *de novo* regeneration.

ABBREVIATIONS AND GLOSSARY

ABA abscisic acid
Callus an intermediate plant tissue that, similar to regenerative blastemas in animals, is an undifferentiated structure that can give rise to new tissues
Cell totipotency the ability of an entire plant to be regenerated from single somatic cells
CIM callus-inducing medium
Dedifferentiation the process by which somatic cells of explant tissues respond to hormonal signals to acquire features similar to meristematic cells
Direct regeneration the induction of *in vitro* organs directly from explant tissues
ECIM embryonic callus-inducing medium
Explant a small piece of plant somatic tissue that can reproduce a new tissue or growth structure during plant regeneration
GFP green fluorescent protein
Indirect regeneration the formation of a *de novo* organ from callus, an intermediate tissue
NPA naphthylphthalamic acid
RAM root apical meristem
RIM root-inducing medium
SAM shoot apical meristem
SEIM somatic embryo-inducing medium
SEs somatic embryos
SIM shoot-inducing medium
Transdifferentiation the plant regeneration process in which cells directly transform into cell types different from their already established differentiation paths
YFP yellow fluorescent protein

1. INTRODUCTION

Plant regeneration involves the *in vitro* culture of cells, tissues, and organs under defined physical and chemical conditions. Critical for *in vitro* plant propagation and biotechnology, this phenomenon is also applicable to studies of plant developmental regulatory mechanisms. Regeneration has long been known to occur in plants, with more recent discovery in animals. With the exception of stem cells, animal cells generally lose the ability to produce other cell types upon differentiation. In plants, however, differentiated cells are able to regenerate into the full array of tissues under appropriate culture conditions (Birnbaum & Sánchez Alvarado, 2008).

As classically defined, plant regeneration refers to regeneration of a growth structure lost by injury, for example, regeneration of an excised root or leaf tip in *Arabidopsis* (Sugimoto, Gordon, & Meyerowitz, 2010). Alternatively, a small piece of plant somatic tissue—an explant—can reproduce a new tissue or growth structure not present before injury. In "cell totipotency," an entire plant can even be regenerated from a single somatic cell (Haberlandt, 1902). However, the mechanisms underlying this totipotency remain elusive (Birnbaum & Sánchez Alvarado, 2008; Vogel, 2005).

In 1902, Haberlandt predicted that someday "one could successfully cultivate embryos from vegetative cells" under correct *in vitro* culture conditions (Haberlandt, 1902; Krikorian & Berquam, 1969). Effective plant regeneration techniques were established three decades later. In 1939, regeneration using larger explant tissues from carrot and other species was successfully carried out in culture medium containing the critical phytohormone indole-3-aceticacid (auxin) (Gautheret, 1985). Auxins, the first discovered plant hormone, are small compounds containing an aromatic ring and a carboxylic acid group. Cytokinin is another phytohormone with a structure resembling adenine. An important advance in the study of plant regeneration was the identification of the major effect of auxin/cytokinin ratios on regenerated tissue type. In 1957, Skoog and Miller found that treating tobacco pith with high auxin/cytokinin ratios led to root formation. In contrast, high cytokinin/auxin ratios induced shoot regeneration. When high concentrations of both hormones were added to explants, a mass of growing cells known as a "callus" was induced. This pioneering work provided the conceptual framework for the role of plant hormones and their interactions in establishing distinct regeneration paths for plant tissue cultures. Widespread success using different culture conditions has since led to the production of a large variety of plant tissues and much information regarding plant regeneration.

Regeneration can involve direct or indirect organogenesis (Hicks, 1980). In direct regeneration, *in vitro* organs are directly induced from explant tissues; in indirect regeneration, a *de novo* organ is typically formed from an intermediate tissue, the callus. Plant calli, like regenerative blastemas in animals, are undifferentiated structures that can give rise to new tissues (Birnbaum & Sánchez Alvarado, 2008). Plant leaves, shoots, roots, and embryos can variously be elicited from a growing callus by treating it with different ratios of hormones (Gautheret, 2003; Skoog & Miller, 1957; Street, 1977). In 1986, Feldmann and Marks established the indirect two-step *Arabidopsis* organ regeneration method (Feldmann & Marks, 1986), one of the most widely used *in vitro* systems. The first step in this procedure entails callus formation from explants incubated on auxin-rich callus-inducing medium (CIM). Shoots and roots can subsequently be induced by culturing on shoot-inducing medium (SIM) or root-inducing medium (RIM), respectively, with different ratios of auxin to cytokinin (Fig. 2.1).

Following the development of the *Arabidopsis* regeneration technique, many studies have focused on hormonal regulation of regeneration in hundreds of plant species (An, Li, Su, & Zhang, 2004; Guan, Zhu, Li, & Zhang, 2006; Hicks & McHughen, 1974; Li, Li, Bai, Lu, & Zhang, 2002; Lu, 2002, 2003; Lu, Enomoto, Fukunaga, & Kuo, 1988; Tran Thanh Van, 1973; Xu et al., 2004). In *in vitro* floral organogenesis of *Hyacinthus orientalis*, high levels of cytokinin and auxin trigger the formation of tepals from explants (Lu et al., 1988). After transfer to medium containing low levels of both hormones, ovaries and ovules can be induced from regenerated floral buds. In leaf protoplast culture of alfalfa, cells grown on medium containing different auxin concentrations develop into either embryogenic or nonembryogenic cell types (Feher, Pasternak, Otvos, Miskolczi, & Dudits, 2002; Pasternak et al., 2002). Different shoot and root regeneration frequencies from *Arabidopsis* inflorescence stem explants have recently been induced from cultures grown on different media containing 216 combinations of exogenous auxin and cytokinin (Zhao et al., 2013). In addition to auxin and cytokinin, other hormones, such as gibberellins, ethylene, and abscisic acid (ABA), affect *in vitro* tissue and organ growth. Ethylene, a colorless, flammable gas, is a hydrocarbon with carbon–carbon double bonds. Inhibition of either its activity by $AgNO_3$ or its production by $CoCl_2$

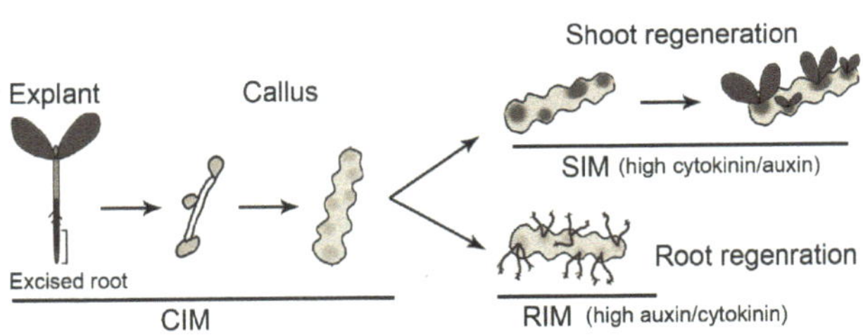

Figure 2.1 Schematic drawing of plant regeneration. *Arabidopsis in vitro* shoot or root regeneration. Callus is induced from root explants with the first auxin-rich hormonal treatment (CIM). Then the subsequent culture of callus on different media causes the cells to be specified to form new organs. Shoots can be induced by culturing on SIM with high ratios of cytokinin to auxin, and roots can be induced on RIM with high ratios of auxin to cytokinin. (For color version of this figure, the reader is referred to the online version of this chapter.)

prevents somatic embryo (SE) formation in *Coffea canephora* leaf cultures (Hatanaka, Sawabe, Azuma, Uchida, & Yasuda, 1995). Ethylene is also necessary for embryonic callus growth and SE maturation in *Medicago sativa* (Kępczyński, McKersie, & Brown, 1992). In addition, SEs can be produced from carrot seedlings cultured on medium containing ABA—the compound responsible for fruit abscission—as the sole exogenous hormone (Nishiwaki, Fujino, Koda, Masuda, & Kikuta, 2000). Taken together, concentrations and types of exogenous hormones are critical to cell fate determination during *in vitro* regeneration.

Plant regeneration patterns depend not only on the specific balance of applied exogenous hormones but also on the response of explant tissues to these hormones (Sugiyama, 1999). Generally, three phases can be recognized throughout plant regeneration. First, somatic cells of explant tissues can respond to hormonal signals to acquire features similar to meristematic cells, a process known as "dedifferentiation." Interestingly, recent work has shown that proliferating callus cells are not dedifferentiated to the fundamental state of meristematic cells, but instead resemble root tissue cells with respect to gene expression patterns during some plant regeneration processes (Atta et al., 2009; Sugimoto, Jiao, & Meyerowitz, 2010). "Transdifferentiation" is thus a better term for such hormone-regulated switches in cell-type identity (Sugimoto, Gordon, et al., 2010). Second, callus cells with organogenic competence are reprogrammed and determined for specific organ formation under the influence of hormone balance. The third regeneration phase, morphogenesis, is independent of exogenously supplied hormones. Thus, exogenous hormone treatment is the critical factor triggering early developmental events in *in vitro* regeneration.

Favorable hormone balances may exist not only in growth media but also in calli. Endogenous hormone production may be induced by various exogenous hormone and stimulating treatments (Peres et al., 1999), or by explantation in the absence of treatment (Peres & Kerbauy, 1999). This implies that endogenous hormonal metabolism and perception are key parameters influencing regeneration (Auer, Cohen, Laloue, & Cooke, 1992; Cary, Uttamchandani, Smets, Van Onckelen, & Howell, 2001; Centeno, Rodríguez, Feito, & Fernández, 1996; Sarul, Vlahova, Ivanova, & Atanassov, 1995; Yoshimatsu & Shimomura, 1994). Recent studies have suggested that exogenous hormones determine the developmental fate of callus cells by regulating biosynthesis and distribution of endogenous hormones, triggering the specialized hormonal signaling required for cell differentiation (Gordon et al., 2007; Su, Liu, & Zhang,

2011; Su et al., 2009; Sugimoto, Gordon, et al., 2010). Based on mutant phenotypes with disrupted hormonal biosynthesis or perception and developments in molecular biology, further understanding of endogenous hormone functions in cell development has been achieved (Sugiyama, 1999). In particular, it is now known that cytokinin regulates cell proliferation and gibberellin promotes cell elongation, while auxin and brassinosteroids—plant hormones structurally similar to animal and insect steroids—are involved in both processes (Hardtke, Dorcey, Osmont, & Sibout, 2007; Nakaya, Tsukaya, Murakami, & Kato, 2002). In contrast, molecular mechanisms underlying endogenous hormonal regulation of *in vitro*-cultured plant organ development still remain to be elucidated. This review describes recent findings that provide insights into endogenous hormone-regulated plant regeneration at the molecular level.

2. SPATIOTEMPORAL PATTERNS OF HORMONAL RESPONSE ARE CRITICAL TO *De novo* REGENERATION

Based on pharmacological and genetic visualization methods, plant hormonal-response signals are asymmetrically distributed across adjacent cells during crucial stages of plant growth and development. Spatiotemporal-response patterns observed for auxin and cytokinin suggest that both hormones control important developmental processes, such as shoot meristem formation and maintenance (Benková et al., 2003; Gordon, Chickarmane, Ohno, & Meyerowitz, 2009; Leibfried et al., 2005; Reinhardt et al., 2003) and embryonic root stem-cell specification (Friml et al., 2003; Müller & Sheen, 2008).

2.1. Auxin-responsive patterns in callus formation and organ regeneration

Auxin is probably the best-known plant hormone exhibiting local accumulation and response in cells and tissues. Dynamic gradients of auxin response are often visualized using reporters such as green fluorescent (GFP) and yellow fluorescent (YFP) proteins under the control of the auxin-responsive DR5 element (Casimiro et al., 2001; Ulmasov, Hagen, & Guilfoyle, 1997). Asymmetrically distributed auxin-response signals are involved in virtually every aspect of *in vivo* plant growth and development, including embryo axis formation (Friml et al., 2003), flower primordium initiation and patterning (Heisler et al., 2005; Reddy, Heisler, & Ehrhardt, 2004; Reddy & Meyerowitz, 2005), vascular tissue differentiation (Mattsson, Ckurshumova, & Berleth, 2003), root meristem maintenance (Sabatini et al., 1999), and tropic growth (Friml, Wisniewska, Benkova, Mendgen, & Palme, 2002). In these different developmental contexts, auxin polar transport mediated by efflux carrier proteins PINFORMEDs (PINs) contributes to the establishment of local auxin-responsive gradients in specific cells of plant tissues (Friml, 2010). PIN1, for example, plays an important role in initiating and maintaining auxin-responsive gradients within various plant tissues (Friml et al., 2003; Heisler et al., 2005).

Although auxin-response patterning during plant *in vivo* development is well understood, little is known regarding its role in *de novo* regeneration of plant tissues in culture. The distribution of auxin-responsive signals is determined in the process of callus formation from *Arabidopsis* root explants on auxin-rich CIM (Gordon et al., 2007). In root explants harvested from 2-week-old seedlings, a DR5-visualized auxin response occurs in some root pericycle, lateral root progenitor, and columellar root cap cells. After induction in CIM, auxin-responsive signals are initially present in clusters of small cells proliferating to form callus, then later diminish within the callus, suggesting that auxin response is only required for early cell proliferation during callus induction (Gordon et al., 2007). After the callus is transferred to cytokinin-rich SIM, expression of *WUSCHEL* (*WUS*), required for stem-cell formation and maintenance in shoot apical meristem (SAM), is upregulated in the center of *de novo*-regenerated SAM. Interestingly, auxin-responsive signals are low or undetectable in areas of SAM initiation, but are strong in surrounding regions. Spatial patterns of auxin response are also clearly shown in pistil-induced shoot regeneration (Cheng et al., 2013). Auxin-responsive signals are uniformly detected at the edges of mature callus on CIM; after shoot induction on SIM, signals are translocated to outermost cell layer regions surrounding the *WUS* expression domain (Fig. 2.2A). Auxin-responsive signals thus accumulate in regions of SAM initiation. As a contributor to the spatially restricted auxin response, PIN1 exhibits polarized membrane localization at future sites of SAM initiation. This polarization can be induced by SIM incubation (Cheng et al., 2013). Applications of the auxin transport inhibitor naphthylphthalamic acid (NPA) on SIM disrupt spatiotemporal auxin response and shoot regeneration, indicating that auxin polar transport and asymmetric distribution of auxin response are required for *de novo* SAM initiation.

Auxin-responsive gradients are also essential for *Arabidopsis* somatic embryogenesis (Su et al., 2009). Endogenous auxin-responsive signals are not detected at the edges of embryonic callus incubated on auxin-rich

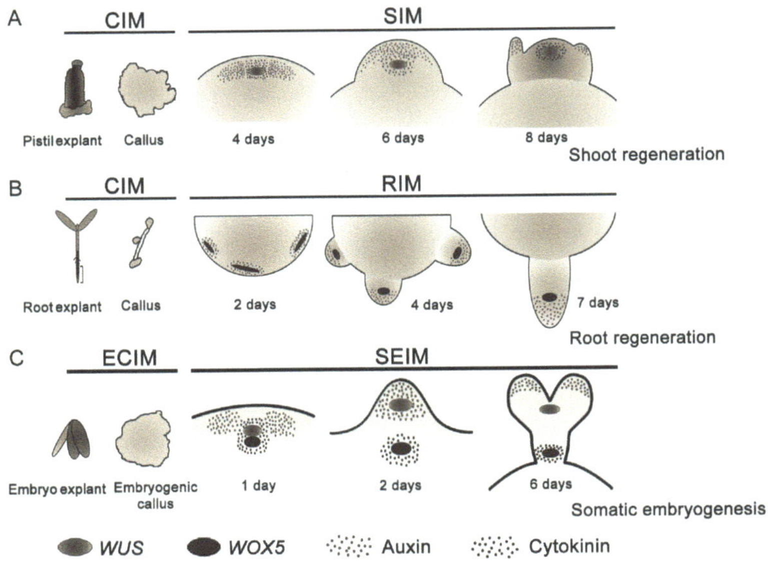

Figure 2.2 Hormone-responsive patterns in shoot, root, and SE regeneration. Auxin- and cytokinin-responsive patterns in shoot (A), root (B), and SE (C) regeneration. (A) Pistils (explants) are cut and transferred to CIM to induce callus formation (Cheng et al., 2013). Calli cultured on CIM for 20 days are transferred onto SIM for shoot induction. *WUS* expression is induced at 4 days in the organizing center (OC) of the initiated SAM. At this time, auxin response is observed in areas surrounding *WUS* expression, whereas cytokinin-responsive signals are concentrated in the center of SAM, the region of *WUS* expression. When the shoot primordium emerged at 6 days, both auxin and cytokinin responses occur at the top of the primordium. (B) During *de novo* root tissue formation, auxin response exhibits a regional distribution pattern for root induction on RIM at about 2 days, corresponding to the *WOX5* expression domain. The *WOX5* signal is subsequently localized in the quiescent center (QC) of the regenerated roots, with auxin-responsive signals in the root apex. (C) Embryonic callus is incubated on auxin-rich embryonic callus-inducing medium (ECIM) for 14 days. After 1 day of SE induction on somatic embryo-inducing medium (SEIM), auxin-responsive gradients are established in edge regions surrounding areas of *WUS* expression, which marked the OC of promeristems (Su, Cheng, Su, & Zhang, 2010; Su et al., 2009). Auxin-responsive signals are later redistributed to the top of the promeristems for *de novo* formation of somatic proembryos. Cytokinin-response signals are asymmetrically distributed over the areas of *WOX5* expression. (For color version of this figure, the reader is referred to the online version of this chapter.)

embryonic callus-inducing medium (ECIM). When exogenous auxin is eliminated from ECIM to induce SE, endogenous auxin-responsive gradients are established in edge regions surrounding the areas of *WUS* expression (Fig. 2.2C), which marks the organizing centers (OCs) of promeristems that develop into somatic proembryos (Su et al., 2009, 2010). Auxin-responsive signals are later redistributed to upper promeristem regions for *de novo* formation of somatic proembryos (Fig. 2.2C). PIN1 polar localization is identified in groups of cells located just above the *WUS* expression, further indicating the importance of auxin polar transport in promeristems during SE regeneration (Su et al., 2009).

Recently, we also examined auxin response during *de novo* root regeneration from root explants. Following incubation on CIM for 4 days, callus was induced for root regeneration by transfer to auxin-rich RIM (Che, Lall, Nettleton, & Howell, 2006). *DR5::YFP* signals were first uniformly identified in edge regions of the callus on CIM (Fig. 2.3A). Root induction on RIM for 2–4 days induced a restricted distribution of auxin-responsive signals corresponding to expression patterns of the root meristem-specific *WUSCHEL-RELATED-HOMEOBOX 5* (*WOX5*) gene (Figs. 2.2B and 2.3B and C). PIN1 expression also exhibited polarized membrane localization at future sites of root apical meristem (RAM) initiation (Fig. 2.3D–F). Our results suggest that the establishment of auxin-responsive gradients is correlated with *de novo* RAM induction and root regeneration.

2.2. Cytokinin-responsive patterns in callus formation and organ regeneration

Cytokinin is another important factor in regulating plant growth and development. The role of the cytokinin response in plant organogenesis has been evaluated using reporters controlled by TCS, a synthetic cytokinin-responsive promoter having activity consistent with cytokinin action (Gordon et al., 2009; Müller & Sheen, 2008). At the early globular stage of embryogenesis, cytokinin-responsive signals are detected in the embryo hypophysis (Müller & Sheen, 2008). By the transition stage, when the hypophysis has undergone asymmetrical cell division, the signals are retained in the apical lens-shaped cell of the embryonic root. Using *TCS::GFP* reporters, cytokinin response can also be visualized during floral meristem development, showing that the localized response domain is similar to that of *WUS* expression (Gordon et al., 2009). *ARABIDOPSIS RESPONSE REGULATOR 5* (*ARR5*), whose expression is correlated with cytokinin content in various tissues, can also be used to demonstrate spatial distribution of cytokinin response (Aloni, Langhans, Aloni, & Ullrich, 2004; Leibfried et al., 2005).

The role of the cytokinin response in *de novo* regeneration of cultured plant tissues has been investigated. During callus formation from root explants, ARR5-visualized cytokinin-responsive signals are detected mainly

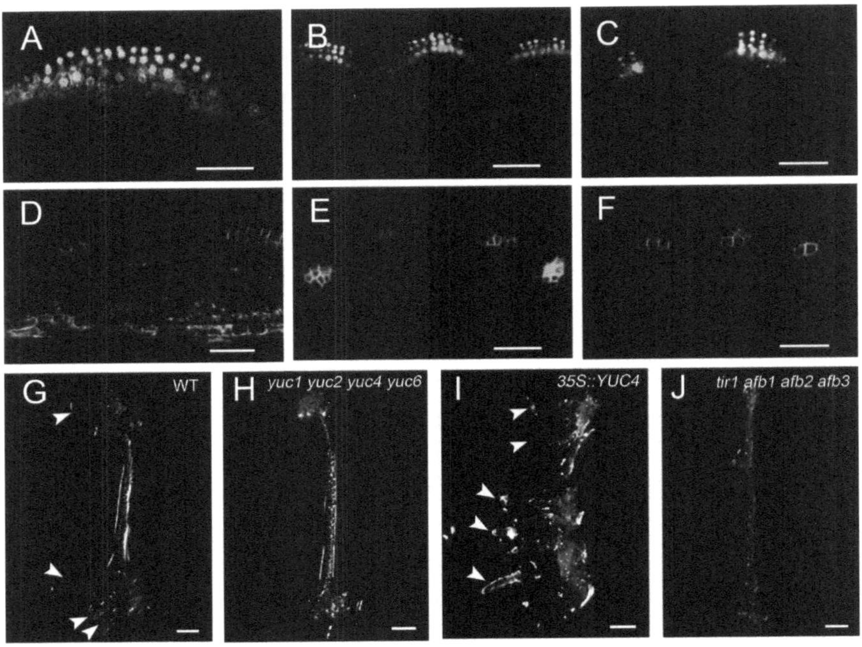

Figure 2.3 Auxin response, polar transport, and biosynthesis in *Arabidopsis* root regeneration. *Arabidopsis* seedlings (ecotype Columbia) are grown on MS medium (Murashige & Skoog, 1962) for 10 days. Root segments (5 mm) are cut and preincubated on CIM (Che et al., 2006) for 6 days and then transferred to RIM (Che et al., 2006) for root induction. (A–C) *DR5rev::YFP* signals (yellow) and *PWOX5::GFP* signals (green) at the edges of callus incubated on CIM for 6 days (A) and on RIM for 2 days (B) and 4 days (C). (D–F) *PIN1::GFP* signals (green) at the edges of callus incubated on CIM for 6 days (D) and on RIM for 2 days (E) and 4 days (F). (G–J) Regenerated roots from callus of wild-type (WT) plants (G), *yuc1 yuc2 yuc4 yuc6* mutants (H), *35S::YUC4* (I), and *tir1 afb1 afb2 afb3* mutants (J) grown on RIM for 10 days. Arrowheads indicate regenerated roots. Scale bars = 150 μm (A–F) and 500 μm (G–J). (See color plate.)

in root explant vasculature, and are strongly distributed in proliferating callus cells incubated on CIM (Gordon et al., 2007). After shoot induction on SIM, cytokinin-responsive patterns are reestablished in areas of SAM initiation, and later within developing SAM. These patterns are distinct from the auxin response ones that occur during *de novo* SAM initiation. Spatiotemporal distribution of cytokinin response revealed by *TCS::GFP* reporters also corresponds to *WUS* expression and SAM formation in *Arabidopsis* pistil-derived *de novo* shoot regeneration (Cheng et al., 2013; Fig. 2.2A). In mature callus (shoot noninduced callus) derived from pistils cultured on CIM, cytokinin response occurs in edge regions, similar to auxin response. *De novo* shoot initiation on SIM induces regional redistribution of cytokinin-responsive signals in areas of SAM initiation and *WUS* expression (Fig. 2.2A). Moreover, spatial expression of the cytokinin-responsive gene *ARR5* in *in vitro*-induced root organs implies that root regeneration is accompanied by a localized cytokinin response in the callus (Pernisová et al., 2009).

The distribution of cytokinin response during SE induction was examined using GFP reporters driven by the *ARR7* promoter (Fig. 2.4A–C), another A-type *ARR* gene that is cytokinin-inducible (Zhao et al., 2010). Interestingly, unlike auxin-responsive distribution in SE promeristems, cytokinin-response signals were asymmetrically distributed over areas of *WOX5* expression associated with embryonic root meristem (Figs. 2.2C and 2.4A–C). These results imply that cytokinin is extremely important in RAM establishment during SE initiation.

3. HORMONAL BIOSYNTHESIS CONTRIBUTES TO THE DISTRIBUTION OF THE HORMONAL RESPONSE AND *De novo* REGENERATION

Differential distribution of hormonal response is essential for plant development during *de novo* regeneration. Multiple hormonal regulation pathways, such as those involved in biosynthesis, transport, perception, and signaling, contribute to the maintenance of optimal hormonal-response patterns within tissues. Elucidation of the molecular mechanisms underlying hormonal biosynthesis will greatly increase our understanding of plant developmental regulation.

3.1. Auxin biosynthesis functions in plant regeneration

Three aspects of auxin likely contribute to its *de novo* production and action: (1) creation of an auxin biosynthesis source, (2) polar transport of synthesized auxin to generate a localized accumulation, and (3) the effect of local auxin response on plant development (Chandler, Cole, Flier, & Werr, 2009). Consequently, regional patterns of auxin response can result from both local auxin biosynthesis and dynamic auxin transport. The molecular components of auxin biosynthesis have already been identified (Zhao et al., 2001; Zhao, 2008), with the *YUCCA* (*YUC*) gene family encoding flavin monooxygenases the best characterized. YUC-mediated auxin biosynthesis is required for establishment of embryonic basal body regions and initiation of embryonic and postembryonic organs (Cheng, Dai, & Zhao, 2007).

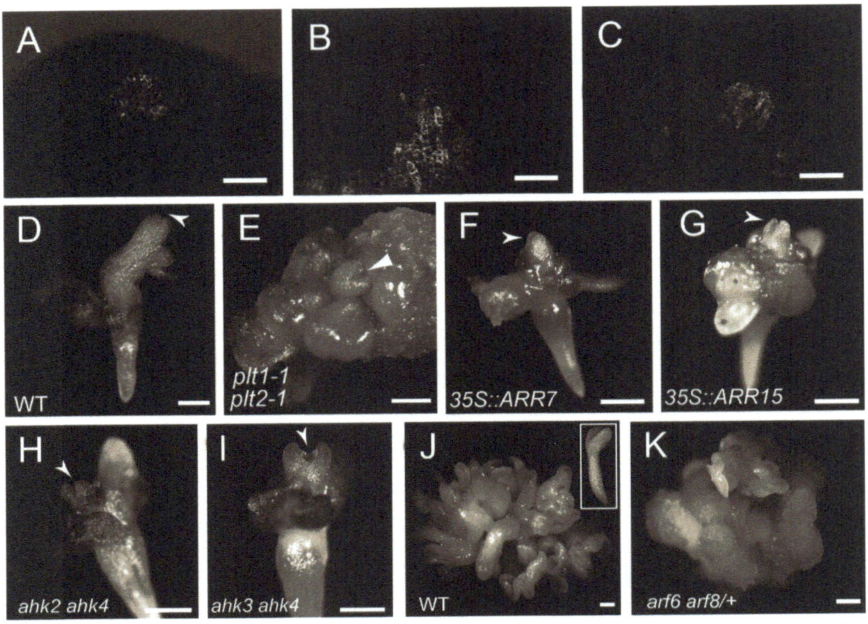

Figure 2.4 Cytokinin and auxin response in *Arabidopsis* somatic embryogenesis. The process of *Arabidopsis* somatic embryogenesis has been reported by Su et al. (2009). Green primary somatic embryos (PSEs) can be induced from explants (zygotic embryos, ecotype Columbia) cultured on medium containing 2,4-D after 10 days. PSEs are then transferred into liquid medium containing 2,4-D (ECIM) for 14 days to form embryonic calli. The resulting calli are transferred into 2,4-D free liquid medium (SEIM) to induce secondary somatic embryos (SSEs). (A–C) *PARR7::GFP* signals (green) and *PWOX5::RFP* signals (red) at the edges of embryonic callus incubated on SEIM for 1 day (A), 2 days (B), and 3 days (C). (D–I) Phenotypes of PSE induction from explants of WT (D), *plt1 plt2* mutants (E), *35S::ARR7* (F), *35S::ARR15* (G), *ahk2 ahk4* mutants (H), and *ahk3 ahk4* mutants (I) grown on solid medium for 10 days. Arrowheads indicate PSEs. (J–K) Phenotypes of SSE induction from calli of WT (J) and *arf6 arf8/+* mutants (K) grown on SEIM for 8 days. Scale bars = 60 μm (A–C), 0.4 mm (D–I), and 1.2 mm (J–K). (See color plate.)

yuc multiple mutants show impaired local auxin distribution and severe defects in floral patterning, vascular formation, and formation of hypocotyls or root meristem (Cheng, Dai, & Zhao, 2006, Cheng et al., 2007).

During *de novo* regeneration in many species, treatment with high levels of exogenous auxin stimulates root regeneration and inhibits shoot regeneration. However, little is known regarding the role of endogenous auxin biosynthesis in *de novo* organ regeneration. Endogenous auxin biosynthesis mediated by YUCs has been recently observed during shoot regeneration from *Arabidopsis* pistils (Cheng et al., 2013). During shoot induction,

YUC1 and *YUC4* expression is upregulated, with both genes showing localized expression patterns within the callus. Following incubation on CIM, *YUC1* and *YUC4* expression signals are not detected, whereas shoot induction on SIM induces regional transcriptional signals from both genes at future SAM initiation sites. *YUC4* signals are detected around regions of initiated shoot promeristems marked by *WUS* expression, similar to patterns observed with the dynamic distribution of auxin response. These results indicate the important role of auxin biosynthesis in shoot regeneration. The regeneration ability is evaluated in dominant gain-of-function, auxin overproducing *yuc1* mutants (*yuc1D*). These mutants regenerate roots in their cotyledon or hypocotyl explants under hormone-free *in vitro* culture conditions (Iwase et al., 2011; Zhao et al., 2001). In rice, increased auxin production caused by *OsYUCCA1* overexpression inhibits shoot regeneration from explants of crown roots, resulting in the regeneration of abundant hairy roots (Yamamoto, Kamiya, Morinaka, Matsuoka, & Sazuka, 2007). Therefore, although localized endogenous auxin biosynthesis is indispensable for shoot regeneration, overproduced endogenous auxin can inhibit shoot induction, similar to the effects of exogenous auxin treatment. Although root regeneration was inhibited from root explants of quadruple mutant *yuc1 yuc2 yuc4 yuc6* (Fig. 2.3G and H), *YUC4* overexpression driven by the 35S promoter enhanced *de novo* root formation (Fig. 2.3G and I), suggesting that endogenous auxin biosynthesis is critical for root regeneration, taking on a function similar to exogenous auxin treatment.

The induction of SEs in *Arabidopsis* also requires local *YUC* expression (Bai, Su, Yuan, & Zhang, 2013). Treatment with high levels of exogenous auxin (2,4-D) induces embryonic callus formation, whereas removal of 2,4-D from the medium stimulates SE initiation and enhances *YUC*-mediated endogenous auxin biosynthesis. Spatial expression patterns of *YUC4* and *YUC1* demonstrate that localized auxin biosynthesis occurs early in promeristem initiation sites along the edges of the embryonic callus, and later in regions of somatic proembryo formation (Bai et al., 2013). In addition, SE production is severely inhibited in the quadruple mutant *yuc1 yuc4 yuc10 yuc11*, suggesting an essential role for auxin biosynthesis in this process.

3.2. Cytokinin biosynthesis functions in plant regeneration

Cytokinin plays critical regulatory roles during cell proliferation, cell differentiation, and numerous other developmental processes *in vivo* (Mok & Mok, 2001). Endogenous cytokinin homeostasis is spatially and temporally regulated by the balance between synthesis and catabolism. Many studies have been conducted to isolate and characterize enzymes that function in plant cytokinin biosynthesis. In *Arabidopsis*, the first step of the cytokinin biosynthetic pathway is catalyzed by ATP/ADP isopentenyltransferases (AtIPTs) (Kakimoto, 2001; Takei, Sakakibara, & Sugiyama, 2001). The *atipt1 atipt3 atipt5 atipt7* quadruple mutant accordingly demonstrates severely reduced cytokinin levels and reduced shoot meristem size and flower numbers (Miyawaki et al., 2006; Werner et al., 2003).

Cytokinins play pivotal roles in *de novo* regeneration. Treatment with exogenous cytokinin induces cell proliferation and stimulates shoot induction from calli (Skoog & Miller, 1957). Expression of the *Agrobacterium ipt* gene to increase endogenous cytokinin levels in the callus can also induce cell division and initiate shoot formation (Ebinuma, Sugita, Matsunaga, & Yamakado, 1997; Kunkel, Niu, Chan, & Chua, 1999). Endogenous cytokinin biosynthesis mediated by *AtIPT* genes in *Arabidopsis* are analyzed during shoot regeneration (Cheng et al., 2013). *AtIPT3*, *AtIPT5*, and *AtIPT7* transcription is upregulated during shoot initiation, suggesting that cytokinin biosynthesis is enhanced during this process. In addition, *AtIPT5* exhibits spatiotemporal expression patterns during shoot induction. In mature callus grown on CIM, *AtIPT5* expression is detected at low levels around callus edges, while shoot induction on SIM induces strong but restricted *AtIPT5* signal distribution at future shoot initiation sites. Patterns of cytokinin biosynthesis appear to be similar to those of cytokinin response, indicating that localized *AtIPT*-mediated cytokinin biosynthesis contributes to the spatiotemporal distribution of cytokinin response for *de novo* shoot regeneration.

Genetic analysis reveals that both the *atipt5 atipt7* double mutant and the *atipt3 atipt5 atipt7* triple mutant show much less shoot regeneration than the wild type (Cheng et al., 2013). In contrast, *AtIPT4* overexpression causes shoot formation from callus incubated on medium containing auxin rather than cytokinin, although root formation is induced on wild-type callus in such a situation (Kakimoto, 2001). Similarly, the gain-of-function *AtIPT8* mutation results in *de novo* shoot formation from root-inducing callus on medium lacking cytokinin (Sun et al., 2003). *IPT* genes from maize (*ZmIPT*) have similar functions to those of *AtIPTs* in *Arabidopsis* (Brugière, Humbert, Rizzo, Bohn, & Habben, 2008). On medium containing only auxin, endogenous cytokinin overproduction mediated by *ZmIPT2*, *ZmIPT7*, or *ZmIPT8* overexpression induces shoot regeneration in *Arabidopsis* hypocotyl-derived calli, whereas calli transformed with the

35S::GUS construct regenerate roots. In the absence of exogenous cytokinin application, elevated endogenous cytokinin levels by overexpression of cytokinin biosynthetic gene can thus stimulate shoot regeneration from calli. In addition, treatment with exogenous cytokinin negatively regulates auxin-induced root induction from hypocotyl explants, which functions through endogenous cytokinin signaling (Pernisová et al., 2009). Furthermore, decreases in endogenous cytokinin attributed to the overexpression of cytokinin oxidase/dehydrogenase genes (*AtCKX2* and *AtCKX3*) enhance root regeneration competence (Pernisová et al., 2009). These results suggest that cytokinin response and endogenous cytokinin biosynthesis contribute to cytokinin-induced shoot induction *in vitro*.

3.3. Biosynthesis of endogenous ethylene and ABA in plant regeneration

Ethylene is an important hormone in many *in vitro* culture systems (Hatanaka et al., 1995; Mantiri et al., 2008; Meskaoui, Desjardins, & Tremblay, 2000). In plants, ethylene synthesis can be rapidly induced by various biotic and abiotic stresses, including explant excision during tissue culture processing (Biddington, 1992; Bleecker & Kende, 2000; Johnson & Ecker, 1998; Li & Guo, 2007). During SE initiation and development in leaf cultures of *C. canephora*, inhibition of ethylene production by $CoCl_2$ treatment prevents SE formation (Hatanaka et al., 1995). High levels of ethylene are produced in embryonic callus during *M. sativa* somatic embryogenesis (Kępczyńska, Ruduś, & Kępczyński, 2009). Analysis of *Medicago truncatula* somatic embryogenesis shows that ethylene biosynthesis is required for SE induction (Mantiri et al., 2008). Conversely, downregulation of ethylene biosynthesis is essential for SE initiation in *Arabidopsis* (Bai et al., 2013). Transcriptional analyses reveal that three genes (*ACS2*, *ACS6*, and *ACS8*) encoding 1-aminocyclopropane-1-carboxylate synthases—the enzymes in a rate-limiting step of ethylene biosynthesis—are downregulated by SE induction. Ethylene production, as detected by rate of ethylene release in embryonic calli, progressively decreases after SE induction, consistent with expression patterns of *ACS* genes. However, enhancement of endogenous ethylene biosynthesis in embryonic calli by adding the precursor 1-aminocyclopropane-1-carboxylic acid (ACC) inhibits SE *de novo* formation (Bai et al., 2013). Mutation of *ETHYLENE-OVERPRODUCTION1* (*ETO1*), a negative regulator of ethylene production (Guzmán & Ecker, 1990; Wang, Yoshida, Lurin, & Ecker, 2004), leads to similar suppression phenotypes. Therefore, endogenous ethylene biosynthesis is repressed following the removal of exogenous auxin during SE induction in *Arabidopsis*.

ABA is an important plant growth regulator mediating various physiological and developmental processes, such as zygotic embryo morphogenesis, storage protein synthesis, and desiccation tolerance (Finkelstein, Gampala, & Rock, 2002; Koornneef & Karssen, 1994; Nambara & Marion-Poll, 2005; Rock & Quatrano, 1995). Because endogenous ABA increases in response to various stress treatments, it is believed to play a role in plant regeneration under stress conditions (Fedina, Tsonev, & Guleva, 1994; Jimnez, Guevara, Herrera, & Bangerth, 2005; Saab, Sharp, & Pritchard, 1992), especially somatic embryogenesis (Karami, Aghavaisi, & Pour, 2009; Karami & Saidi, 2010). Application of exogenous ABA promotes SE formation when shoot apical tips of carrot are used as explants (Kikuchi, Sanuki, Higashi, Koshiba, & Kamada, 2006; Nishiwaki et al., 2000). In *Arabidopsis* somatic embryogenesis, treatment with fluridone, a potent inhibitor of *de novo* ABA synthesis, inhibits SE initiation (Su, Su, Liu, & Zhang, 2013). Mutation of the ABA biosynthetic gene *ABA2* reduces SE formation ability compared with wild type, suggesting that ABA biosynthesis is involved in SE induction.

4. HORMONAL SIGNALING IN *De novo* PLANT REGENERATION

Plant growth and development are controlled by both external environmental cues and intrinsic growth regulators such as hormones. Environmental cues target the biosynthesis and perception of endogenous hormones, conveying environmental inputs to developmental programs. Once synthesized, the endogenous hormone binds to the receptor protein, resulting in activation of a signal transduction pathway that ultimately leads to cell-type-specific responses. Many investigations have been conducted *in vivo* on hormonal perception and signaling mechanisms in plant development, including those involved in embryo development, stem-cell control of root and shoot meristems, vascular tissue differentiation, root and shoot growth and branching, and seed development (Müller & Sheen, 2007; Paciorek & Friml, 2006). These analyses have thoroughly examined the contribution of relevant processes, such as hormonal signal transduction and spatiotemporal regulation of hormonal response, to plant growth and patterning.

4.1. Auxin signaling in *de novo* plant regeneration

Over the past decades, auxin receptors and downstream signaling components have been identified (Fig. 2.5). Cellular response to auxin is mediated by receptors such as the F-box protein TRANSPORT INHIBITOR RESPONSE 1 (TIR1) and its homologs, AUXIN BINDING F-BOX PROTEINs (AFBs) (Dharmasiri, Dharmasiri, & Estelle, 2005; Kepinski & Leyser,

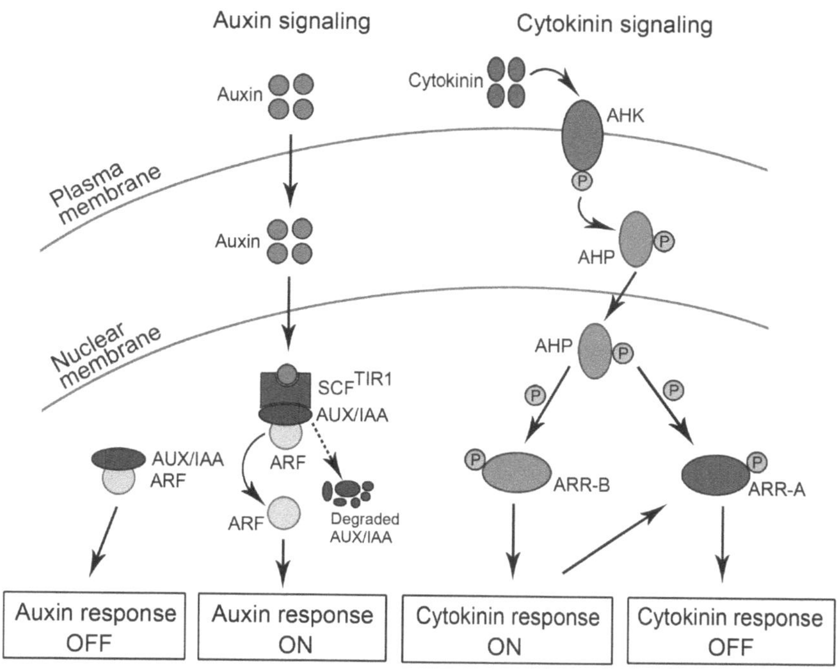

Figure 2.5 A model of auxin and cytokinin signaling. At low auxin concentrations in the cell, Aux/IAAs heterodimerize with ARF transcription factors to repress transcription of auxin-regulated genes (auxin response OFF). Auxin can flow across the plasma membrane. When auxin concentrations in the cell are high, auxin binds to the TIR1 receptor, stimulating the interaction of Aux/IAAs with the SCF^{TIR1} ubiquitin–ligase complex. This interaction promotes the degradation of Aux/IAAs, releasing ARFs to transcribe auxin-regulated genes (auxin response ON). Cytokinin is perceived by cytokinin receptors AHKs at the plasma membrane, activating a multistep phosphorelay. Cytokinin binding to AHKs activates their autophosphorylation, with a phosphate group (P) subsequently transferred to AHPs. AHPs can translocate into the nucleus to transfer the P to type-A or type-B ARRs (cytokinin primary response genes). Type-B ARRs act as transcription factors, and their phosphorylation activates transcription of cytokinin-regulated genes, including type-A ARRs (cytokinin response ON). Phosphorylated type-A ARRs negatively regulate cytokinin signaling (cytokinin response OFF). (For color version of this figure, the reader is referred to the online version of this chapter.)

2005). These receptors are integral components of the SCF^{TIR1} ubiquitination E3 complex involved in proteasome-mediated protein degradation of the AUXIN/INDOLE ACETIC ACID (AUX/IAA) family (Benjamins & Scheres, 2008; Paciorek & Friml, 2006; Vanneste & Friml, 2009). Aux/IAA degradation is a key event in auxin signaling (Ulmasov, Murfett, Hagen, & Guilfoyle, 1997), as it releases activating AUXIN RESPONSE FACTOR (ARF) proteins, a class of transcription factors that mediate auxin-dependent transcriptional regulation (Paciorek & Friml, 2006; Ulmasov, Murfett, et al., 1997).

Global analysis of gene expression events during *in vitro* shoot regeneration in *Arabidopsis* reveals a role for auxin signaling during this process (Che, Gingerich, Lall, & Howell, 2002). Many Aux/IAA genes, such as *IAA1*, *IAA5*, *IAA9*, and *IAA11*, are upregulated during preincubation on CIM and downregulated during shoot induction on SIM. However, expression levels of some Aux/IAA genes, such as *IAA17*, increase dramatically during early incubation on SIM and then decrease rapidly, suggesting different functions for these genes. Effects of the auxin receptor *TIR1* on *Arabidopsis* root-induced shoot regeneration have recently been investigated (Qiao, Zhao, & Xiang, 2012). *TIR1* expression is upregulated in callus upon transfer to SIM after preincubation on CIM. During CIM incubation, *TIR1* transcriptional signals are detected throughout the entire callus. After shoot induction on SIM, signals are enhanced in proliferated callus, and then concentrated in shoot initiation sites. Shoot regeneration efficiency is reduced by mutations of *TIR1* but significantly enhanced by its overexpression, suggesting that *TIR1* positively regulates shoot regeneration (Qiao et al., 2012). *tir1-1* mutants also lose the ability to undergo somatic embryogenesis, which requires auxin as the sole hormone in embryonic callus induction (Su et al., 2009). We further explored the functions of *TIR1* and its homologs during root regeneration, demonstrating that root explants of the *tir1 afb1 afb2 afb3* quadruple mutant can neither induce callus formation nor stimulate *de novo* root formation (Fig. 2.3J).

In pistil-induced shoot regeneration, *ARF3* is upregulated by SIM incubation, consistent with the results of a transcriptomic screening using roots as explants (Che et al., 2006; Cheng et al., 2013). *ARF3* transcription signals are evenly distributed at the edges of calli on CIM, whereas SIM incubation spatially restricts *ARF3* expression, similar to the effects of DR5 auxin-responsive signals (Cheng et al., 2013). In addition, *ARF3* mutations significantly reduce shoot regeneration, indicating that this gene mediates auxin response during *in vitro* shoot induction (Cheng et al., 2013). During somatic embryogenesis, we observed that two other *ARF* genes, *ARF6* and *ARF8*, function redundantly in SE induction. *arf6-2*, *arf8-3*, and *arf6-2 arf8-3/+* mutants produced substantially fewer SEs than wild type (Fig. 2.4J and K), and SE regeneration frequency was lower in the *arf6-2 arf8-3/+* mutant than in *arf6-2* and *arf8-3* single mutants. These results indicate that *ARF6* and *ARF8* mediate auxin-induced gene activation during SE induction. Cellular response to auxin mediated by receptors and auxin-responsive factors is thus required in *de novo* regeneration.

4.2. Cytokinin signaling in *de novo* plant regeneration

Components of the cytokinin signal transduction pathway have been identified during the past few years (Fig. 2.5), including sensor histidine kinases (AHKs) (Hwang & Sheen, 2001; Inoue et al., 2001; Riefler, Novak, Strnad, & Schmülling, 2006; To & Kieber, 2008), histidine phosphotransmitters (AHPs) and response regulators (ARRs) (Ferreira & Kieber, 2005; Heyl & Schmülling, 2003; Kakimoto, 2003). Cytokinin receptor AHKs are autophosphorylated during initial cytokinin perception, and then transfer phosphate groups to AHPs. AHPs subsequently translocate to the nucleus where they phosphorylate type-A or type-B ARRs (Ferreira & Kieber, 2005; Heyl & Schmülling, 2003; Kakimoto, 2003). Genetic studies have demonstrated that type-B ARRs positively regulate the expression of cytokinin-induced genes (Mason et al., 2005; Yokoyama et al., 2007), whereas type-A ARRs repress the cytokinin signaling pathway (To et al., 2004; To & Kieber, 2008).

As cytokinin plays a major role in directing plant shoot development, its effects on *de novo* shoot regeneration should be most apparent after induction on cytokinin-rich SIM. Indeed, previous results have shown that cytokinin signaling is critical for *de novo* stem-cell initiation and SAM establishment in *Arabidopsis* (Che et al., 2002; Cheng, Zhu, Gao, & Zhang, 2010; Gordon et al., 2007). Expression profiles of shoot regeneration from root explants show significant changes in expression of genes involved in cytokinin signaling pathway (Che et al., 2002). For example, the cytokinin receptor *AHK4/CRE1* is rapidly induced after transfer onto SIM. Similar results are observed with the hybrid His kinase-encoding gene *CYTOKININ-INDEPENDENT1* (*CKI1*) implicated in cytokinin responses. Although the function of *CKI1* in cytokinin signaling remains unknown, its overexpression stimulates *in vitro* shoot formation independently of cytokinin, both in calli derived from hypocotyl segments of seedling (Kakimoto, 1996) and in proliferating tissues derived from SAM of the seedlings

(Hwang & Sheen, 2001). Calli from hypocotyls of the *ahk3-1* and *ahk4-1* single mutants and the *ahk2-1 ahk3-1* double mutant exhibit a cytokinin-insensitive phenotype, showing weak stimulation of cell proliferation and shoot induction even at high exogenous cytokinin concentrations (Nishimura et al., 2004). These results confirm that *AHK* genes function as positive signal transduction molecules in the cytokinin signaling pathway during shoot regeneration.

Expression of two cytokinin primary type-A ARR genes, *ARR4* and *ARR5*, is also highly increased during *de novo* shoot induction (Che et al., 2002), although their upregulation may simply be a response to exogenous cytokinin upon callus transfer to the high cytokinin-containing SIM medium. *ARR5* exhibits spatial expression patterns during callus formation and shoot induction (Che et al., 2002; Gordon et al., 2007), suggesting a role for cytokinin signaling during *de novo* shoot initiation. *In vitro* shoot formation has also been studied using root explants with variously altered A-type *ARR* expressions. Overexpression of *ARR7* and *ARR15* severely suppresses shoot regeneration (Buechel et al., 2009). The *arr7* and *arr15* single mutants strongly promote cell proliferation during callus development and shoot formation; this effect is further enhanced in *arr3,4,5,6,7,8,9* septuple mutants. These results suggest that A-type *ARRs*, which are negative regulators of cytokinin signaling, may function as suppressors of shoot regeneration. Interestingly, type-A ARRs *ARR4* and *ARR8* exhibit opposite ectopic expression effects during shoot regeneration from root tissues (Osakabe et al., 2002). *ARR4* functions as a positive regulator of *in vitro* shoot formation, whereas *ARR8* is a negative regulator, suggesting different response roles to cytokinin signal transduction under tissue culture conditions. Although expression profile analysis demonstrates that most type-B *ARRs* are not induced during *de novo* shoot formation (Che et al., 2002), overexpression of type-B ARR *ARR2* promotes cell proliferation and shoot regeneration from SAM of seedlings in the absence of exogenous cytokinin (Hwang & Sheen, 2001). In contrast, the *arr1 arr10 arr12* triple mutant exhibits reduced callus formation and enhanced root induction compared with wild-type plants, even at high exogenous cytokinin concentrations, when hypocotyl segments are used as explants (Mason et al., 2005). Increased root regeneration is also detected from hypocotyls of single or double *AHK* mutants compared with wild type, revealing a negative role for cytokinin signaling in the modulation of *de novo* root formation (Pernisová et al., 2009).

We have further detected a role for cytokinin signaling in SE induction. As shown in Fig. 2.4D–I, embryonic calli of *ARR7*- and *ARR15*-overexpressing plants and *ahk2 ahk4* and *ahk3 ahk4* double mutants produce abnormal SEs with very few hypocotyls or radicles, similar to phenotypes of the RAM-deficient mutant *plt1-1 plt2-1*. Therefore, cytokinin signaling regulates correct pattern formation for embryonic root meristem initiation.

4.3. Conclusions regarding hormonal signaling in plant regeneration

Shoot regeneration usually requires incubation on cytokinin-rich SIM. It is thus reasonable to suggest that exogenous cytokinin mainly functions in regulating shoot induction. Consistent with this proposal, cytokinin signaling, which is responsible for exogenous cytokinin signal input, plays a positive role in shoot regeneration. Auxin signaling components, however, such as the auxin receptor *TIR1* and response gene *ARF3*, are also required for *de novo* shoot formation (Cheng et al., 2013; Qiao et al., 2012). This suggests that the effects of exogenous cytokinin on shoot induction are mediated not only by cytokinin signaling but at least partially by endogenous auxin signaling as well. Root regeneration requires exogenous auxin in RIM as well as preincubation in auxin-rich CIM. Suppression of auxin signaling consequently results in failure of both callus formation and root induction (Fig. 2.3J). In contrast, explants carrying mutations of cytokinin signaling genes also induce root formation, even in the presence of high exogenous cytokinin concentrations (Mason et al., 2005). This indicates that the balance between endogenous auxin and cytokinin signaling is critical to organ regeneration. In somatic embryogenesis, which requires high levels of exogenous auxin, auxin signaling is essential for SE formation. Interestingly, we determined that cytokinin signaling has an important role in initiation of SE root meristem. In addition, constitutive ethylene signaling caused by the mutation of *CONSTITUTIVE TRIPLE RESPONSE 1* (*CTR1*), a negative regulator of ethylene response (Guo & Ecker, 2004; Ju et al., 2012; Kieber, Rothenberg, Roman, Feldmann, & Ecker, 1993; Zhao & Guo, 2011), arrests SE initiation (Bai et al., 2013). This indicates that ethylene signaling has negative effects on SE induction.

5. HORMONE INTERACTIONS DURING PLANT REGENERATION

Previous studies have revealed the functions of individual hormones in different developmental processes. In recent years, however, it has become apparent that plant hormones rarely act alone; plant developmental output is instead regulated by a complex network of interlocking hormonal signaling pathways. Many reviews have summarized the molecular basis of hormonal interactions and their regulatory networks in developmental processes such as root and shoot meristem development (Su et al., 2011; Vanstraelen & Benková, 2012), shoot branching (Shimizu-Sato, Tanaka, & Mori, 2009), lateral root formation (Fukaki & Tasaka, 2009), seed germination (Vanstraelen & Benková, 2012), and vascular differentiation (Dettmer, Elo, & Helariutta, 2009).

In addition to these *in vivo* plant developmental processes, *de novo* organogenesis also requires the regulation of plant hormones. Pioneering work has shown that the exogenous hormone balance used in culture conditions determines the types of organs regenerated (Skoog & Miller, 1957). Furthermore, organ regeneration induced by an exogenous hormone also requires another hormone signaling type. For example, auxin signaling plays important roles in cytokinin-induced shoot regeneration, suggesting hormone cross talk during *de novo* organogenesis. Several synergistic or antagonistic interactions between various plant hormones have currently been identified during plant regeneration, but the molecular mechanisms underlying these interactions are largely unknown.

5.1. Interactions of auxin and cytokinin during plant regeneration

A high auxin/cytokinin ratio induces root regeneration, whereas a low ratio promotes shoot induction (Skoog & Miller, 1957). Auxin and cytokinin thus appear to be the first key phytohormones recognized to interact in regulation of organ regeneration. During shoot induction from *Arabidopsis* root tissue, incubation on auxin-rich CIM leads to upregulation of the cytokinin receptor gene *AHK4*, which is required for *WUS* induction during SAM initiation on SIM (Gordon et al., 2009). On modified CIM containing auxin as the sole hormone, calli can also be induced by upregulated expression of the cytokinin-responsive gene *ARR5* in proliferated cells (Gordon et al., 2007). These results suggest that auxin pretreatment on CIM enhances cytokinin signaling during callus formation, which is essential for later shoot induction following SIM incubation. Cytokinin also regulates auxin response during shoot formation on cytokinin-rich SIM. Treatment with exogenous cytokinin leads to auxin-responsive signals primarily in the surrounding regions of SAM initiation in the callus, while *PIN1* is upregulated at sites of SAM initiation (Gordon et al., 2007). Direct interaction between auxin and cytokinin during shoot regeneration has recently been revealed using pistils as explants (Cheng et al., 2013). Shoot meristem initiation requires spatially restricted distributions of both auxin and cytokinin in callus. Cytokinin response takes place in the center region of the meristem, overlapping *WUS* expression, while auxin response is restricted to the region surrounding the location of cytokinin-response signal expression (Cheng et al., 2013). Therefore, a mutually exclusive distribution of auxin and cytokinin responses exists in the SAM initiation region. Application of the auxin transport inhibitor NPA disrupts the restricted distribution of the cytokinin response, indicating a role for interaction of auxin and cytokinin response in SAM initiation. Direct evidence shows that ARF3, an auxin-response mediator, negatively regulates expression of the cytokinin biosynthetic gene *AtIPT5* by directly binding to its promoter. This suggests that auxin modulates cytokinin-induced *de novo* shoot regeneration through the direct control of localized cytokinin biosynthesis. On the other hand, cytokinin influences auxin-induced root regeneration via regulation of auxin efflux-mediated auxin polar transport (Pernisová et al., 2009). Exogenous cytokinin treatment affects the restricted distribution patterns of auxin response in regenerated root primordium, which resembles the effects of NPA treatment. While endogenous cytokinins are required for maintaining expression of PIN auxin efflux carriers in root tips, which regulates the formation of local auxin-response maxima and root meristem development (Pernisová et al., 2009). Other hormones also regulate plant regeneration, but their effects are generally attributed to auxin or cytokinin because of a lack of information (Gazzarini & McCourt, 2001; Pullman, Mein, Johnson, & Zhang, 2005).

5.2. Interactions of auxin and ethylene during plant regeneration

The role of ethylene has been examined in various organ regeneration processes. Ethylene inhibits auxin-induced root regeneration from cultured tomato leaf discs (Coleman, Huxter, Reid, & Thorpe, 1980), while the suppression of endogenous ethylene activity significantly stimulates auxin-induced *de novo* root initiation in explants. On the other hand, ethylene production is positively regulated by increased exogenous auxin concentrations during root induction (Coleman et al., 1980). Based on these observations, we suggest that ethylene is involved in auxin function during root regeneration. Although the associated molecular mechanisms are not well understood, ethylene has important roles in SE initiation and development (Hatanaka et al., 1995; Mantiri et al., 2008; Meskaoui et al., 2000). In a more recent investigation, ethylene has been found to interact with auxin in *Arabidopsis* somatic embryogenesis (Bai et al., 2013). In that study, excessive ethylene produced by ACC treatment or by the *ETO1* mutation negatively regulates SE initiation. Local auxin biosynthesis mediated by *YUC1* and *YUC4* expression is disrupted in ACC-treated embryonic calli and the calli of the *eto1-1* mutant. Another finding is that the expression patterns of *YUC* genes are disturbed in *CTR1*-mutated calli with constitutive ethylene signaling (Bai et al., 2013). Therefore, constitutive ethylene biosynthesis and responses inhibit SE induction by interfering with local auxin biosynthesis and subsequent auxin responses. On the other hand, auxin is involved in endogenous ethylene biosynthesis (Abel, Nguyen, Chow, & Theologis, 1995; Aharoni & Yang, 1983; Eklund & Little, 1994; Ohmiya & Haji, 2002). Exogenous auxin stimulates expression of ethylene biosynthetic genes *ACSs* in many plant tissues (Abel et al., 1995; Abeles, Morgan, & Saltveit, 1992; Che et al., 2006; Tsuchisaka & Theologis, 2004). Its removal for SE initiation in *Arabidopsis* downregulates endogenous ethylene biosynthesis and responses, which is required for local auxin biosynthesis (Bai et al., 2013). *Arabidopsis* SE initiation therefore requires mutual regulation between auxin and ethylene.

5.3. Interactions of ABA and auxin during plant regeneration

The role of ABA during regeneration has not been extensively studied, but several studies have shown that ABA prevents callus induction in various plants (Fazeli-nasab, Omidi, & Amiritokaldani, 2012; Kovalenco & Kurchii, 1998; Nadina, Martinez, Castillo, & Gonzalez, 2001). Combined effects of exogenous ABA and hormones such as auxin and cytokinin have also been investigated (Ella & Zapata, 1991; Fernando & Gamage, 2000; Ghanati & Rahmati Ishka, 2009; Maggon & Deo Singh, 1995). In rice, exogenous ABA inhibits shoot regeneration in the presence of both exogenous auxin and cytokinin, but has almost no effect in the absence of exogenous auxin (Xing, Huang, Shiragami, & Unno, 1995). These results imply that interactions exist between ABA and auxin during shoot regeneration. The molecular mechanisms behind these processes have rarely been studied. ABA acts as an inducer in somatic embryogenesis of many plant species, including carrot and coconut (Fernando & Gamage, 2000; Kikuchi et al., 2006; Nadina et al., 2001; Nishiwaki et al., 2000). Recent studies demonstrate that endogenous ABA biosynthesis is required for SE initiation in *Arabidopsis*, and that ABA functions are correlated with auxin activity

(Su et al., 2013). Inhibition of endogenous ABA biosynthesis suppresses localized expression of *YUC* genes and polar localization of PIN1. ABA may mediate both auxin biosynthesis and polar transport to establish the auxin-response pattern required for SE induction. These investigations thus shed light on interactions of ABA and auxin in *Arabidopsis* SE initiation.

6. CONCLUDING REMARKS AND PERSPECTIVES

Although it is widely believed that any living plant cell can maintain totipotency, the mechanisms enabling such high plasticity remain to be investigated. Recent studies have focused on hormone-regulated developmental fates of regenerating tissue cells. In the commonly used *Arabidopsis* regeneration systems, hormonal-response patterns in callus are established (Fig. 2.2). During initial shoot regeneration, auxin response is distributed in the areas surrounding *WUS* expression, whereas cytokinin-responsive signals are localized in the central region corresponding to *WUS* expression (Fig. 2.2A). Both local auxin biosynthesis and transport contribute to auxin response in the surrounding regions, and this local auxin response defines the spatiotemporal distribution of cytokinin response through ARF3-regulated suppression of *AtIPT* (Fig. 2.6). Cytokinin-initiated *WUS* expression gives rise to stem cells, and subsequently the SAM. During root regeneration, auxin is the major factor initiating RAM formation, as shown in Fig. 2.2B. Auxin response, rather than that of cytokinin, is located in the expression domain of *WOX5* within the callus. Of particular interest is the role of auxin and cytokinin responses in SE induction. Auxin response appears to induce embryonic SAM initiation, whereas cytokinin response is involved in *de novo* formation of embryonic RAM, implying a distinct role for these two hormones in the axis establishment of SEs (Fig. 2.2C).

In animal systems, morphological patterning is dependent on morphogens, which are formative substances secreted by source cells. Spatial patterns of gene expression occur along resulting morphogen concentration gradients in target tissues (Gurdon & Bourillot, 2001; Tabata & Takei, 2004). Although criteria-defining morphogens are different in animals than in plants, auxin may be considered as a morphogen *in planta* due to the asymmetric distribution of its response during plant cell fate determination (Dubrovsky et al., 2008). Local hormonal responses during plant regeneration might accordingly serve as instructive factors for cell specification, similar to the function of morphogen gradients in animal organogenesis.

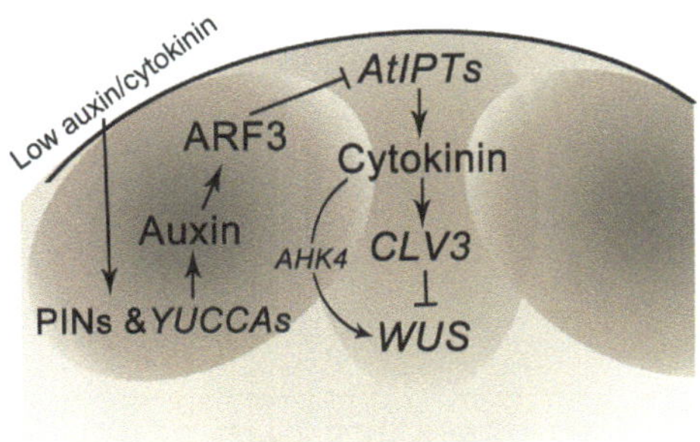

Figure 2.6 Interactions of auxin and cytokinin contribute to form a specific hormone-responsive pattern during SAM initiation. Local auxin biosynthesis and transport mediated by YUCs and PINs regulate the local auxin response in regions surrounding SAM, which negatively regulates expression of cytokinin biosynthetic genes *IPTs* through the direct binding of ARF3 to their promoters. In the center region of SAM, cytokinin signaling may partially regulate *WUS* expression through a CLV-dependent pathway as shown *in planta* (Gordon et al., 2009). Cytokinin-induced increase of *WUS* transcript levels is mediated primarily through an *AHK4*-dependent pathway. Red color indicates auxin-response regions, blue color indicates cytokinin-response regions, and yellow color indicates callus. (For interpretation of the references to color in this figure legend, the reader is referred to the online version of this chapter.)

Although the fundamental model for hormone-regulated *de novo* organogenesis under culture conditions has been outlined, some interesting questions remain to be investigated. These questions include:

(i) How do exogenous hormones control *de novo* formation of various types of organs or SEs? Future work will focus on mechanisms of exogenously supplied hormones regulating endogenous hormone biosynthesis and response during *de novo* regeneration.

(ii) During shoot regeneration, how does cytokinin signaling induce *WUS* expression in the central region of *de novo*-initiated SAM? Is there a feedback regulation loop between cytokinin signaling and *WUS* expression, similar to that seen during shoot development *in planta*?

(iii) Even at high exogenous cytokinin concentrations, enhanced root regeneration has been detected in multiple mutations of type-B *ARRs* or *AHKs* (Mason et al., 2005; Pernisová et al., 2009). These results demonstrate that absence of cytokinin signaling in calli inhibits *WUS* expression but induces *WOX5* expression. What are the mechanisms of cytokinin action on *WOX5* expression during RAM establishment in root regeneration? Are there auxin–cytokinin interactions during this process?

(iv) Epigenetic modification of hormone-regulated organ regeneration, a completely new field, may determine the cellular origin of new organs. Progress in this area should elucidate the mechanisms of specific cell formation and hormonal spatiotemporal response during *in vitro* organogenesis.

In the future, molecular and genetic approaches will be employed to analyze gene regulatory mechanisms involved in cellular origin of regenerated organs or SEs under hormonal regulation. In addition, high-resolution data sets from live imaging and histological analysis will be used to validate the cytological basis of specific cells in organ regeneration under regulation of distinct hormonal response. The principles revealed by such approaches may be critical to an understanding of hormone-regulated plant regeneration processes, and should assist in the study of *in vivo* plant development.

ACKNOWLEDGMENTS

We are grateful to all members for their assistance in our laboratory. We also thank Yu Bo Liu and Jia Yuan for their support in the figures of this manuscript. This research was supported by grants from the National Natural Science Foundation of China (90917015, 91217308, 31000652, and 31170272).

REFERENCES

Abel, S., Nguyen, M. D., Chow, W., & Theologis, A. (1995). ACS4, a primary indoleacetic acid-responsive gene encoding 1-aminocyclopropane-1-carboxylate synthase in Arabidopsis thaliana. Structural characterization, expression in Escherichia coli, and expression characteristics in response to auxin. *The Journal of Biological Chemistry*, *270*, 19093–19099.

Abeles, F. B., Morgan, P. W., & Saltveit, J. M. E. (1992). *Ethylene in plant biology*. San Diego: Academic Press.

Aharoni, N., & Yang, S. F. (1983). Auxin-induced ethylene production as related to auxin metabolism in leaf discs of tobacco and sugar beet. *Plant Physiology*, *73*, 598–604.

Aloni, R., Langhans, M., Aloni, E., & Ullrich, C. I. (2004). Role of cytokinin in the regulation of root gravitropism. *Planta*, *220*, 177–182.

An, Y. R., Li, X. G., Su, H. Y., & Zhang, X. S. (2004). Pistil induction by hormones from callus of Oryza sativa in vitro. *Plant Cell Reports*, *23*, 448–452.

Atta, R., Laurens, L., Boucheron-Dubuisson, E., Guivarc'h, A., Carnero, E., Giraudat-Pautot, V., et al. (2009). Pluripotency of *Arabidopsis* xylem pericycle underlies shoot regeneration from root and hypocotyl explants grown *in vitro*. *The Plant Journal*, *57*, 626–644.

Auer, C. A., Cohen, J. D., Laloue, M., & Cooke, T. J. (1992). Comparison of benzyladenine metabolism in two Petunia lines differing in shoot organogenesis. *Plant Physiology*, *98*, 1035–1041.

Bai, B., Su, Y. H., Yuan, J., & Zhang, X. S. (2013). Induction of somatic embryos in Arabidopsis requires local YUCCA expression mediated by the down-regulation of ethylene biosynthesis. *Molecular Plant*, *6*, 1247–1260. http://dx.doi.org/10.1093/mp/sss154.

Benjamins, R., & Scheres, B. (2008). Auxin: The looping star in plant development. *Annual Review of Plant Biology, 59*, 443–465.

Benková, E., Michniewicz, M., Sauer, M., Teichmann, T., Seifertova, D., Jurgens, G., et al. (2003). Local, efflux-dependent auxin gradients as a common module for plant organ formation. *Cell, 115*, 591–602.

Biddington, N. L. (1992). The influence of ethylene in plant tissue culture. *Plant Growth Regulation, 11*, 173–187.

Birnbaum, K. D., & Sánchez Alvarado, A. (2008). Slicing across kingdoms: Regeneration in plants and animals. *Cell, 132*, 697–710.

Bleecker, A. B., & Kende, H. (2000). Ethylene: A gaseous signal molecule in plants. *Annual Review of Cell and Developmental Biology, 16*, 1–18.

Brugière, N., Humbert, S., Rizzo, N., Bohn, J., & Habben, J. E. (2008). A member of the maize isopentenyl transferase gene family, Zea mays isopentenyl transferase 2 (ZmIPT2), encodes a cytokinin biosynthetic enzyme expressed during kernel development. Cytokinin biosynthesis in maize. *Plant Molecular Biology, 67*, 215–229.

Buechel, S., Leibfried, A., To, J. P., Zhao, Z., Andersen, S. U., Kieber, J. J., et al. (2009). Role of A-type *ARABIDOPSIS RESPONSE REGULATORS* in meristem maintenance and regeneration. *European Journal of Cell Biology, 89*, 279–284.

Cary, A., Uttamchandani, S. J., Smets, R., Van Onckelen, H. A., & Howell, S. H. H. (2001). Arabidopsis mutants with increased organ regeneration in tissue culture are more competent to respond to hormonal signals. *Planta, 213*, 700–707.

Casimiro, I., Marchant, A., Bhalerao, R. P., Beeckman, T., Dhooge, S., Swarup, R., et al. (2001). Auxin transport promotes Arabidopsis lateral root initiation. *Plant Cell, 13*, 843–852.

Centeno, M. L., Rodríguez, A., Feito, I., & Fernández, B. (1996). Relationship between endogenous auxin and cytokinin levels and morphogenic responses in Actinidia deliciosa tissue cultures. *Plant Cell Reports, 16*, 58–62.

Chandler, J. W., Cole, M., Flier, A., & Werr, W. (2009). BIM1, a bHLH protein involved in brassinosteroid signalling, controls Arabidopsis embryonic patterning via interaction with DORNROESCHEN and DORNROESCHEN-LIKE. *Plant Molecular Biology, 69*, 57–68.

Che, P., Gingerich, D. J., Lall, S., & Howell, S. H. (2002). Global and hormone-induced gene expression changes during shoot development in Arabidopsis. *Plant Cell, 14*, 2771–2785.

Che, P., Lall, S., Nettleton, D., & Howell, S. H. (2006). Gene expression programs during shoot, root, and callus development in Arabidopsis tissue culture. *Plant Physiology, 141*, 620–637.

Cheng, Y. F., Dai, X. H., & Zhao, Y. D. (2006). Auxin biosynthesis by the YUCCA flavin monooxygenases controls the formation of floral organs and vascular tissues in Arabidopsis. *Genes & Development, 20*, 1790–1799.

Cheng, Y. F., Dai, X. H., & Zhao, Y. D. (2007). Auxin synthesized by the YUCCA flavin monooxygenases is essential for embryogenesis and leaf formation in Arabidopsis. *Plant Cell, 19*, 2430–2439.

Cheng, Z. J., Wang, L., Sun, W., Zhang, Y., Zhou, C., Su, Y. H., et al. (2013). Pattern of auxin and cytokinin responses for shoot meristem induction results from the regulation of cytokinin biosynthesis by AUXIN RESPONSE FACTOR3. *Plant Physiology, 161*, 240–251.

Cheng, Z. J., Zhu, S. S., Gao, X. Q., & Zhang, X. S. (2010). Cytokinin and auxin regulates WUS induction and inflorescence regeneration in vitro in Arabidopsis. *Plant Cell Reports, 29*, 927–933.

Coleman, W. K., Huxter, T. J., Reid, D. M., & Thorpe, T. A. (1980). Ethylene as an endogenous inhibitor of root regeneration in tomato leaf discs cultured in vitro. *Physiologia Plantarum, 48*, 519–525.

Dettmer, J., Elo, A., & Helariutta, Y. (2009). Hormone interactions during vascular development. *Plant Molecular Biology, 69*, 347–360.

Dharmasiri, N., Dharmasiri, S., & Estelle, M. (2005). The F-box protein TIR1 is an auxin receptor. *Nature, 435*, 441–445.

Dubrovsky, J. G., Sauer, M., Napsucialy-Mendivil, S., Ivanchenko, M. G., Friml, J., Shishkova, S., et al. (2008). Auxin acts as a local morphogenetic trigger to specify lateral root founder cells. *Proceedings of the National Academy of Sciences of the United States of America, 105*, 8790–8794.

Ebinuma, H., Sugita, K., Matsunaga, E., & Yamakado, M. (1997). Selection of marker-free transgenic plants using the isopentenyl transferase gene. *Proceedings of the National Academy of Sciences of the United States of America, 94*, 2117–2121.

Eklund, L., & Little, C. H. A. (1994). Interaction between indole-3-acetic acid and ethylene in the control of tracheid production in detached shoots of Abies balsamea. *Tree Physiology, 15*, 27–34.

Ella, E. S., & Zapata, F. J. (1991). Effect of abscisic acid and zeatin on plant regeneration from Scutellum-derived callus of rice. *Philippine Journal of Crop Science, 16*, 3–6.

Fazeli-nasab, B., Omidi, M., & Amiritokaldani, M. (2012). callus induction and plant regeneration of wheat mature embryos under abscisic acid treatment. *International Journal of Agriculture and Crop Sciences, 4*, 17–23.

Fedina, I. S., Tsonev, T. D., & Guleva, E. I. (1994). ABA as modulator of the response of Pisum sativum to salt stress. *Journal of Plant Physiology, 143*, 245–249.

Feher, A., Pasternak, T., Otvos, K., Miskolczi, P., & Dudits, D. (2002). Induction of embryogenic competence in somatic plant cells: A review. *Biologia, 57*, 5–12.

Feldmann, K. A., & Marks, M. D. (1986). Rapid and efficient regeneration of plants from explants of Arabidopsis thaliana. *Plant Science, 47*, 63–69.

Fernando, S. C., & Gamage, C. K. (2000). Abscisic acid induced somatic embryogenesis in immature embryo explants of coconut (Cocos nucifera L.). *Plant Science, 151*, 193–198.

Ferreira, F. J., & Kieber, J. J. (2005). Cytokinin signalling. *Current Opinion in Plant Biology, 8*, 518–525.

Finkelstein, R. R., Gampala, S. S., & Rock, C. D. (2002). Abscisic acid signaling in seeds and seedlings. *Plant Cell, 14*, S15–S45.

Friml, J. (2010). Subcellular trafficking of PIN auxin efflux carriers in auxin transport. *European Journal of Cell Biology, 89*, 231–235.

Friml, J., Vieten, A., Sauer, M., Weijers, D., Schwarz, H., Hamann, T., et al. (2003). Efflux-dependent auxin gradients establish the apical-basal axis of Arabidopsis. *Nature, 426*, 147–153.

Friml, J., Wisniewska, J., Benkova, E., Mendgen, K., & Palme, K. (2002). Lateral relocation of auxin efflux regulator PIN3 mediates tropism in Arabidopsis. *Nature, 415*, 806–809.

Fukaki, H., & Tasaka, M. (2009). Hormone interactions during lateral root formation. *Plant Molecular Biology, 69*, 437–449.

Gautheret, R. J. (1985). History of plant tissue and cell culture: A personal account. In: Vasil, I. K. (Ed.), *Cell Culture and Somatic Cell Genetics of Plants* (Vol. 2, pp. 1–59). New York: Academic Press.

Gautheret, R. J. (2003). *Plant tissue culture: The history. Plant Tissue Culture: 100 Years since Gottlieb Haberlandt.* New York: SpringWien.

Gazzarini, S., & McCourt, P. (2001). Genetic interaction between ABA, ethylene and sugarcane signaling pathways. *Current Opinions in Plant Biology, 4*, 387–391.

Ghanati, F., & Rahmati Ishka, M. (2009). Investigation of the interaction between abscisic acid (ABA) and excess benzyladenine (BA) on the formation of shoot in tissue culture of tea (*Camellia sinensis* L.). *International Journal of Plant Production, 3*, 7–14.

Gordon, S. P., Chickarmane, V. S., Ohno, C., & Meyerowitz, E. M. (2009). Multiple feedback loops through cytokinin signaling control stem cell number within the Arabidopsis shoot meristem. *Proceedings of the National Academy of Sciences of the United States of America, 106*, 16529–16534.

Gordon, S. P., Heisler, M. G., Reddy, G. V., Ohno, C., Das, P., & Meyerowitz, E. M. (2007). Pattern formation during de novo assembly of the Arabidopsis shoot meristem. *Development, 134*, 3539–3548.

Guan, C. M., Zhu, S. S., Li, X. G., & Zhang, X. S. (2006). Hormone-regulated inflorescence induction and TFL1 expression in Arabidopsis callus in vitro. *Plant Cell Reports, 25*, 1133–1137.

Guo, H., & Ecker, J. R. (2004). The ethylene signaling pathway: New insights. *Current Opinion in Plant Biology, 7*, 40–49.

Gurdon, J. B., & Bourillot, P. Y. (2001). Morphogen gradient interpretation. *Nature, 413*, 797–803.

Guzmán, P., & Ecker, J. R. (1990). Exploiting the triple response of Arabídopsís to identify ethylene-related mutants. *Plant Cell, 2*, 513–523.

Haberlandt, G. (1902). Culturversuche mit isolierten pflanzenzellen. *Sitzungsber. Akademie der Wissenschaften Wien. Mathematisch Naturwissenschaftliche Klasse, 111*, 69–92.

Hardtke, C. S., Dorcey, E., Osmont, K. S., & Sibout, R. (2007). Phytohormone collaboration: Zooming in on auxin-brassinosteroid interactions. *Trends in Cell Biology, 17*, 485–492.

Hatanaka, T., Sawabe, E., Azuma, T., Uchida, N., & Yasuda, T. (1995). The role of ethylene in somatic embryogenesis from leaf disks of *Coffea canephora*. *Plant Science, 107*, 199–204.

Heisler, M. G., Ohno, C., Das, P., Sieber, P., Reddy, G. V., Long, J. A., et al. (2005). Patterns of auxin transport and gene expression during primordium development revealed by live imaging of the Arabidopsis inflorescence meristem. *Current Biology, 15*, 1899–1911.

Heyl, A., & Schmülling, T. (2003). Cytokinin signal perception and transduction. *Current Opinion in Plant Biology, 6*, 480–488.

Hicks, G. S. (1980). Patterns of organ development in plant tissue culture and the problem of organ determination. *The Botanical Review, 46*, 1–23.

Hicks, G. S., & McHughen, A. (1974). Altered morphogenesis of placental tissues of tobacco in vitro. Stigmatoid and carpelloid outgrowths. *Planta, 121*, 193–196.

Hwang, I., & Sheen, J. (2001). Two-component circuitry in Arabidopsis cytokinin signal transduction. *Nature, 413*, 383–389.

Inoue, T., Higuchi, M., Hashimoto, Y., Seki, M., Kobayashi, M., Kato, T., et al. (2001). Identification of CRE1 as a cytokinin receptor from *Arabidopsis*. *Nature, 409*, 1060–1063.

Iwase, A., Mitsuda, N., Koyama, T., Hiratsu, K., Kojima, M., Arai, T., et al. (2011). The AP2/ERF transcription factor WIND1 controls cell dedifferentiation in Arabidopsis. *Current Biology, 21*, 508–514.

Jimnez, V. M., Guevara, E., Herrera, J., & Bangerth, F. (2005). Evolution of endogenous hormone concentration in embryogenic cultures of carrot during early expression of somatic embryogenesis. *Plant Cell Reports, 23*, 567–572.

Johnson, P. R., & Ecker, J. R. (1998). The ethylene gas signal transduction pathway: A molecular perspective. *Annual Review of Genetics, 32*, 227–254.

Ju, C., Yoon, G. M., Shemansky, J. M., Lin, D. Y., Ying, Z. I., Chang, J., et al. (2012). CTR1 phosphorylates the central regulator EIN2 to control ethylene hormone signaling from the ER membrane to the nucleus in Arabidopsis. *Proceedings of the National Academy of Sciences of the United States of America, 109*, 19486–19491.

Kakimoto, T. (1996). CKI1, a histidine kinase homolog implicated in cytokinin signal transduction. *Science, 274*, 982–985.

Kakimoto, T. (2001). Identification of plant cytokinin biosynthetic enzymes as dimethylallyl diphosphate: ATP/ADP isopentenyltransferases. *Plant & Cell Physiology, 42*, 677–685.

Kakimoto, T. (2003). Perception and signal transduction of cytokinins. *Annual Review of Plant Biology, 54*, 605–627.

Karami, O., Aghavaisi, B., & Pour, A. M. (2009). Molecular aspects of somatic-to-embryogenic transition in plants. *Journal of Chemical Biology, 2*, 177–190.

Karami, O., & Saidi, A. (2010). The molecular basis for stress-induced acquisition of somatic embryogenesis. *Molecular Biology Reports, 37*, 2493–2507.

Kępczyńska, E., Ruduś, I., & Kępczyński, J. (2009). Endogenous ethylene in indirect somatic embryogenesis of Medicago sativa L. *Plant Growth Regulation, 59*, 63–73.

Kępczyński, J., McKersie, B. D., & Brown, D. C. W. (1992). Requirement of ethylene for growth of callus and somatic embryogenesis in *Medicago sativa* L. *Journal of Experimental Botany, 43*, 1199–1202.

Kepinski, S., & Leyser, O. (2005). The Arabidopsis F-box protein TIR1 is an auxin receptor. *Nature, 435*, 446–451.

Kieber, J. J., Rothenberg, M., Roman, G., Feldmann, K. A., & Ecker, J. R. (1993). CTR1, a negative regulator of the ethylene response pathway in Arabidopsis, encodes a member of the Raf family of protein kinases. *Cell, 72*, 427–441.

Kikuchi, A., Sanuki, N., Higashi, K., Koshiba, T., & Kamada, H. (2006). Abscisic acid and stress treatment are essential for the acquisition of embryogenic competence by carrot somatic cells. *Planta, 223*, 637–645.

Koornneef, M., & Karssen, C. M. (1994). *Seed dormancy and germination*. New York: Cold Spring Harbor, 313–334.

Kovalenco, P. G., & Kurchii, B. (1998). A using of abbcisic acid in the plant tissue culture of licorice Glycyrrhiza Glabra L. electroporated protoplast. In *Abstracts of II International Symposium on Plant Biotechnology, Kyiv, Ukraine* (p. 65).

Krikorian, A. D., & Berquam, D. L. (1969). Plant cell and tissue cultures: The role of haberlandt. *The Botanical Review, 35*, 59–88.

Kunkel, T., Niu, Q. W., Chan, Y. S., & Chua, N. H. (1999). Inducible isopentenyl transferase as a high-efficiency marker for plant transformation. *Nature Biotechnology, 17*, 916–919.

Leibfried, A., To, J. P. C., Busch, W., Stehling, S., Kehle, A., Demar, M., et al. (2005). WUSCHEL controls meristem function by direct regulation of cytokinin-inducible response regulators. *Nature, 438*, 1172–1175.

Li, H., & Guo, H. (2007). Molecular basis of the ethylene signaling and response pathway in Arabidopsis. *Journal of Plant Growth Regulation, 26*, 106–117.

Li, Q. Z., Li, X. G., Bai, S. N., Lu, W. L., & Zhang, X. S. (2002). Isolation of HAG1 and its regulation by plant hormones during in vitro floral organogenesis in Hyacinthus orientalis L. *Planta, 215*, 533–540.

Lu, W. L. (2002). The direct regeneration of inflorescence from callus in *Dracaena fragrans cv. Massangeana Hort. Acta Botanica Sinica, 44*, 113–116.

Lu, W. L. (2003). Control of in vitro regeneration of individual reproductive and vegetative organs in *Dracaena fragrans cv. Massangeana Hort.*-regularities of the direct regeneration of individual organs *in vitro*. *Acta Botanica Sinica, 45*, 1453–1464.

Lu, W. L., Enomoto, K., Fukunaga, Y., & Kuo, C. (1988). Regeneration of tepals stamens and ovules in explants from perianth of Hyacinthus orientalis L. importance of explant age and exogenous hormones. *Planta, 175*, 478–484.

Maggon, R., & Deo Singh, B. (1995). Promotion of adventitious bud regeneration by ABA in combination with BAP in epicotyl and hypocotyl explants of sweet orange (*Citrus sinensis* L. Osbeck). *Scientia Horticulturae, 63*, 123–128.

Mantiri, F. R., Kurdyukov, S., Lohar, D. P., Sharopova, N., Saeed, N. A., Wang, X. D., et al. (2008). The transcription factor MtSERF1 of the ERF subfamily identified by transcriptional profiling is required for somatic embryogenesis induced by auxin plus cytokinin in Medicago truncatula. *Plant Physiology, 146*, 1622–1636.

Mason, M. G., Mathews, D. E., Argyros, D. A., Maxwell, B. B., Kieber, J. J., Alonso, J. M., et al. (2005). Multiple type-B response regulators mediate cytokinin signal transduction in Arabidopsis. *Plant Cell*, *17*, 3007–3018.

Mattsson, J., Ckurshumova, W., & Berleth, T. (2003). Auxin signaling in Arabidopsis leaf vascular development. *Plant Physiology*, *131*, 1327–1339.

Meskaoui, A. E., Desjardins, Y., & Tremblay, F. M. (2000). Kinetics of ethylene biosynthesis and its effects during maturation of white spruce somatic embryos. *Physiologia Plantarum*, *109*, 333–342.

Miyawaki, K., Tarkowski, P., Matsumoto-Kitano, M., Kato, T., Sato, S., Tarkowska, D., et al. (2006). Roles of *Arabidopsis* ATP/ADP isopentenyltransferases and tRNA isopentenyltransferases in cytokinin biosynthesis. *Proceedings of the National Academy of Sciences of the United States of America*, *103*, 16598–16603.

Mok, D. W. S., & Mok, M. C. (2001). Cytokinin metabolism and action. *Annual Review of Plant Physiology and Plant Molecular Biology*, *52*, 89–118.

Müller, B., & Sheen, J. (2007). Advances in cytokinin signaling. *Science*, *318*, 68–69.

Müller, B., & Sheen, J. (2008). Cytokinin and auxin interaction in root stem-cell specification during early embryogenesis. *Nature*, *453*, 1094–1097.

Murashige, T., & Skoog, F. (1962). A revised medium for rapid growth and bioassays with tobacco tissue cultures. *Physiologia Plantarum*, *15*, 473–497.

Nadina, N., Martinez, M. E., Castillo, R., & Gonzalez, O. (2001). Effect of abscisic acid and jasmonic acid on partial desiccation of encapsulated somatic embryos of sugarcane. *Plant Cell, Tissue and Organ Culture*, *65*, 15–21.

Nakaya, M., Tsukaya, H., Murakami, N., & Kato, M. (2002). Brassinosteroids control the proliferation of leaf cells of Arabidopsis thaliana. *Plant & Cell Physiology*, *43*, 239–244.

Nambara, E., & Marion-Poll, A. (2005). Abscisic acid biosynthesis and metabolism. *Annual Review of Plant Biology*, *56*, 165–185.

Nishimura, C., Ohashi, Y., Sato, S., Kato, T., Tabata, S., & Ueguchi, C. (2004). Histidine kinase homologs that act as cytokinin receptors possess overlapping functions in the regulation of shoot and root growth in *Arabidopsis*. *Plant Cell*, *16*, 1365–1377.

Nishiwaki, M., Fujino, K., Koda, Y., Masuda, K., & Kikuta, Y. (2000). Somatic embryogenesis induced by the simple application of abscisic acid to carrot (Daucus carota L.) seedlings in culture. *Planta*, *211*, 756–759.

Ohmiya, A., & Haji, T. (2002). Promotion of ethylene biosynthesis in peach mesocarp discs by auxin. *Plant Growth Regulation*, *36*, 209–214.

Osakabe, Y., Miyata, S., Urao, T., Seki, M., Shinozaki, K., & Yamaguchi-Shinozaki, K. (2002). Overexpression of Arabidopsis response regulators, ARR4/ATRR1/IBC7 and ARR8/ATRR3, alters cytokinin responses differentially in the shoot and in callus formation. *Biochemical and Biophysical Research Communications*, *293*, 806–815.

Paciorek, T., & Friml, J. (2006). Auxin signaling. *Journal of Cell Science*, *119*, 1199–1202.

Pasternak, T. P., Prinsen, E., Ayaydin, F., Miskolczi, P., Potters, G., Asard, H., et al. (2002). The role of auxin, pH, and stress in the activation of embryogenic cell division in leaf protoplast-derived cells of alfalfa. *Plant Physiology*, *129*, 1807–1819.

Peres, L. E. P., Amar, S., Kerbauy, G. B., Salatino, A., Zaffari, G. R., & Mercier, H. (1999). Effects of auxin, cytokinin and ethylene treatments on the endogenous ethylene and auxin-to-cytokinin ratio related to direct root tip conversion of Catasetum fimbriatum Lindl. (Orchidaceae) into buds. *Journal of Plant Physiology*, *155*, 551–555.

Peres, L. E. P., & Kerbauy, G. B. (1999). High cytokinin accumulation following root tip excision changes the endogenous auxin to cytokinin ratio during root-to-shoot conversion in Catasetum fimbriatum Lindl. (Orchidaceae). *Plant Cell Reports*, *18*, 1002–1006.

Pernisová, M., Klima, P., Horak, J., Valkova, M., Malbeck, J., Soucek, P., et al. (2009). Cytokinins modulate auxin-induced organogenesis in plants via regulation of the auxin

efflux. *Proceedings of the National Academy of Sciences of the United States of America, 106*, 3609–3614.
Pullman, G. S., Mein, J., Johnson, S., & Zhang, Y. (2005). Gibberellin inhibitors improve embryogenic tissue initiation in conifers. *Plant Cell Reports, 23*, 596–605.
Qiao, M., Zhao, Z.-J., & Xiang, F.-N. (2012). Arabidopsis thaliana in vitro shoot regeneration is impaired by silencing of TIR1. *Biologia Plantarum, 56*, 409–414.
Reddy, G. V., Heisler, M. G., & Ehrhardt, D. W. (2004). Real-time lineage analysis reveals oriented cell divisions associated with morphogenesis at the shoot apex of Arabidopsis thaliana. *Development, 131*, 4225–4237.
Reddy, G. V., & Meyerowitz, E. M. (2005). Stem-cell homeostasis and growth dynamics can be uncoupled in the Arabidopsis shoot apex. *Science, 310*, 663–667.
Reinhardt, D., Pesce, E. R., Stieger, P., Mandel, T., Baltensperger, K., Bennett, M., et al. (2003). Regulation of phyllotaxis by polar auxin transport. *Nature, 426*, 255–260.
Riefler, M., Novak, O., Strnad, M., & Schmülling, T. (2006). *Arabidopsis* cytokinin receptor mutants reveal functions in shoot growth, leaf senescence, seed size, germination, root development, and cytokinin metabolism. *Plant Cell, 18*, 40–54.
Rock, C. D., & Quatrano, R. S. (1995). The role of hormones during seed development. In P. J. Davies (Ed.), *Plant hormones: Physiology, biochemistry and molecular biology* (2nd ed., pp. 671–697). Dordrecht: Kluwer.
Saab, I. N., Sharp, R. E., & Pritchard, J. (1992). Effect of inhibition of abscisic acid accumulation on spatial distribution of elongation in the primary root and mesocotyl of maize at low water potentials. *Plant Physiology, 99*, 26–33.
Sabatini, S., Beis, D., Wolkenfelt, H., Murfett, J., Guilfoyle, T., Malamy, J., et al. (1999). An auxin-dependent distal organizer of pattern and polarity in the Arabidopsis root. *Cell, 99*, 463–472.
Sarul, P., Vlahova, M., Ivanova, A., & Atanassov, A. (1995). Direct shoot formation in spontaneously occurring root pseudonodules of alfalfa (Medicago sativa L.). *In Vitro Cellular and Developmental Biology, 31*, 21–25.
Shimizu-Sato, S., Tanaka, M., & Mori, H. (2009). Auxin-cytokinin interactions in the control of shoot branching. *Plant Molecular Biology, 69*, 429–435.
Skoog, F., & Miller, C. O. (1957). Chemical regulation of growth and organ formation in plant tissues cultured in vitro. *Symposia of the Society for Experimental Biology, 54*, 118–130.
Street, H. E. (Ed.), (1977). *Plant tissue and cell culture*. Oxford: Blackwell Scientific Publications.
Su, Y. H., Cheng, Z. J., Su, Y. X., & Zhang, X. S. (2010). Pattern analysis of stem cell differentiation during *in vitro Arabidopsis* organogenesis. *Frontiers of Biology, 5*, 464–470.
Su, Y. H., Liu, Y. B., & Zhang, X. S. (2011). Auxin-cytokinin interaction regulates meristem development. *Molecular Plant, 4*, 616–625.
Su, Y. H., Su, Y. X., Liu, Y. G., & Zhang, X. S. (2013). Abscisic acid is required for somatic embryo initiation through mediating spatial auxin response in Arabidopsis. *Plant Growth Regulation, 69*, 167–176.
Su, Y. H., Zhao, X. Y., Liu, Y. B., Zhang, C. L., O'Neill, S. D., & Zhang, X. S. (2009). Auxin-induced *WUS* expression is essential for embryonic stem cell renewal during somatic embryogenesis in *Arabidopsis*. *The Plant Journal, 59*, 448–460.
Sugimoto, K., Gordon, S. P., & Meyerowitz, E. M. (2010). Regeneration in plants and animals: Dedifferentiation, transdifferentiation, or just differentiation? *Trends in Cell Biology, 21*, 212–218.
Sugimoto, K., Jiao, Y. L., & Meyerowitz, E. M. (2010). *Arabidopsis* regeneration from multiple tissues occurs via a root development pathway. *Developmental Cell, 18*, 463–471.
Sugiyama, M. (1999). Organogenesis in vitro. *Current Opinion in Plant Biology, 2*, 61–64.

Sun, J., Niu, Q.-W., Tarkowski, P., Zheng, B., Tarkowska, D., Sandberg, G., et al. (2003). The Arabidopsis AtIPT8/PGA22 gene encodes an isopentenyl transferase that is involved in de novo cytokinin biosynthesis. *Plant Physiology, 131*, 167–176.

Tabata, T., & Takei, Y. (2004). Morphogens: Their identification and regulation. *Development, 131*, 703–712.

Takei, K., Sakakibara, H., & Sugiyama, T. (2001). Identification of genes encoding adenylate isopentenyltransferase, a cytokinin biosynthesis enzyme, in *Arabidopsis thaliana*. *The Journal of Biological Chemistry, 276*, 26405–26410.

To, J. P., Haberer, G., Ferreira, F. J., Deruere, J., Mason, M. G., Schaller, G. E., et al. (2004). Type-A Arabidopsis response regulators are partially redundant negative regulators of cytokinin signaling. *Plant Cell, 16*, 658–671.

To, J. P. C., & Kieber, J. J. (2008). Cytokinin signaling: Two-components and more. *Trends in Plant Science, 13*, 85–92.

Tran Thanh Van, M. (1973). Direct flower neoformation from superficial tissue of small explants of Nicotiana tabacum L. *Planta, 115*, 87–92.

Tsuchisaka, A., & Theologis, A. (2004). Unique and overlapping expression patterns among the Arabidopsis 1-amino-cyclopropane-1-carboxylate synthase gene family members. *Plant Physiology, 136*, 2982–3000.

Ulmasov, T., Hagen, G., & Guilfoyle, T. J. (1997). ARF1, a transcription factor that binds to auxin response elements. *Science, 276*, 1865–1868.

Ulmasov, T., Murfett, J., Hagen, G., & Guilfoyle, T. J. (1997). Aux/IAA proteins repress expression of reporter genes containing natural and highly active synthetic auxin response elements. *Plant Cell, 9*, 1963–1971.

Vanneste, S., & Friml, J. (2009). Auxin: A trigger for change in plant development. *Cell, 136*, 1005–1016.

Vanstraelen, M., & Benková, E. (2012). Hormonal interactions in the regulation of plant development. *Annual Review of Cell and Developmental Biology, 28*, 463–487.

Vogel, G. (2005). Deriving 'controversy-free' ES cells is controversial. *Science, 310*, 416–417.

Wang, K. L., Yoshida, H., Lurin, C., & Ecker, J. R. (2004). Regulation of ethylene gas biosynthesis by the Arabidopsis ETO1 protein. *Nature, 428*, 945–950.

Werner, T., Motyka, V., Laucou, V., Smets, R., Van Onckelen, H., & Schmulling, T. (2003). Cytokinin-deficient transgenic *Arabidopsis* plants show multiple developmental alterations indicating opposite functions of cytokinins in the regulation of shoot and root meristem activity. *Plant Cell, 15*, 2532–2550.

Xing, X.-H., Huang, M., Shiragami, N., & Unno, H. (1995). Effect of abscisic acid on shoot regeneration from rice (*Oryza sativa* L.) callus. *Plant Tissue Culture Letters, 12*, 125–130.

Xu, H. Y., Li, X. G., Li, Q. Z., Bai, S. N., Lu, W. L., & Zhang, X. S. (2004). Characterization of HoMADS 1 and its induction by plant hormones during in vitro ovule development in Hyacinthus orientalis L. *Plant Molecular Biology, 55*, 209–220.

Yamamoto, Y., Kamiya, N., Morinaka, Y., Matsuoka, M., & Sazuka, T. (2007). Auxin biosynthesis by the YUCCA genes in rice. *Plant Physiology, 143*, 1362–1371.

Yokoyama, A., Yamashino, T., Amano, Y. I., Tajima, Y., Imamura, A., Sakakibara, H., et al. (2007). Type-B ARR transcription factors, ARR10 and ARR12, are implicated in cytokinin-mediated regulation of protoxylem differentiation in roots of *Arabidopsis thaliana*. *Plant & Cell Physiology, 48*, 84–96.

Yoshimatsu, K., & Shimomura, K. (1994). Plant regeneration on cultured root segments of Cephalis ipecacuanha A. Richard. *Plant Cell Reports, 14*, 98–101.

Zhao, Z., Andersen, S. U., Ljung, K., Dolezal, K., Miotk, A., Schultheiss, S. J., et al. (2010). Hormonal control of the shoot stem-cell niche. *Nature, 465*, U1089–U1154.

Zhao, Y., Christensen, S. K., Fankhauser, C., Cashman, J. R., Cohen, J. D., Weigel, D., et al. (2001). A role for flavin monooxygenase-like enzymes in auxin biosynthesis. *Science, 291*, 306–309.

Zhao, Q., & Guo, H. W. (2011). Paradigms and paradox in the ethylene signaling pathway and interaction network. *Molecular Plant*, *4*, 626–634.
Zhao, Y. (2008). The role of local biosynthesis of auxin and cytokinin in plant development. *Current Opinion in Plant Biology*, *11*, 16–22.
Zhao, X. Y., Su, Y. H., Zhang, C. L., Wang, L., Li, X. G., & Zhang, X. S. (2013). Differences in capacities of *in vitro organ* regeneration between two *Arabidopsis* ecotypes Wassilewskija and Columbia. *Plant Cell, Tissue and Organ Culture*, *112*, 65–74.

CHAPTER THREE

Histone Modifications and Regeneration in the Planarian *Schmidtea mediterranea*

Sofia M.C. Robb[*,†,‡], **Alejandro Sánchez Alvarado**[**,1]
[*]Department of Plant Pathology & Microbiology, University of California-Riverside, California, USA
[†]Institute for Integrative Genome Biology, University of California-Riverside, California, USA
[‡]Department of Botany & Plant Sciences, University of California-Riverside, California, USA
[**]Howard Hughes Medical Institute & Stowers Institute for Medical Research, Kansas City, Missouri, USA
[1]Corresponding author: e-mail address: asa@stowers.org

Contents

Abstract

The freshwater planarian *Schmidtea mediterranea* has emerged as a powerful model system for studying regeneration and adult stem cell (ASC) biology. This is largely due to the developmental plasticity of these organisms and the abundant distribution and experimental accessibility of their ASCs. Techniques such as whole mount *in situ* hybridization, dsRNA-mediated interference, halogenated thymidine analogs for defining cell lineages, and fluorescence-activated cell sorting among other methods, have allowed researchers to interrogate the biology and attendant pluripotent stem cells of these animals in great detail. Therefore, it has now become possible to interrogate and define the roles that epigenetic states may play in regulating ASCs, and by extension, regeneration proper. Here, we provide a primer on the types and number of histone families found in *S. mediterranea*, known as epigenetic marks of these molecules and a survey of epigenetic modifying enzymes encoded by the planarian genome. We also review experimental evidence indicating that such modifications may in fact play key roles in determining the activities of planarian stem cells.

Current Topics in Developmental Biology, Volume 108
ISSN 0070-2153
http://dx.doi.org/10.1016/B978-0-12-391498-9.00004-8

1. INTRODUCTION

Because regeneration and tissue homeostasis are processes that are strongly affected by environmental factors (e.g., amputation and starvation), could epigenetics play a role in regulating these processes and key cellular effectors such as adult stem cells? One approach to answer this important question is to focus efforts on defining the dynamics of modifications present in the basic unit of chromatin (a complex of 147 base pairs of DNA wrapped around an octamer of histones, two each of histones H3, H4, H2A, and H2B). In particular, by defining and establishing cause/effect relationships between marks introduced into histone tails (the 20–30 amino acids of the amino terminus of these molecules) by environmental changes and specific cellular functions.

Specific modifications have been observed to correspond with specific transcriptional and cellular events. For example, histone H3 serine 10 becomes phosphorylated (H3P) during the G2 to M transition of the cell cycle, trimethylation of histone H3 lysine 9 (H3K9-Me3) is typically found in regions of the genome which are transcriptionally inactive or silenced, while histone H3 lysine 9, lysine 14 acetylation (H3K9K14-Ac) is found in transcriptionally active regions of the genome (Peterson & Laniel, 2004). Such histone modifications make up the present foundation of epigenetics, that is, the interaction of environmental and genetic factors resulting in an observable characteristic or trait (phenotype).

Most studies of epigenetic regulation in stem cells have been carried out *in vitro*, particularly in embryonic stem (ES) cells or induced pluripotent stem cells. Studies in ES cells, for example, indicate that epigenetic regulation has a very important role in ES cell function. When ES cells are exposed to retinoic acid, specific changes in gene expression and differentiation into neuronal cells are observed. Four days after exposure to retinoic acid the mRNA levels of Hoxb1 rise, and by day 10 such expression is significantly downregulated. The acetylation levels at the promoter region of this gene have a similar pattern. The increase in histone H3 lysine 9 acetylation (H3K9-Ac) and histone H3 lysine 4 methylation (H3K4-Me) is coupled with the increase in Hoxb1 expression, and a decrease in these modifications is found when gene expression drops. The inverse is seen with H3K9-Me. These observations indicate that histone posttranslational modifications or chromatin organization is closely associated with biological processes such as differentiation (Chambeyron & Bickmore, 2004).

It has also been demonstrated that a specific pattern of modifications found in ES cells, termed "bivalent domains," help to maintain pluripotency. This bivalent domain consists of a pairing of H3 lysine 27 trimethylation (H3K27-Me3) and H3 lysine 4 trimethylation (H3K4-Me3), which silences developmental genes in ES cells, but leaves them "poised" for expression during differentiation (Bernstein et al., 2006).

However, ES cells are the product of the collection and culturing of cells from the inner cell mass (ICM) of the mammalian blastocyst. As such, are cultured ES cells and the *in vivo* ICM cells experimentally the same? Can the data collected from ES cell experiments be transposed onto *in vivo* studies? A study comparing the chromatin states of ES and ICM cells has demonstrated that *in vitro* data is not directly applicable to *in vivo* biology (O'Neill, VerMilyea, & Turner, 2006). Typically, *in vivo* ICM cells are too few for conventional chromatin immunoprecipitation (ChIP), but modifications of the ChIP methodology can be used that mitigate the limitations on cell number. Histone posttranslational modification levels of a set of genes known to be expressed in ICM and ES cells (*Nanog* and *Oct4*) and genes not expressed in these cell types (*Cdx2, Hhex, Gapdh*) were compared. The two cell types have many differences in posttranslational modifications especially in H3K4-Me. There is as much as a fivefold increase in H3K4-Me in the population of *in vitro* cultured ES cells in regions of *Nanog*. In ES cells, H3K4-Me is present on regions of *Cdx2*, a gene expressed in intestinal epithelium, but is undetected in ICM cells. These results alone clearly indicate that fundamental differences are likely to exist between cultured ES cells and the cells found *in vivo* in the ICM, and that the *in vitro* data should not be used blindly to infer *in vivo* states.

How different are ES cells and ASCs? Can information learned from ES cells be used interchangeably with ASCs? ASCs are not as commonly used to study epigenetics as ES cells are due to the small numbers, inaccessibility, and difficulties of long-term culturing of these cells (Eilertsen, Floyd, & Gimble, 2008). It is becoming increasingly evident that the two types of stem cells are very different. For instance, dissimilarity in the importance of specific genes to stem cell self-renewal and maintenance has been identified. Specifically, *Nanog* and *Oct4* while necessary for ES cell maintenance are not important for ASCs. Conversely, *Bmi-1* has been shown to be necessary for adult stem cell self-renewal, but not for ES cells (Molofsky, Pardal, & Morrison, 2004).

Planarians are well known for their regenerative capacities. Understanding the biological principles guiding this remarkable attribute is likely to shed light on our understanding of how cellular pluripotentiality may be regulated. Thus, planarians offer a singular experimental paradigm to address these issues. Planarians lend themselves to studies that circumvent the common issues associated with ASC studies. For instance, the worm has a large population of neoblasts, or planarian ASCs, that are distributed throughout their body. This allows for *in vivo* studies and overcomes issues associated with small cell numbers or artifacts of culturing conditions that are commonplace in other organisms. In addition, planarians afford the study of lineage progression from undifferentiated stem cells due to the identification of molecular markers for stem cells and their progeny (Eisenhoffer, Kang, & Sánchez Alvarado, 2008). An additional benefit of using *Schmidtea mediterranea* as a model organism for studying epigenetics is that there is an excellent collection of bioinformatic tools to make many studies possible. These tools include the genome assembly, annotations (Cantarel et al., 2008), genome browser (Robb, Ross, & Alvarado, 2007), and the ability to develop custom bioinformatic tools as needed.

2. HISTONE FAMILIES AND VARIANTS IN *S. MEDITERRANEA*

All eukaryotic organisms have and use highly conserved histone proteins to organize DNA into chromatin (Malik & Henikoff, 2003), but the ways in which these proteins are regulated are different (Marino-Ramirez, Jordan, & Landsman, 2006). How many copies of each of the core histone are present? Which histone variants exist in the genome, and how many? Do histone genes in planarians form clusters in the genome as they do in other organisms?

Due to the high degree of amino acid identity for histone proteins across species (Mariño-Ramírez, Hsu, Baxevanis, & Landsman, 2005), it is straightforward to identify genome annotations for all four core histones, variants, and the linker histone, H1, in *S. mediterranea* (Tables 3.1 and 3.2). The numbers of histones found in the *S. mediterranea* genome are also comparable to numbers identified in other organisms (Table 3.2). Additionally, histone variants of histone H3 and histone H2A can be identified. The H3 variant present in the planarian genome is H3.3. This variant replaces canonical H3 during transcription. Both H2A.X and H2A.Z are also present. H2A.X has a serine in the C-terminal tail that is phosphorylated in response to double-stranded breaks. H2A.Z, on the other hand, is often found at heterochromatic boundaries, and it is thought that this histone

Table 3.1 Histone numbers in *Schmidtea mediterranea*

Histone	**Count**
H1	6
H2A	247
H2B	18
H3	13
H4	9
H3.3	2
H2A.X	1
H2A.Z	2

The total unique nucleotide sequences of each of the four core histones, the linker histone H1, and histone variants are totaled in the column-labeled "Count."

Table 3.2 Comparative species histone gene numbers

	S. mediterranea	***H. sapiens***	***D. melanogaster***	***C. elegans***
H1	6	26	24	8
H2A	249	116	26	19
H2B	18	105	23	17
H3	15	184	49	28
H4	9	50	25	16

Total number of histone sequences identified in the genomes of *S. mediterranea*, *H. sapiens*, *D. melanogaster*, and *C. elegans* with the use of Blast and custom scripts.

enables the DNA to resist condensation and keeps heterochromatin from spreading into euchromatic regions (Malik & Henikoff, 2003).

Moreover, many of the histone genes identified are organized in clusters in the planarian genome (Table 3.3; Fig. 3.1), a feature likely to be the result of evolutionary conservation (Rooney, Piontkivska, & Nei, 2002). Many species have histones in clusters, though the chromosomal organization is not always the same (Sittman et al., 1981). In *S. mediterranea*, nine histone gene clusters can be identified. The largest cluster is made up of five genes, two H3.3 sequences, two H2B sequences, and one H2A sequence. The other clusters are smaller and are made up of only two genes each. In the examination of the genome, neither H4 nor the linker histone H1 was found

Table 3.3 Histone clusters in *Schmidtea mediterranea*

Genomic contig ID	Histone name	MAKER ID
v31.000056	H2A	mk4.000056.00.01
	H2A	mk4.000056.09.01
v31.000114	H2A	mk4.000114.12.01
	H2A	mk4.000114.12.02
v31.000124	H2B	mk4.000124.04.01
	H2A	mk4.000124.05.01
v31.000503	H3	mk4.000503.06.01
	H2A	mk4.000503.19.01
v31.000688	H3.3	mk4.000688.02.01
	H2A	mk4.000688.04.01
	H2B	mk4.000688.06.01
	H2B	mk4.000688.07.01
	H3.3	mk4.000688.08.01
v31.000696	H2A	mk4.000696.08.01
	H2A	mk4.000696.10.01
v31.004793	H2A	mk4.004793.01.01
	H2A	mk4.004793.01.02
v31.11360	H2A	mk4.011360.00.01
	H2B	mk4.011360.01.01
v31.013070	H2B	mk4.013070.00.01
	H2A	mk4.013070.01.01

Catalog of identifiable histone clusters with the genomic contig ID in which the cluster was found, the histones present in each cluster, and the MAKER ID for each histone.

to be part of any histone cluster. Whether such genomic organization reflects on complex but specific regulatory programs modulating the temporal and/or spatial regulation of histones in planarians remains to be determined. However, it is known that specific histone variants are primarily expressed in planarian neoblasts, in particular H4 (Rouhana, Vieira, Roberts-Galbraith, & Newmark, 2012) and H2B (Solana et al., 2012).

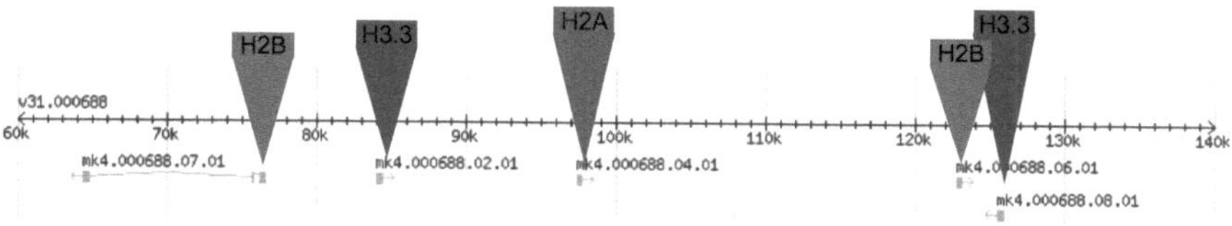

Figure 3.1 Largest identifiable *S. mediterranea* histone gene cluster. This is a screen capture from SmedGD of genomic contig v31.000688. It is the largest identifiable histone cluster and contains two H2B genes, two H3.3 genes, and one H2A gene. Only the first exon of the MAKER transcript mk4.000688.07.01 is the single exon H2B gene. (See color plate.)

3. A SURVEY OF HISTONE MODIFICATIONS IN *S. MEDITERRANEA*

Because of extensive sequence conservation, many commercially available antibodies generated against epitopes of histone posttranslational modifications in other species cross-react with planarian proteins. *S. mediterranea* Histone H4 is identical to human H4, and histone H3 has only two amino acid differences (Fig. 3.2, top). Antibodies against various histones tested by Western blots (Fig. 3.2, bottom), and in whole animals appear to work specifically (Table 3.4). For example, the antibody against H3Pser-10 works robustly in whole animals (Fig. 3.3A), and antibodies against H3K9K14-Ac and H2K9-Me3 when tested in whole-mount fluorescent immunohistochemistry assays produced discrete, cell-specific patterns throughout the animals (Fig. 3.3B and C).

4. DISTRIBUTION OF ACETYLATION MARKS IN PLANARIAN CELLS

Combining microscopy to simultaneously identify cell types and histone modification such as acetylation (AC) via immunofluorescence allows for measurements of epigenetic states in planarians. For instance, cells collected from whole worm dissociations and the levels of histone H3K9K14-Ac were measured by a combination of immunocytochemistry and fluorescent microscopy. Different cell types collected from whole worm dissociations can be identified by visual morphology (Fig. 3.4, top), and their relative levels of acetylation (in this case H3K9K14-Ac) can be measured. The majority of the cells were found to have low levels of AC/DAPI (Fig. 3.4A, B, and E–L). However, high levels of AC/DAPI were present in a few cells (Fig. 3.4C, D, and M). This becomes readily apparent when the levels of acetylation versus DNA

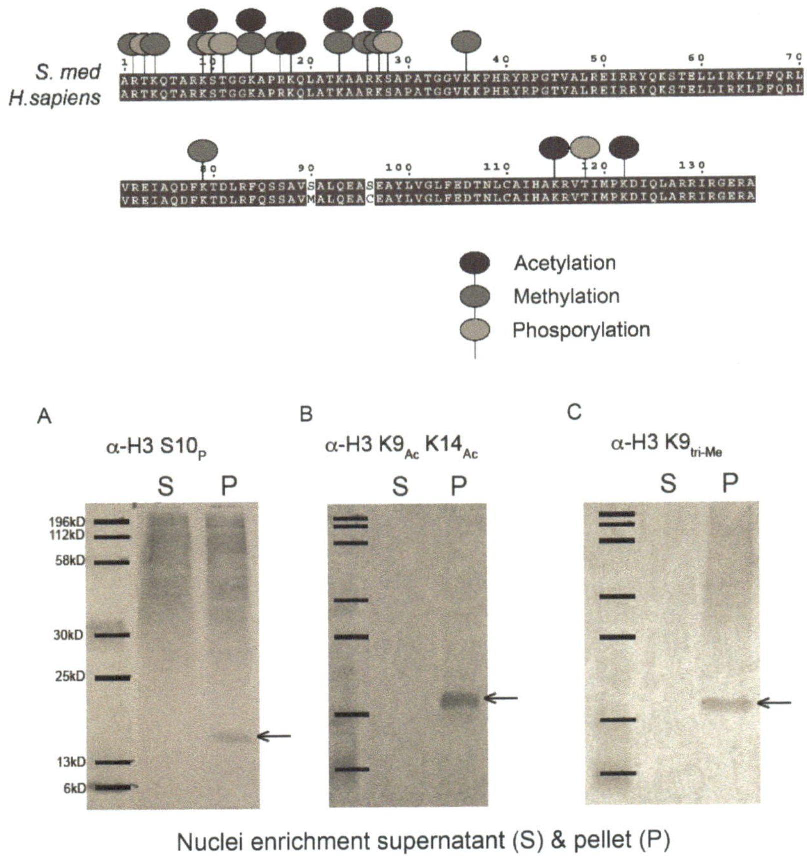

Figure 3.2 (Top) Amino acid alignment of histone H3 in *H. sapiens* and *S. mediterranea*. This is an alignment created with ClustalW (Thompson, Higgins, & Gibson, 1994), Planarian H3 is on top and human below. Locations of select posttranslational modifications are marked with colored lollipops. There are only two amino acid differences between *H. sapiens* and *S. mediterranea*. (Bottom) Histone antibodies cross-react with planarian epitopes. Nuclei were enriched and the supernatant (S) and pellet (P) were run on a 12% SDS polyacrylamide gel, transferred to nitrocellulose and immunostained. Bands of the appropriate size were obtained (arrows). (For interpretation of the references to color in this figure legend, the reader is referred to the online version of this chapter.)

content (DAPI intensity) are plotted (Fig. 3.4, bottom). The three outliers are three cells with a very high AC/DAPI ratio (Fig. 3.4C, D, and M).

Additionally, cell types with different AC/DAPI ratios but with similar morphologies can also be identified (Fig. 3.5). Each of these cells may be a neoblast given their large nuclear to cytoplasmic ratio and small cell size (5–8 μm). The majority of the neoblast-like cells have a very low AC/DAPI

Table 3.4 List of antibodies against histones, histone modification, and DNA methylation tested in whole animals

Antibody	Catalog #	1° Ab dilution	Positive for signal
H3K9K14-Ac	Upstate 06-599	1:500	Yes
H3K27-Me3	Upstate 07-449	1:100	No
H3K4-Me3	Abcam ab8580	1:500	No
H3K4-Me2	Upstate 07-030	1:100	Yes
H3K9-Me3	Upstate 07-442	1:500	Yes
H3S10-P	Upstate 05-817	1:200	Yes
H3.3S31-P	Upstate 07-679	1:500	No
H3	Upstate 06-755	1:500	Yes
H4	Abcam ab10158	1:100	No
H2A	Upstate 07-146	1:100	No
H2BK12-Ac	Cell signaling #2575	1:100	No
H2B	Abcam ab1790	1:100	No
H1	Upstate 05-629	1:500	Yes
5-Methylcytidine	Eurogentec BI-MECY	1:500	Yes
HDAC1	Upstate 06-720	1:100	Yes
HDAC2	Upstate 05-814	1:100	Yes
HDAC3	Upstate 05-813	1:100	Yes
HP1	Abcam ab9057	1:100	No

ratio of 0.1; however, some have comparably higher ratios suggesting the existence of tightly regulated histone acetylation and deacetylation mechanisms whose precise function in stem cell activities remains to be elucidated.

5. EPIGENETIC MODIFYING ENZYMES IN PLANARIANS

Because enzymes ultimately effect modifications of histones, we used annotated protein collections from other species to generate a list of candidate genes coding for epigenetic modifying enzymes in order to identify and clone their respective homologues in planaria (Table 3.5). Expression patterns for many of these genes have been defined (Fig. 3.6) and illustrate a

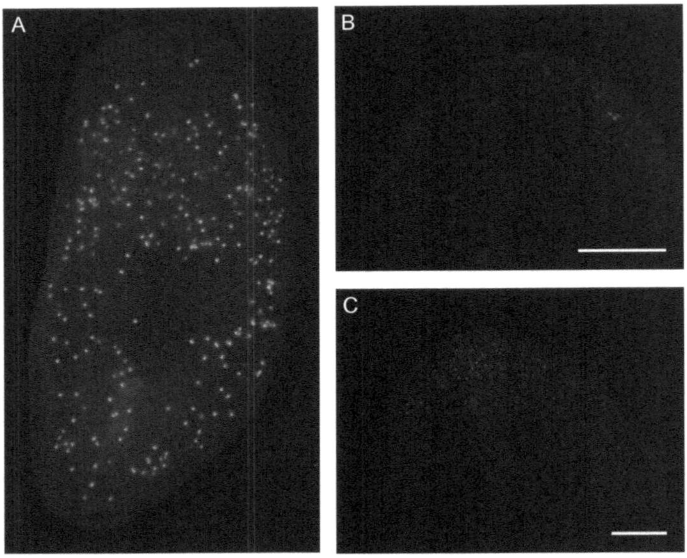

Figure 3.3 Whole-mount immunohistochemistry. (A) Animals immunostained with antibodies against (A) H3P, (B) H3K9K14-Ac, and (C) H3K9-Me3. (See color plate.)

variety of discrete patterns, and an absence of ubiquitously expressed genes. Some expression patterns are shared by multiple classes of enzymes. For example, a stem cell-like pattern and a nervous system-like pattern are readily apparent (Fig. 3.6, magenta rectangle). The stem cell-like pattern is characterized by the expression of mRNA between the branches of the gut and an absence of expression anterior to the photoreceptors and in the pharynx. This pattern was found for a histone deacetylase (*Smed-HDAC1-1*), a histone methyltransferase (*Smed-SETD8-1*), and two putative DNA demethylase (*Smed-NHP2L1 and Smed-RL7A*).

The nervous system-like expression pattern is characterized by mRNA accumulation in the brain, down the ventral cords, and at the distal tip of the pharynx (Fig. 3.6, green rectangle). This pattern is found in three of the histone deacetylases (*Smed-HDAC5-1*, *Smed-HDAC8-1*, and *Smed-Sin3-2*), five of the histone acetyltransferases (*Smed-CBP-5*, *Smed-CBP-6*, *Smed-Ep300-3*, *Smed-Ep300-1*, and *Smed-CBP-4*), two of the histone demethylases (*Smed-JARID-2* and *Smed-JARID-3*), and three of the histone methyltransferases (*Smed-TRR-1*, *Smed-DOT1L-1*, and *Smed-SUV92-2*). Altogether, the discrete spatial distribution of expression of the genes coding for enzymes capable of modifying histones suggests that complex regulation of such marks is occurring simultaneously and in many different cell types

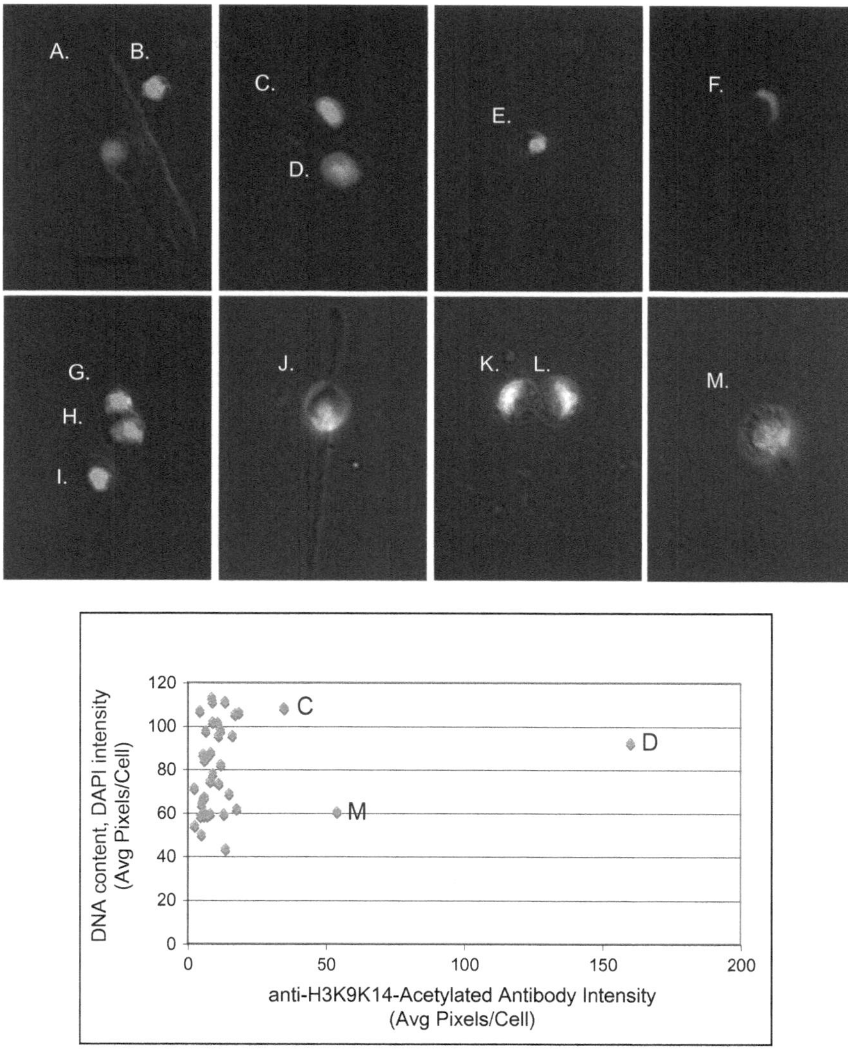

Figure 3.4 Sampling of cell types. (Top) Sampling of cell types by morphology. The scale bar is 10 μm. (A) A muscle cell (B, C, D, E, F, I, and M), potential stem cells (G, H, K, and L), dividing stem cells. (J) A neuron. (Bottom) Plot of H3K9K14-Ac versus DNA content. The letters labeling the three outliers refer to the cells identified in Figure 3.5 top.

during normal physiological homeostasis. Recent functional studies aimed at identifying chromatin regulators in planarian stem cells support this hypothesis. For example, components of six chromatin complexes, nucleosome remodeling and deacetylase, CAF-1, BRG1/Brm-associated factor,

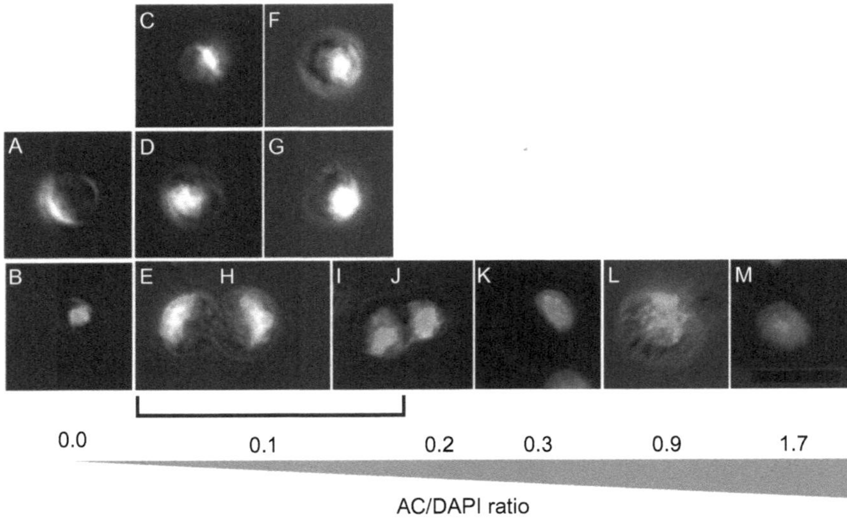

Figure 3.5 Stem cells and AC/DAPI ratios. The scale bar is 10 μm. (A, B) Neoblasts with AC/DAPI ratio of 0.0. (C–I) Most neoblasts have a ratio of 0.1. (J–K) Neoblasts with a ratio of 0.2 and 0.3. (L–M) Neoblasts with a much higher ratio, 0.9 and 1.7.

Table 3.5 Classes and counts of epigenetic modifying enzyme genes in *Schmidtea mediterranea*

Genes	**Found in genome**	**Cloned and sequenced**
DNA de/methylases	11	11
Histone phosphorylases	3	3
Histone acetyltransferases	26	25
Histone methyltransferases	24	23
Histone deacetylases	14	14
Histone demethylases	12	11
Histone deubiquinases	1	0
Count	91	87

Counts of classes of enzymes found in the genome and the counts of the genes that were positively cloned and sequenced.

facilitates chromatin transcription, Cdk-activating kinase, and minichromosome maintenance complex have been shown to participate in planarian stem cell functions and regeneration (Zeng et al., 2013). Particularly, heterochromatin Ppotein 1 (SMED-HP1-1), a molecule known to

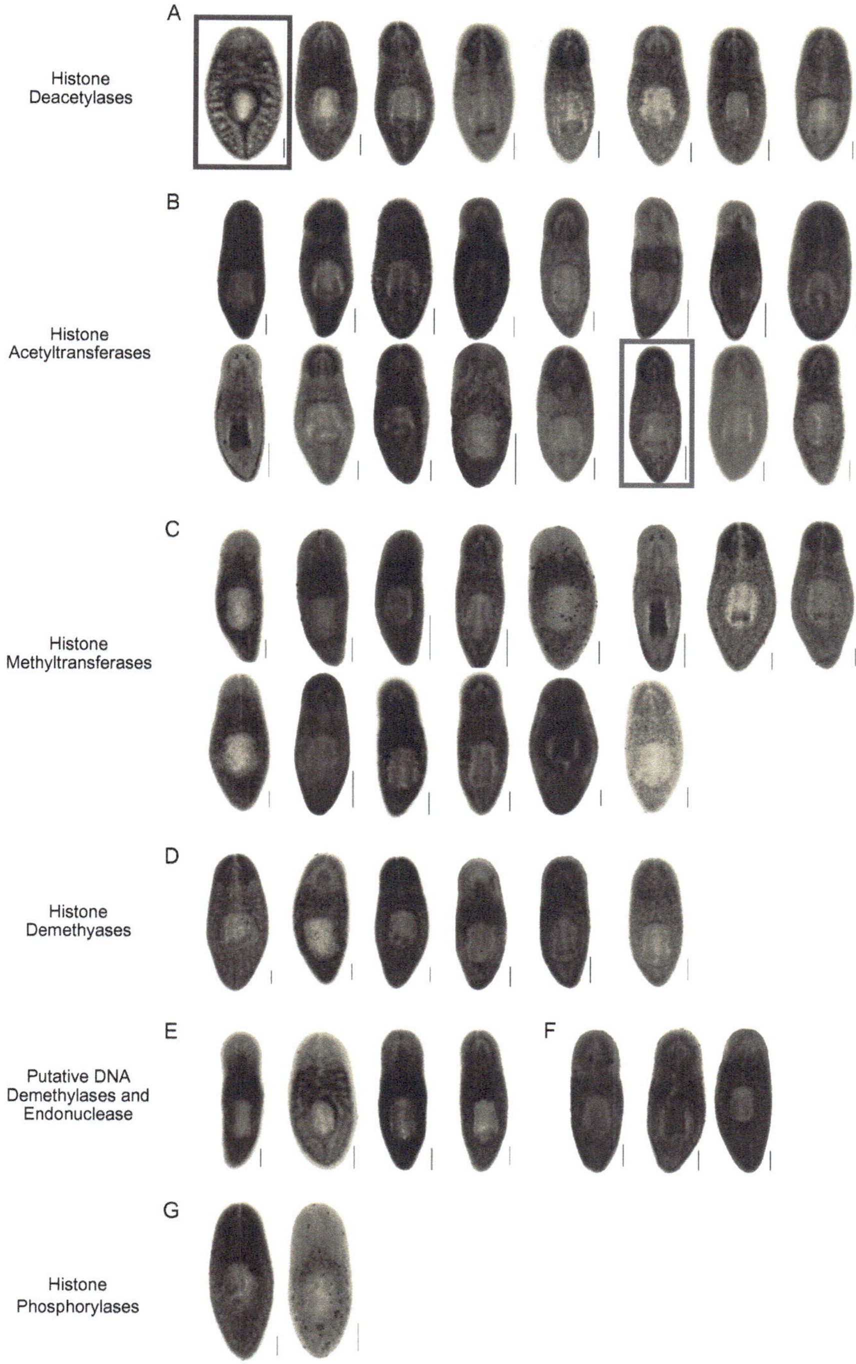

Figure 3.6 See figure caption on next page.

interact with genomic DNA modifications such as lysine 9-methylated histone H3 tails (Bannister et al., 2001; Lachner, O'Carroll, Rea, Mechtler, & Jenuwein, 2001) has been shown to be exclusively expressed in planarian neoblasts and to play a key role in planarian stem cell self-renewal (Zeng et al., 2013). Unraveling the functional relationships of chromatin modifications deployed in planarians should significantly inform our understanding of how a stem cell population and its diverse progeny may coordinate self-renewal and the production of the right number and types of cells necessary to maintain the form and function of the organism.

6. HISTONE MODIFICATIONS, STEM CELLS, AND CELL LINEAGES

Smed-HDAC1-1 is specifically expressed in the planarian ASCs and loss of gene function by RNAi of this gene results in a neoblasts deficiency phenotype, that is, regression of cephalic structures, ventral curling, and an inability to launch a regenerative response after amputation (Eisenhoffer et al., 2008; Reddien, Bermange, Murfitt, Jennings, & Sánchez Alvarado, 2005). Given its expression and RNAi phenotype, *Smed-HDAC1-1* provides an interesting opportunity to measure the consequences of perturbing a discrete epigenetic process (deacetylation) on stem cell lineage in both unamputated (Fig. 3.7) and regenerating organisms (Fig. 3.8).

In unamputated animals, the expression of the stem cell marker *smedwi-1* under *Smed-HDAC1-1(RNAi)* conditions begins to noticeably decrease by

Figure 3.6—Cont'd Whole-mount *in situ* hybridization expression patterns of epigenetic modifying enzymes. (A) Histone deacetylases; *Smed-HDAC1-1, Smed-HDA2-1, Smed-HDAC5-1, HDAC8-1, Smed-Sin3-1, Smed-SAP18-1, Smed-SAP180-1,* and *HDAC1-3.* (B) Histone acetyltransferases; *Smed-ELP3-1, Smed-NCOAT-2, Smed-CBP-2, Smed-YEATS4-1, Smed-CBP-4, Smed-NCOAT-1, Smed-MYST3-1, Smed-MYST4-2, Smed-MYST2-2, Smed-Ep300-2, Smed-MYST4-1, Smed-HAT-1, Smed-CBP-5, Smed-CBP-6, Smed-Ep300-3,* and *Smed-Tip60-1.* (C) Histone methyltransferases; *Smed-SETD8-1, Smed-SUV42-1, Smed-SETB1-1, Smed-CARM1-1, Smed-NSD2-2, Smed-SUV92-1, Smed-TRR-1, Smed-DOT1L-1, Smed-ANM8-1, Smed-ANMX-1, Smed-ANM1-1, Smed-ASH2L-1, Smed-SETB1-2,* and *Smed-SUV92-2.* (D) Histone demethylases; *Smed-JARID-3, Smed-JmjC-1, Smed-JARID-2, Smed-JmjC-6, Smed-JmjC-4,* and *Smed-JmjC-7.* (E) Putative DNA demethylases; *Smed-RL7A, Smed-NHP2L1, Smed-RPS12,* and *Smed-RPL30.* (F) Endonucleases; Smed-XPG-1, Smed-XPG-2, and Smed-FEN1. (G) Histone phosphorylases; *Smed-TAF1-2 and Smed-TAF1-3.* Magenta square indicates stem-cell like expression pattern. Green square, central nervous system expression pattern. (See color plate.)

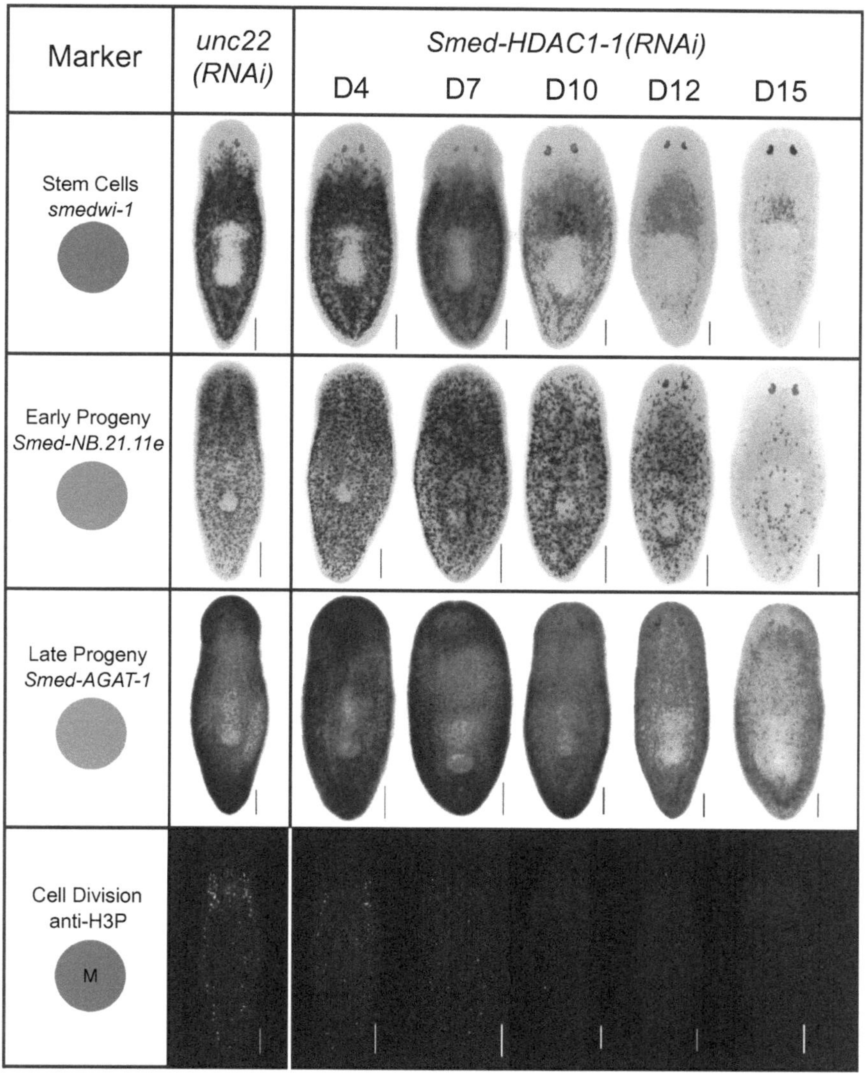

Figure 3.7 Stem cell lineage analysis of intact *Smed-HDAC1-1(RNAi)* animals. Intact *Smed-HDAC1-1(RNAi)* animals were fixed on D4, D7, D10, D12, and D15 and tested for the presence of the four markers *smedwi-1*, *Smed-NB.21.11e*, *Smed-AGAT-1*, and anti-H3P. *unc22(RNAi)* animals were used as a control. (See color plate.)

D7 after RNAi treatment. There is a detectable reduction by D10 of the early division progeny marker *Smed-NB.21.11e*, followed by a decrease at D12 of the late division progeny marker *Smed-AGAT-1* (Fig. 3.7). In contrast, these markers decrease rapidly when the animals are amputated.

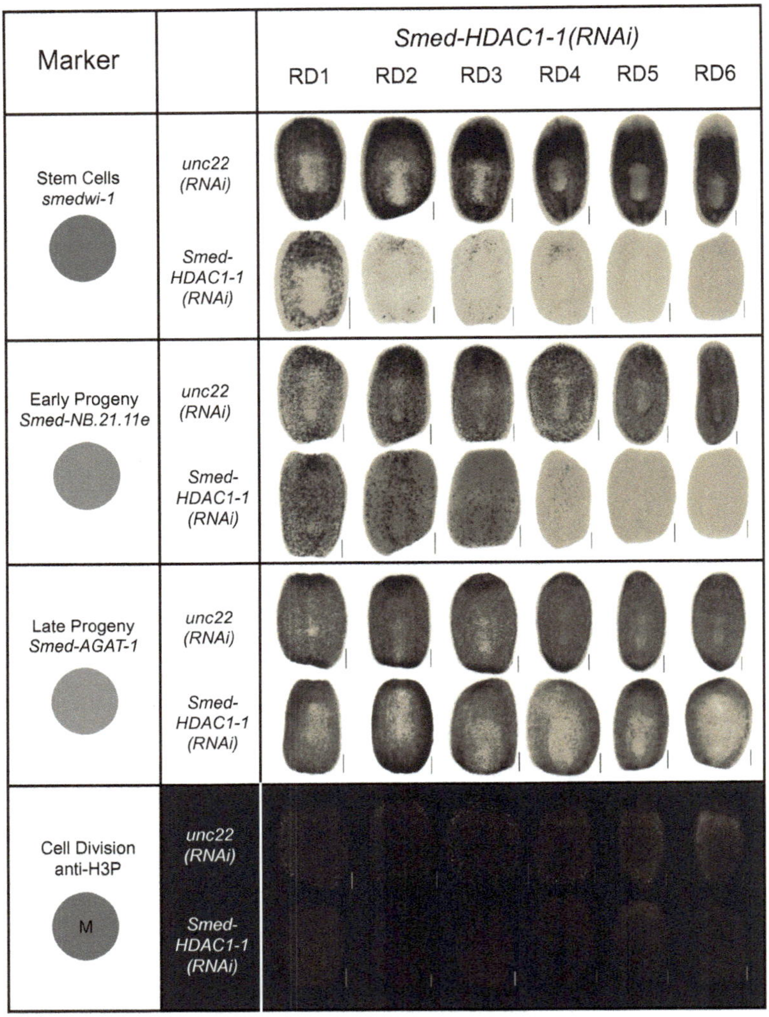

Figure 3.8 Stem cell lineage analysis of *Smed-HDAC1-1(RNAi)* intact and amputated animals. Intact *Smed-HDAC1-1(RNAi)* animals were fixed on D4, D7, and D10 and amputated. *Smed-HDAC1-1(RNAi)* animals (labeled "e" for experimental) were fixed on D11 through D16 (RD1 through RD6) and tested for the presence of the four markers *smedwi-1*, *Smed-NB.21.11e*, *Smed-AGAT-1*, and anti-H3P. *unc22(RNAi)* animals were used as a control for intact animals and for amputated animals (labeled "c" for control). (See color plate.)

smedwi-1 is almost completely depleted in regeneration day 2 (RD2) animals, while *Smed-NB.21.11e* is almost undetectable by RD4, and *Smed-AGAT-1* expression is severely depressed by RD6 (Fig. 3.8). Such changes in gene expression in both conditions (unamputated and amputated) are accompanied by a reduction in mitotic figures (Fig. 3.9A). Interestingly, quantification of stem cells, early progeny and late progeny cells of *Smed-HDAC1-1(RNAi)* animals on D12 (see Fig. 3.7) by flow cytometry indicates that stem cells are still present in the animal in numbers that are indistinguishable from control (Fig. 3.9B). This is a time after RNAi treatment when the animals appear healthy, yet the stem cell marker is severely depressed, the early progeny marker has begun to disappear, and the late progeny marker is beginning to be downregulated. As such, it is likely that acetylation in planarians plays an important role in effecting mechanisms associated with the way in which stem cells and their progeny interact with their environment under normal physiological conditions, after injury, and during regeneration.

7. CONCLUSIONS

Adult stem cells are an understudied population of cells, and *S. mediterranea* is an ideal model that makes the investigation of ASCs much more amenable. The abundance and experimental accessibility of ASCs in this organism should help advance our understanding of the importance and variety of roles epigenetics and epigenetic modifying enzymes have in stem cell function. Planarians provide a unique opportunity and an unexplored paradigm for studying epigenetics in ASCs and the role this biological process plays in regeneration and tissue homeostasis. Planarians possess large numbers of core histones, histone clusters, histone variants, and many of the widely studied posttranslational modifications, which makes it abundantly clear that intricate regulatory choreographies of histone gene expression and posttranslational modulation await elucidation.

There are many well-studied histone posttranslational modifications, which are also represented in planarians and typically correspond to a specific cellular event, such as double-stranded DNA breaks, transcription, and stages of the cell cycle. They are dynamic and possess a predictable nature. The use of antibodies against these modifications may be an informative method for separating the mixed staged cycling stem cell population, for example. H3K9K14-Ac levels compared to DNA content may be a sliding scale indicator to the phase of the cell cycle (Fig. 3.5). Analyses of H3K9K14-Ac, H3K9-Me3, and H3K27-Me3 in planarian cells show very promising results for identification of specific cell populations and specific stages of the cell cycle. Examination of H3K9K14-Ac in a mixed cell population indicates that cells with the visual characteristics of stem cells have

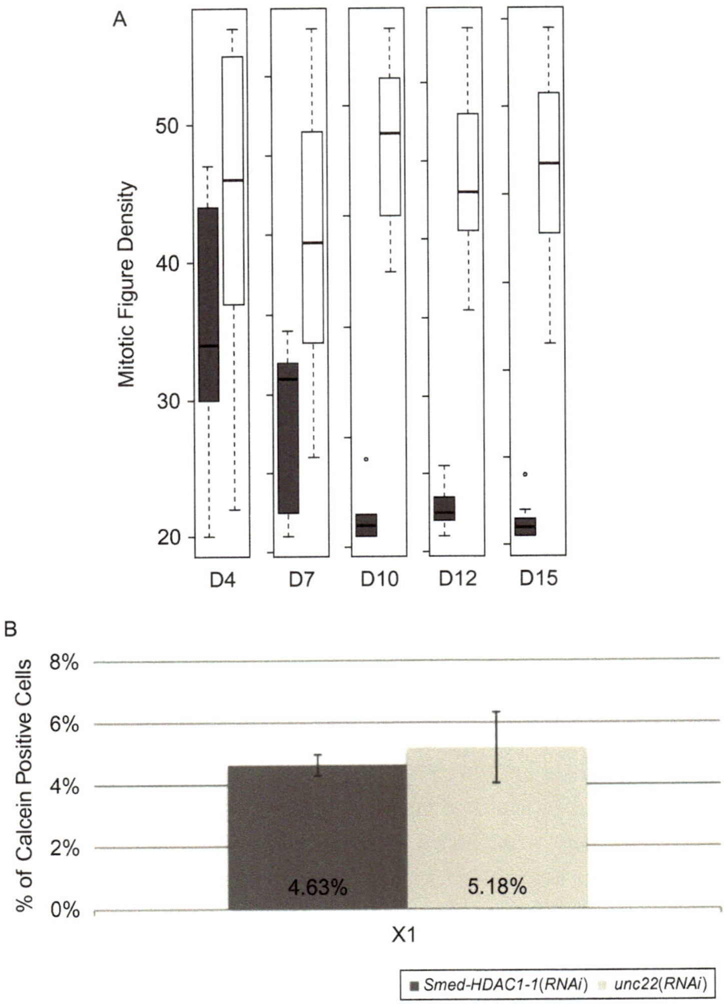

Figure 3.9 (A) Mitotic stem cell density in *Smed-HDAC1-1(RNAi)* animals. For *Smed-HDAC1-1(RNAi)* (blue) and *unc22(RNAi)* (white) animals on D4, D7, D10, D12, and D15 mitotic figures were identified, counted, and compared to the volume of the worm to produce a mitotic figure density. This graph is a boxplot, the box itself contains the middle 50% of the data. The upper edge of the box indicates the 75th percentile of the data set, and the lower edge indicates the 25th percentile. The line in the box indicates the median value of the data. The ends of the vertical lines indicate the minimum and maximum data values, and the points outside the ends of the vertical lines are outliers or suspected outliers. (B) Stem cells are still present in D12 *Smed-HDAC1-1 (RNAi)* animals. D12 *Smed-HDAC1-1(RNAi)* animals and unc22(RNAi) animals were dissociated and the stem cells were sorted and counted. The numbers reported are an average of a 220,000 total events over three separate sorts for *Smed-HDAC1-1(RNAi)* and for *unc22(RNAi)* animals. (For color version of this figure, the reader is referred to the online version of this chapter.)

different levels of acetylation (Fig. 3.5). This is interesting because we could take advantage of this observation not only to label the stem cells as they transition from G2 to M with anti-H3P (Van Hooser, Goodrich, Allis, Brinkley, & Mancini, 1998), but to further separate the mixed-staged cycling stem cell population by detecting H3K9K14-Ac (Fig. 3.5) and H3K9-Me3 levels.

Additionally, a combination of histone mark detections will allow for higher resolution studies of cell cycle dynamics because specific modifications are known to occur during specific stages of the cycle. For example, H3K9K14-Ac is present in areas of the genome that are being actively transcribed (Berger, 2007). Therefore when levels of this modification are high, it should indicate times during the cell cycle when many genes are being transcribed. Transcription increases from G1 through S and peaks during G2 followed by an almost complete cessation of transcription during mitosis (White, Gottlieb, Downes, & Jackson, 1995). H3K9-Me3 is found in areas of the genome that are transcriptionally silent (Berger, 2007), and its levels have been shown to increase during mitosis (Park et al., 2011). Use of antibodies against these three sets of modifications, H3P, H3K9K14-Ac, and H3K9-Me3 should make it possible to identify cell cycle stage specific stem cells.

Equally interesting are the expression patterns uncovered for proteins known to effect histone posttranslational modifications. Two common patterns are a nervous system and stem cell-like pattern. Four genes resemble a stem cell-like pattern, *Smed-HDAC1-1* a histone deacetylase, *Smed-SETD8-1* a histone methyltransferase, and two putative DNA demethylases *Smed-NHP2L1* and *Smed-RL7A*. The other common expression pattern was a nervous system-like pattern and was found in thirteen genes of four enzyme classes. When the function of each of the genes were perturbed, six genes in three different epigenetic modifying classes produced a stem cell-like deficient phenotype. Four are the same genes that gave stem cell-like expression patterns and the remaining two are in the same family as *Smed-RL7A* and *Smed-NHP2L1*, putative DNA demethylases.

The number of genes with expression patterns in the nervous system may reflect multiple roles for epigenetics in the regulation of neuronal circuits. It has been shown that epigenetic regulation of gene expression is indispensable for neural lineage differentiation, memory, and learning (Feng, Fouse, & Fan, 2007; Taniura, Sng, & Yoneda, 2007). Therefore, changes in the epigenetic state may contribute to phenotypes associated with neural aging and neurodegenerative disorders, such as Alzheimer's disease

(Peleg et al., 2010; Rumbaugh & Miller, 2011; Stilling & Fischer, 2011). It would be interesting to determine if the activities detected in the planarian nervous system may reflect similar changes and serve, perhaps, as experimental paradigms to study these processes.

The functional studies described here of *Smed-HDAC1-1*, a class I histone deacetylase, are also intriguing. Class I histone deacetylases are usually found in the nucleus, are expressed in most tissues and cell lines in humans (Verdin, Dequiedt, & Kasler, 2003), and cause embryonic lethality when eliminated in mice (Dovey, Foster, & Cowley, 2010). The HDAC1 homolog in *C. elegans*, *hda-1*, is also ubiquitously expressed and it too results in embryonic lethality when mutated in these animals (Dufourcq et al., 2002). HDAC inhibitors have been shown to result in cell cycle arrest in cancer cell lines (Tonelli et al., 2006). In marked contrast, *Smed-HDAC1-1* is not ubiquitously expressed in planarians, and the loss of this protein is likely causing cell cycle arrest. In *Smed-HDAC1-1(RNAi)* animals, the stem cell marker decreases first, followed by a concomitant decrease in the early progeny first, and the late progeny second.

Closer examination of the stem cells by FACS in these RNAi-treated animals at D12 (Fig. 3.9B), a day in which the stem cell marker was almost completely absent, showed that the stem cells are in fact still present in normal numbers. This is accompanied by a loss of mitotic figures, measured by the absence of anti-H3P labeling (Fig. 3.9A). This indicates that the loss of *Smed-HDAC1-1* causes the cell cycle to arrest. Since there is a loss of phosphorylation of the H3 serine 10, the cell cycle is not progressing from G2 to M. And because one of our criteria for identifying stem cells by FACS is an increase in the concentration of DNA content (>2n), the cells are likely to be arrested somewhere between G1 and G2, or in one of the phases of mitosis. *Smed-HDAC1-1* is likely responsible for the regulation of genes that are required for the decision of stem cells to self-renew and/or to differentiate. The loss of *Smed-HDAC1-1*, the stem cells are halted; they are no longer dividing and no longer differentiating into early progeny. The lack of stem cell differentiation is proposed because they are not being depleted even as there is a subsequent reduction in the expression of *Smed-NB.21.11e* and *Smed-AGAT-1*, presumably due to their differentiation or death.

With many methodologies needed for studying chromatin state in place, and many more growing rapidly in complexity and sensitivity, great strides can now be made in the field of ASC epigenetics in planaria. Large-scale studies such as ChIP-Seq could be used to discover genes differentially modified in ASC and their progeny. Whole genome tiling arrays using chromatin immunoprecipitated DNA as well as ChiP seq is an excellent technique for the identification of regulatory regions of genes that are essential for stem cell maintenance, self-renewal, and differentiation. There are many proteomic approaches that can be used to better characterize epigenetics in ASCs as well. Mass spectrometry analysis of histone proteins from ASCs and other cell types could yield novel histone posttranslational modifications or combinations of modifications that are unique to ASCs. Hence we anticipate planarians being able to contribute in significant ways to our understanding of epigenomics and the role it may or may not play in stem cell regulation *in vivo*, cell differentiation, regeneration, and plasticity.

ACKNOWLEDGMENTS

This work was supported by NIH Genetics Training Grant T32GM007464 to SMCR and by NIH Grant R37GM057260 to A. S. A. A. S. A. is an Investigator of the Howard Hughes Medical Institute and the Stowers Institute for Medical Research.

REFERENCES

Bannister, A. J., Zegerman, P., Partridge, J. F., Miska, E. A., Thomas, J. O., Allshire, R. C., et al. (2001). Selective recognition of methylated lysine 9 on histone H3 by the HP1 chromo domain. *Nature, 410*, 120–124.

Berger, S. L. (2007). The complex language of chromatin regulation during transcription. *Nature, 447*, 407–412.

Bernstein, B. E., Mikkelsen, T. S., Xie, X., Kamal, M., Huebert, D. J., Cuff, J., et al. (2006). A bivalent chromatin structure marks key developmental genes in embryonic stem cells. *Cell, 125*, 315–326.

Cantarel, B. L., Korf, I., Robb, S. M., Parra, G., Ross, E., Moore, B., et al. (2008). MAKER: An easy-to-use annotation pipeline designed for emerging model organism genomes. *Genome Research, 18*, 188–196.

Chambeyron, S., & Bickmore, W. A. (2004). Chromatin decondensation and nuclear reorganization of the HoxB locus upon induction of transcription. *Genes and Development, 18*, 1119–1130.

Dovey, O. M., Foster, C. T., & Cowley, S. M. (2010). Histone deacetylase 1 (HDAC1), but not HDAC2, controls embryonic stem cell differentiation. *Proceedings of the National Academy of Sciences of the United States of America, 107*, 8242–8247.

Dufourcq, P., Victor, M., Gay, F., Calvo, D., Hodgkin, J., & Shi, Y. (2002). Functional requirement for histone deacetylase 1 in Caenorhabditis elegans gonadogenesis. *Molecular Cell Biology, 22*, 3024–3034.

Eilertsen, K. J., Floyd, Z., & Gimble, J. M. (2008). The epigenetics of adult (somatic) stem cells. *Critical Reviews in Eukaryotic Gene Expression, 18*, 189–206.

Eisenhoffer, G. T., Kang, H., & Sánchez Alvarado, A. (2008). Molecular analysis of stem cells and their descendants during cell turnover and regeneration in the planarian Schmidtea mediterranea. *Cell Stem Cell, 3*, 327–339.

Feng, J., Fouse, S., & Fan, G. (2007). Epigenetic regulation of neural gene expression and neuronal function. *Pediatric Research, 61*, 58R–63R.

Lachner, M., O'Carroll, D., Rea, S., Mechtler, K., & Jenuwein, T. (2001). Methylation of histone H3 lysine 9 creates a binding site for HP1 proteins. *Nature, 410*, 116–120.

Malik, H. S., & Henikoff, S. (2003). Phylogenomics of the nucleosome. *Nature Structural Biology, 10*, 882–891.

Mariño-Ramírez, L., Hsu, B., Baxevanis, A. D., & Landsman, D. (2005). The histone database: A comprehensive resource for histones and histone fold-containing proteins. *Proteins: Structure, Function, and Bioinformatics, 62*, 838–842.

Marino-Ramirez, L., Jordan, I. K., & Landsman, D. (2006). Multiple independent evolutionary solutions to core histone gene regulation. *Genome Biology, 7*, R122.

Molofsky, A. V., Pardal, R., & Morrison, S. J. (2004). Diverse mechanisms regulate stem cell self-renewal. *Current Opinion in Cell Biology, 16*, 700–707.

O'Neill, L. P., VerMilyea, M. D., & Turner, B. M. (2006). Epigenetic characterization of the early embryo with a chromatin immunoprecipitation protocol applicable to small cell populations. *Nature Genetics, 38*, 835–841.

Park, J. A., Kim, A. J., Kang, Y., Jung, Y. J., Kim, H. K., & Kim, K. C. (2011). Deacetylation and methylation at histone H3 lysine 9 (H3K9) coordinate chromosome condensation during cell cycle progression. *Molecules and Cells, 31*, 343–349.

Peleg, S., Sananbenesi, F., Zovoilis, A., Burkhardt, S., Bahari-Javan, S., Agis-Balboa, R. C., et al. (2010). Altered histone acetylation is associated with age-dependent memory impairment in mice. *Science, 328*, 753–756.

Peterson, C. L., & Laniel, M. A. (2004). Histones and histone modifications. *Current Biology, 14*, R546–R551.

Reddien, P. W., Bermange, A. L., Murfitt, K. J., Jennings, J. R., & Sánchez Alvarado, A. (2005). Identification of genes needed for regeneration, stem cell function, and tissue homeostasis by systematic gene perturbation in planaria. *Developmental Cell, 8*, 635–649.

Robb, S. M. C., Ross, E., & Alvarado, A. S. (2007). SmedGD: The Schmidtea mediterranea genome database. *Nucleic Acids Research, 36*, D599–D606.

Rooney, A. P., Piontkivska, H., & Nei, M. (2002). Molecular evolution of the nontandemly repeated genes of the histone 3 multigene family. *Molecular Biology and Evolution, 19*, 68–75.

Rouhana, L., Vieira, A. P., Roberts-Galbraith, R. H., & Newmark, P. A. (2012). PRMT5 and the role of symmetrical dimethylarginine in chromatoid bodies of planarian stem cells. *Development, 139*, 1083–1094.

Rumbaugh, G., & Miller, C. A. (2011). Epigenetic changes in the brain: Measuring global histone modifications. *Methods in Molecular Biology, 670*, 263–274.

Sittman, D. B., Chiu, I. M., Pan, C. J., Cohn, R. H., Kedes, L. H., & Marzluff, W. F. (1981). Isolation of two clusters of mouse histone genes. *Proceedings of the National Academy of Sciences of the United States of America, 78*, 4078–4082.

Solana, J., Kao, D., Mihaylova, Y., Jaber-Hijazi, F., Malla, S., Wilson, R., et al. (2012). Defining the molecular profile of planarian pluripotent stem cells using a combinatorial RNAseq, RNA interference and irradiation approach. *Genome Biology, 13*, R19.

Stilling, R. M., & Fischer, A. (2011). The role of histone acetylation in age-associated memory impairment and Alzheimer's disease. *Neurobiology of Learning and Memory, 96*, 19–26.

Taniura, H., Sng, J. C., & Yoneda, Y. (2007). Histone modifications in the brain. *Neurochemistry International, 51*, 85–91.

Thompson, J. D., Higgins, D. G., & Gibson, T. J. (1994). CLUSTAL W: Improving the sensitivity of progressive multiple sequence alignment through sequence weighting, position-specific gap penalties and weight matrix choice. *Nucleic Acids Research, 22*, 4673–4680.

Tonelli, R., Sartini, R., Fronza, R., Freccero, F., Franzoni, M., Dongiovanni, D., et al. (2006). G1 cell-cycle arrest and apoptosis by histone deacetylase inhibition in MLL-AF9 acute myeloid leukemia cells is p21 dependent and MLL-AF9 independent. *Leukemia, 20*, 1307–1310.

Van Hooser, A., Goodrich, D. W., Allis, C. D., Brinkley, B. R., & Mancini, M. A. (1998). Histone H3 phosphorylation is required for the initiation, but not maintenance, of mammalian chromosome condensation. *Journal of Cell Science*, *111*(Pt. 23), 3497–3506.
Verdin, E., Dequiedt, F., & Kasler, H. G. (2003). Class II histone deacetylases: Versatile regulators. *Trends in Genetics*, *19*, 286–293.
White, R. J., Gottlieb, T. M., Downes, C. S., & Jackson, S. P. (1995). Cell cycle regulation of RNA polymerase III transcription. *Molecular Cell Biology*, *15*, 6653–6662.
Zeng, A., Li, Y. Q., Wang, C., Han, X. S., Li, G., Wang, J. Y., et al. (2013). Heterochromatin protein 1 promotes self-renewal and triggers regenerative proliferation in adult stem cells. *The Journal of Cell Biology*, *201*, 409–425.

CHAPTER FOUR

Control of Growth During Regeneration

Gongping Sun, Kenneth D. Irvine[1]
Howard Hughes Medical Institute, Waksman Institute and Department of Molecular Biology and Biochemistry, Rutgers The State University of New Jersey, Piscataway, New Jersey, USA
[1]Corresponding author: e-mail address: irvine@waksman.rutgers.edu

Contents

Abstract

Regeneration is a process by which organisms replace damaged or amputated organs to restore normal body parts. Regeneration of many tissues or organs requires proliferation of stem cells or stem cell-like blastema cells. This regenerative growth is often initiated by cell death pathways induced by damage. The executors of regenerative growth are a group of growth-promoting signaling pathways, including JAK/STAT, EGFR, Hippo/YAP, and Wnt/β-catenin. These pathways are also essential to developmental growth, but in regeneration, they are activated in distinct ways and often at higher strengths, under the regulation by certain stress-responsive signaling pathways, including JNK signaling. Growth suppressors are important in termination of regeneration to prevent unlimited growth and also contribute to the loss of regenerative capacity in nonregenerative organs. Here, we review cellular and molecular growth regulation mechanisms induced by organ damage in several models with different regenerative capacities.

Current Topics in Developmental Biology, Volume 108
ISSN 0070-2153
http://dx.doi.org/10.1016/B978-0-12-391498-9.00003-6

ABBREVIATIONS

AEC apical epithelial cap
BMP bone morphogenetic protein
Ds Dachsous
EB enteroblast
EC enterocyte
ECM extracellular matrix
EGFR epidermis growth factor receptor
FGF fibroblast growth factor
Fj Four-jointed
HGF hepatocyte growth factor
IL-6 interleukin-6
ILK integrin-linked kinase
ISC intestinal stem cell
JNK c-Jun N-terminal kinase
MAPK mitogen-activated protein kinase
MEF mouse embryonic fibroblast
MMP matrix metalloproteinase
PH partial hepatectomy
TGFβ transforming growth factor β
TLR Toll-like receptor
TNFα tumor necrosis factor α

Regeneration enables animals to recover from severe organ damage or amputation by restoring fully formed and functional organs. In 1901, Morgan classified regeneration into two types, epimorphosis and morphallaxis (Morgan, 1901). Morphallaxis is regeneration through remodeling of preexisting cells, while epimorphosis is regeneration in which growth contributes to restoration of missing tissue (Brockes & Kumar, 2008; Reddien & Sanchez Alvarado, 2004). While epimorphic regeneration requires several steps, including wound closure, activation of preexisting stem cells or dedifferentiation of differentiated cells, cell proliferation, patterning of newly formed tissue, and differentiation (Carlson, 2005; Yokoyama, 2008), a distinguishing feature is the reliance on precisely controlled proliferation to restore correctly sized organs. The capacity for regeneration varies from species to species. For example, Planaria and *Hydra* are famous for their incredible capacity for regeneration, being able to restore the whole body from a small part. Amphibians can regenerate several functional organs depending upon the species, such as limbs, tails, and lens. By contrast, mammals have only a very limited capacity to regenerate damaged organs. Here, we focus on cellular and molecular mechanisms that control growth during epimorphic regeneration, discussing examples from several models that differ in regenerative capacity.

1. CELLS PARTICIPATING IN REGENERATIVE GROWTH

A crucial question in epimorphic regeneration is the origin of the cells that participate in regenerative growth. In most cases, robust proliferation is local rather than systemic, and progenitor cells that concentrate around wound sites and are responsible for producing cells needed for regrowth form what is referred to as a blastema. In different systems, blastema cells can come from stem cells residing in injured tissues, differentiated cells that undergo dedifferentiation, or parenchymal cells with proliferation capacity (King & Newmark, 2012; Tanaka & Reddien, 2011). Planarians maintain multipotent stem cells throughout adult life, and their regeneration is achieved mainly through increased proliferation of these stem cells (Reddien, 2013; Reddien & Sanchez Alvarado, 2004). Regeneration of some organs in other species, such as the vertebrate or insect intestine, is also stem cell-dependent, although in intestine there is no specific blastema region formed. In other cases, such as limb or tail regeneration in amphibians, blastema cells form from dedifferentiation of differentiated cells, and then proliferate to replace missing tissue. In tissues with proliferating parenchymal cells, such as *Drosophila* wing imaginal discs or mammalian livers, regeneration relies mainly on accelerating proliferation of the remaining undamaged cells.

Blastema cells are usually the key players in regenerative growth. However, there are also nonproliferating cells that play important roles by influencing blastema cell formation, proliferation, and differentiation. In many species, formation of a blastema requires wound healing and formation of a wound epithelium (specialized epithelial cells that migrate to cover the wound) (Brockes & Kumar, 2008; Campbell & Crews, 2008). Replacement of this wound epithelium with unwounded skin abolishes limb regeneration in newts (Brockes & Kumar, 2008; Carlson, 2005). Innervation also influences blastema cell proliferation (Brockes & Kumar, 2008; Endo, Bryant, & Gardiner, 2004; Yokoyama, 2008). The requirement for these and other cells could reflect the influence of growth factors and cytokines that they secrete on regenerative growth. For example, regeneration of the adult midgut in *Drosophila* involves proliferation of intestinal stem cells (ISCs), but this proliferation is mainly regulated by signaling molecules secreted from differentiated cells called enterocytes (ECs) (Jiang & Edgar, 2011). In mammalian liver regeneration, proliferation of hepatocytes requires growth factors produced by Kupffer cells (Bohm, Kohler, Speicher, & Werner, 2010).

2. RECOGNIZING TISSUE DAMAGE TO INITIATE REGENERATION

When tissues are damaged, the injured cells often undergo apoptosis. Studies in many organisms, including *Hydra*, *Drosophila*, and mouse, have established that apoptosis is not only a way to remove damaged cells, but also an important trigger stimulating the proliferation of nearby healthy cells.

After mid-gastric bisection of *Hydra*, each half regenerates a complete *Hydra* body. *Hydra* regeneration is considered to be morphallactic (Park, Ortmeyer, & Blankenbaker, 1970). Nonetheless, proliferation is induced in the head-regenerating tip after mid-gastric bisection. This proliferation, as well as induction of Wnt3, which is required for head regeneration, is induced by local apoptosis. Blocking apoptosis using caspase inhibitors leads to failure of head regeneration. Moreover, ectopic induction of apoptosis in foot-regenerating tips stimulates cell proliferation and Wnt3 expression, and leads to ectopic head regeneration (Chera et al., 2009). Thus, in *Hydra*, apoptosis is necessary and sufficient to trigger head regeneration.

The imaginal discs of *Drosophila* have been an important model for the characterization of damage-induced cell proliferation. Imaginal discs are the precursors of adult structures, present within the larva as clusters of proliferating cells. Induction of apoptosis by irradiation or expression of proapoptotic genes triggers elevated proliferation near dying cells, which is referred to as compensatory cell proliferation (Fan & Bergmann, 2008a). In wing discs, stimulating apoptosis while blocking the execution of apoptosis leads to sustained compensatory proliferation and, consequently, tissue overgrowth. This can occur in animals mutant for the effector caspase Drice/Dcp-1 (Kondo, Senoo-Matsuda, Hiromi, & Miura, 2006), or when a Drice inhibitor, the bacculovirus P35 protein, is expressed (Huh, Guo, & Hay, 2004; Perez-Garijo, Martin, & Morata, 2004; Ryoo, Gorenc, & Steller, 2004). Conversely, if apoptosis is prevented at an earlier step, by mutation or downregulation of the initiator caspase Dronc, then compensatory cell proliferation does not occur. These observations indicate that it is not simply the elimination of cells that triggers compensatory proliferation, but specific molecular events associated with induction of apoptosis. However, the molecular mechanisms involved can vary in different tissues, for example, in eye imaginal discs, blocking the effector caspase Drice by overexpressing P35 abolishes ectopic proliferation and tissue regeneration (Fan & Bergmann, 2008b), whereas this is not the case in wing imaginal discs.

A requirement for apoptosis and caspase activation in stimulating cell proliferation after tissue damage is also observed in mammals. For example, lethally irradiated mouse embryonic fibroblasts (MEFs) can stimulate proliferation of nearby stem or progenitor cells both *in vitro* and *in vivo*. Effector caspases (caspase 3 and caspase 7) are required for this cell death-induced cell proliferation, as *Casp3*$^{-}$ or *Casp7*$^{-}$ lethally irradiated MEFs are significantly impaired in their stimulation of proliferation compared to wild-type MEFs. Furthermore, *Casp3*$^{-}$ or *Casp7*$^{-}$ mice are deficient in skin wound healing and liver regeneration (Li et al., 2010). This requirement for effector caspases is similar to what has been observed for *Drosophila* eye discs.

3. REGENERATIVE GROWTH IS CONTROLLED BY SIGNALING NETWORKS

After damage is recognized, injury signals need to be translated to proliferating signals to guide the replacement of damaged parts with new tissue. One pathway involved in this link is c-Jun N-terminal kinase (JNK). JNK belongs to the mitogen-activated protein kinase (MAPK) family and is activated by diverse cellular stresses, including wounding, oxidative stress, loss of cell polarity, infection, and apoptosis. It regulates many biological processes including apoptosis, proliferation, cytoskeletal rearrangement, and inflammation (Bogoyevitch, Ngoei, Zhao, Yeap, & Ng, 2010; Chen, 2012). JNK is among the earliest activated signals in response to wounding, and is required for wound healing. The essential role of JNK in triggering proliferation in regeneration is well established in *Drosophila* wing discs, where suppression of JNK activity blocks cell proliferation in response to damage (Bergantinos, Corominas, & Serras, 2010; Mattila, Omelyanchuk, Kyttala, Turunen, & Nokkala, 2005). The activation of proliferation by damage signals in mammalian liver is more complicated, involving multiple pathways including Toll-like receptor (TLR) signaling, JNK, and cytokine signaling (Bohm et al., 2010; Iimuro & Fujimoto, 2010). Though a role for JNK in vertebrate limb regeneration has not yet been reported, as in *Drosophila*, limb amputation or wounding induces the expression of matrix metalloproteinases (MMPs), which are responsible for extracellular matrix (ECM) degradation to allow cell dedifferentiation and blastema formation

(Yang & Bryant, 1994). Inhibition of MMPs causes abnormal tissue regeneration (Vinarsky, Atkinson, Stevenson, Keating, & Odelberg, 2005). In *Drosophila* and *Caenorhabditis elegans*, MMPs have proved to be regulated by JNK signaling (Page-McCaw, 2008). Thus, it is possible that in vertebrate limb regeneration, JNK also functions at this early stage. In *Hydra*, direct evidence for JNK participation in regeneration has not been reported yet, but MMPs are upregulated and required for head regeneration after decapitation (Shimizu et al., 2002; Yan, Leontovich, Fei, & Sarras, 2000). Moreover, another MAPK, extracellular signal-regulated kinase, and its downstream target CREB, are activated and required for apoptosis-induced compensatory cell proliferation during *Hydra* head regeneration (Chera, Ghila, Wenger, & Galliot, 2011), reminiscent of the requirement for JNK during *Drosophila* wing disc regeneration.

After activation of stress-responsive signals, the maintenance of regenerative growth is executed by multiple growth-controlling pathways including but not limited to Wnt/β-catenin, Hippo/Yorkie, JAK/STAT, and epidermis growth factor receptor (EGFR). All these pathways are also required during normal development. However, the manner in which they are regulated can differ in regenerating tissues. Also, the signaling strength required in regenerating tissues is often different from that in developing tissues, as regenerating tissues can be more sensitive to a reduction in dose of some growth-promoting factors.

3.1. Signals promoting growth during stem cell-driven regeneration: The Drosophila midgut

Stem cells maintain homeostasis in adult organs, providing a reservoir of multipotent cells for tissue renewal. When an organ is damaged, stem cell proliferation can be accelerated to speed repair. An example of this that has been intensively investigated over the last several years occurs in the adult midgut of *Drosophila* (Wang & Hou, 2010). Maintenance of midgut epithelium homeostasis relies on ISCs localized on the basal side of the epithelium. Through asymmetric cell division, an ISC can produce a new ISC and a bipotential progenitor cell called an enteroblast (EB), which then differentiates into either an absorptive EC, which makes up the bulk of the midgut epithelium, or a secretory enteroendocrine cell.

Toxins, injury, or pathogen infection of the epithelium can accelerate expansion and differentiation of ISCs to replace the damaged cells (Wang & Hou, 2010). This occurs through induction of JNK activation in ECs in response to damage or infection. This triggers secretion from ECs of both Unpaired (Upd) proteins, which are signals related to vertebrate cytokines that act as ligands for the JAK-STAT signaling pathway, and of ligands for EGFR signaling (Jiang & Edgar, 2011; Jiang, Grenley, Bravo, Blumhagen, & Edgar, 2011) (Fig. 4.1). JNK induction of JAK-STAT and EGFR ligands is mediated through JNK-dependent activation of Yorkie (Yki) in ECs, which then promotes the expression of JAK-STAT and EGFR ligands (Karpowicz, Perez, & Perrimon, 2010; Ren et al., 2010; Shaw et al., 2010; Staley & Irvine, 2010). Moreover, some groups also reported that inactivation of Hippo signaling and the consequent activation of Yki could occur in ISCs in response to damage of the midgut epithelium, and that this also promotes stem cell proliferation (Karpowicz et al., 2010; Ren et al., 2010; Shaw et al., 2010), although the major signals promoting ISC proliferation are produced by the ECs. JNK activation in ECs can also stimulate Wingless expression in EBs and ISCs, which promote Myc-dependent ISC proliferation in stressed midgut (Cordero, Stefanatos, Scopelliti, Vidal, & Sansom, 2012).

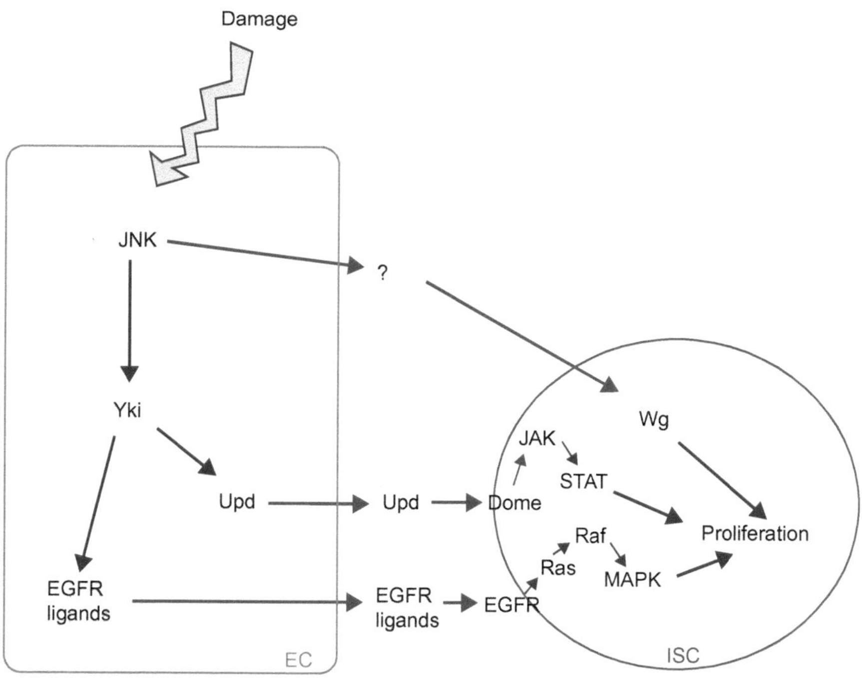

Figure 4.1 Signaling network promoting *Drosophila* midgut regeneration. Schematic of regulatory interactions between components of signaling pathways involved in promoting Drosophila midgut regeneration, see text for details. The mechanism by which JNK activation in ECs stimulates Wg in ISCs is unknown (indicated by ?). (For the color version of this figure, the reader is referred to the online version of this chapter.)

The Upd secreted by ECs activates the kinase JAK in ISCs through binding to its receptor Domeless (Dome). Activated JAK then phosphorylates and promotes nuclear translocation of the downstream transcription factor STAT, which promotes ISC proliferation and differentiation (Jiang et al., 2009). Similarly, EGFR ligands activate EGFRs in ISCs that in turn activate the Ras/Raf/MAPK pathway to promote ISC proliferation (Jiang et al., 2011). Notably, loss of one copy of the *Drosophila* JAK does not affect ISC number under physiological conditions, but it suppresses ISC proliferation caused by EC apoptosis, which indicates that a high level of JAK/STAT signaling is needed for midgut regeneration (Jiang et al., 2009).

These studies of midgut regeneration after induction of damage have revealed that a complex interplay amongst multiple signaling pathways, and involving multiple cell types, is needed to induce and control the cell proliferation needed to repair a damaged intestine.

3.2. Regeneration from proliferating parenchymal cells: Drosophila wing imaginal discs

The *Drosophila* wing imaginal disc is a widely used model to study regulatory mechanisms of cell proliferation during regeneration because of its high regenerative capacity and the ease of genetic manipulation (Bergantinos, Vilana, Corominas, & Serras, 2010). Imaginal disc regeneration can be studied after surgical excision of part of the disc, and such studies have contributed to our understanding of how positional information influences regenerative growth (Haynie & Bryant, 1976). More recently, regeneration has been studied after genetic ablation of a portion of a developing imaginal disc, typically induced by localized, transient expression of proapoptotic genes such as *reaper*, *hid*, or *eiger* (Grusche, Degoutin, Richardson, & Harvey, 2011; Herrera, Martin, & Morata, 2013; Smith-Bolton, Worley, Kanda, & Hariharan, 2009; Sun & Irvine, 2011).

JNK is found among the molecules responding first after tissue damage and is an essential modulator of regenerative growth. JNK activity appears at the leading edge of healing tissues soon after genetic or surgical ablation, and persists and broadens during and after wound healing (Bergantinos, Corominas, et al., 2010; Bosch, Serras, Martin-Blanco, & Baguna, 2005). JNK activation is also essential for apoptosis-induced compensatory cell proliferation (Ryoo et al., 2004). As noted earlier, blocking effector caspase activity by overexpressing P35 blocks the removal of dying cells, but these "undead" cells keep triggering compensatory proliferation, causing hyperplastic growth in discs (Morata, Shlevkov, & Perez-Garijo, 2011; Ryoo et al., 2004). These observations revealed that compensatory cell proliferation is a response to molecular events associated with induction of apoptosis, and consistent with this, JNK activation, which is required for hyperplastic disc growth induced by undead cells, can be induced by activation of the initiator Caspase Dronc (Pahlavan, Feldmann, Zavos, & Kountouras, 2006). There is also a potential for positive feedback between JNK and Dronc (Fig. 4.2), as activation of JNK can promote apoptosis through upregulation of *hid* and *reaper*, which leads to Dronc activation (Moreno, Yan, & Basler, 2002).

Lineage analysis experiments have revealed that a large proportion of the cells that contribute to regeneration after disc damage arise from cells with JNK activation (Bosch, Baguna, & Serras, 2008). Moreover, blocking JNK activity significantly suppresses wing disc regeneration and compensatory proliferation in response to genetic ablation, amputation, or irradiation (Bergantinos, Corominas, et al., 2010; Mattila et al., 2005; Ryoo et al., 2004). As is the case for the intestine, the growth-promoting function of JNK during wing disc regeneration stems from its regulation of other growth control pathways (Fig. 4.2). Activation of JNK leads to the localized elevation of Wingless (Wg, *Drosophila* homolog of Wnt1), Decapentaplegic (Dpp, a *Drosophila* homolog of BMPs), and activation of the Hippo pathway

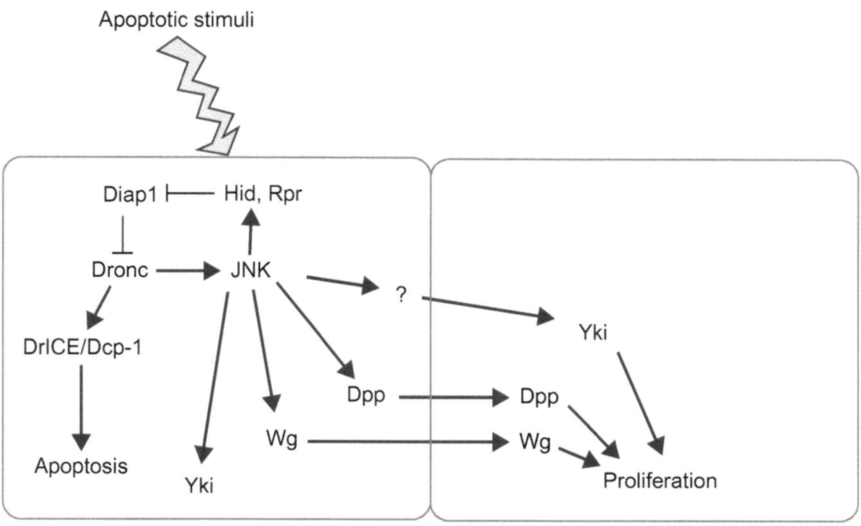

Figure 4.2 Signaling network promoting *Drosophila* wing disc regeneration. Schematic of regulatory interactions between components of signaling pathways involved in promoting *Drosophila* wing disc regeneration, see text for details. (For the color version of this figure, the reader is referred to the online version of this chapter.)

downstream effector Yki (Grusche et al., 2011; Huh, Guo, et al., 2004; Perez-Garijo et al., 2004; Ryoo et al., 2004; Sun & Irvine, 2011). Strong activation of Yki is observed in *Drosophila* wing discs after genetic ablation of the wing pouch, which can be suppressed by blocking JNK activity (Sun & Irvine, 2011). Yki activity is required for cell proliferation during regeneration, and regenerating wing discs require even higher Yki activity than developing discs, as loss of one copy of *yki* severely impairs the regenerative capacity of *Drosophila* wings but does not affect normal wing development (Grusche et al., 2011; Sun & Irvine, 2011).

The roles of Wg and Dpp upregulation in the wing disc damage response are complex and context-dependent. When wing discs expressing P35 were subjected to irradiation or coexpression of proapoptotic genes, ectopic Wg and Dpp expression were detected in both "undead" cells and normal cells nearby (Morata et al., 2011). Both Wg and Dpp are required for the hyperplastic growth caused by "undead" cells (Morata et al., 2011; Perez-Garijo, Shlevkov, & Morata, 2009). However, in the absence of P35 expression, though ectopic expression of Wg and Dpp is observed (Perez-Garijo et al., 2004), it is dispensable for compensatory cell proliferation (Perez-Garijo et al., 2009). When a portion of a disc is ablated by transient expression of proapoptotic genes in the absence of P35, different responses have been observed depending upon the proapoptotic gene employed. When apoptosis is induced by expressing *eiger* or *reaper*, proliferation is stimulated at the edge of the wound, and Wg is required for this compensatory cell proliferation and for wing regeneration (Smith-Bolton et al., 2009). At least in part, this occurs through Wg downregulating Notch, which leads to upregulation of Myc (Smith-Bolton et al., 2009). Intriguingly, when apoptosis is induced by expressing the proapoptotic gene, *hid*, Wg and Dpp expression patterns are not altered (Herrera et al., 2013). Moreover, knocking down *wg* does not block wing regeneration after *hid* expression (Herrera et al., 2013). But unlike the localized elevation of proliferation observed after *egr*- or *rpr*-induced ablation (Smith-Bolton et al., 2009), or proliferation associated with "undead" cells (Ryoo et al., 2004), after *hid*-induced ablation, elevated proliferation is induced uniformly throughout the regenerating disc (Herrera et al., 2013). The basis for these differences in patterns of cell proliferation during regeneration and the requirement of mitogenic signals is not yet clear. They might in part reflect differences in the efficiency of genetic ablation, but it seems that there are multiple, distinct mechanisms that can promote cell proliferation during disc regeneration, which may be preferentially induced depending upon the nature of the tissue damage.

3.3. Regeneration from proliferating parenchymal cells: Mammalian liver

The liver is among the few organs in mammals with regenerative capacity. The original mass and function can be restored even after removal of 75% of the liver through partial hepatectomy (PH). Similar to *Drosophila* wing discs, liver regeneration after PH relies mainly on the proliferation of the major parenchymal cells in liver, hepatocytes, although in recent years, the existence of stem cells in liver and their function in liver regeneration has also been reported (Pahlavan et al., 2006). Hepatocytes are not terminally differentiated, but they normally stay quiescent at the G0 phase of the cell cycle. After PH or chemical injury, hepatocytes are activated and reenter the cell cycle to initiate compensatory cell proliferation. Hepatocyte activation is triggered by other types of cells in liver, such as Kupffer cells and Stellate cells. After PH, TLR is activated in Kupffer cells and, through regulation of the transcription factor NF-κB, induces secretion of tumor necrosis factor α (TNFα) and interleukin-6 (IL-6) (Seki et al., 2005) (Fig. 4.3). TNFα is a ligand for TNF receptor 1 (TNFR1), which is present in hepatocytes and can signal through both JNK and NF-κB. Blocking TNFα signaling using TNFα neutralizing antibodies, or by mutation of *Tnfr1*, impairs liver regeneration after PH (Akerman et al., 1992; Yamada, Kirillova, Peschon, & Fausto, 1997). TNFα can also bind to TNFR1 in Kupffer cells to promote IL-6 secretion. Blocking TNFα signaling leads to suppression of IL-6 upregulation upon PH (Akerman et al., 1992; Yamada et al., 1997). Injection of IL-6 before PH rescues the defect in liver regeneration in TNFR1-deficient mice (Yamada et al., 1997), suggesting that the major function of TNFα and TNFR1 in liver regeneration is to regulate IL-6 expression. Secreted IL-6 activates the JAK/STAT3 pathway in hepatocytes through the IL-6 receptor GP130. Mice deficient in IL-6, GP130, or STAT3 exhibit impaired liver regeneration after PH (Cressman et al., 1996; Li, Liang, Kellendonk, Poli, & Taub, 2002; Wuestefeld et al., 2003).

JNK, the stress-response pathway that is essential for tissue regeneration in *Drosophila*, is also important in mammalian liver regeneration. The activity of JNK and one of its targets, the AP-1 transcription factor subunit c-Jun, is markedly stimulated after PH, possibly by TNF and EGF (Akerman et al., 1992; Schwabe et al., 2003; Westwick, Weitzel, Leffert, & Brenner, 1995; Yamada et al., 1997). Liver-specific deletion of c-Jun reduced hepatocyte proliferation, increased cell death, and inhibited cell cycle progression, leading to impaired liver regeneration (Behrens et al., 2002). Treatment with a JNK-specific inhibitor, SP600125, reduced hepatocyte proliferation and

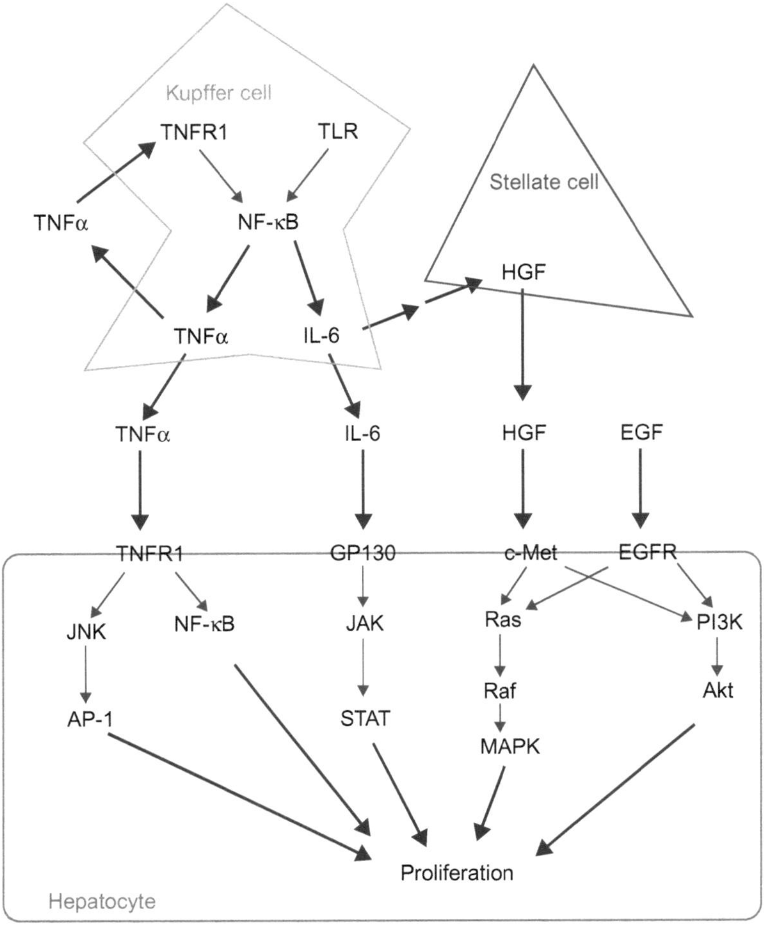

Figure 4.3 Signaling network promoting mammalian liver regeneration. Schematic of regulatory interactions between components of signaling pathways involved in promoting mammalian liver regeneration, see text for details. (For the color version of this figure, the reader is referred to the online version of this chapter.)

cyclin D1 expression in rats after PH (Schwabe et al., 2003). These studies identify JNK as a necessary positive regulator of liver regeneration. However, the role of JNK in liver regeneration is more complicated. There are two isoforms of JNKs in liver, JNK1 and JNK2. Distinct roles of JNK1 and JNK2 were unveiled using gene-specific knock-out mice. Impaired liver regeneration has been reported in *Jnk1*$^{-/-}$ mice, whereas *Jnk2*$^{-/-}$ mice showed no effect on regeneration, or even faster regeneration

(Das, Garlick, Greiner, & Davis, 2011; Sabapathy et al., 2004). The distinct roles of JNK1 and JNK2 in liver regeneration may be explained by differential regulation of c-Jun by these two isoforms. JNK1 promotes cell proliferation by phosphorylation of c-Jun, whereas JNK2 negatively regulates cell proliferation by inducing c-Jun degradation (Sabapathy et al., 2004).

IL-6 also induces expression of hepatocyte growth factor (HGF) (Liu, Michalopoulos, & Zarnegar, 1994), a major mitogen in liver. HGF can stimulate hepatocyte proliferation both *in vitro* and *in vivo* (Block et al., 1996; Sakata et al., 1996). HGF acts on hepatocytes in a paracrine way, through its receptor c-Met. HGF upregulation and activation of its receptor, c-Met, are observed after PH (Michalopoulos, 2007). *c-Met* conditional knock-out mice exhibit impaired regeneration after PH or necrogen CCl4 treatment (Huh, Factor, et al., 2004).

The EGFR pathway has also been intensively studied for its role in promoting growth during liver regeneration. Several ligands for EGFR such as TGFα, EGF, and Amphiregulin are increased in murine liver after PH (Michalopoulos, 2007). Each of these EGFR ligands can promote hepatocyte proliferation *in vitro*, and transgenic mice with overexpression of TGFα, heparin-binding EGF (HB-EGF), or Amphiregulin have higher hepatocyte proliferation rates after PH (Kiso et al., 2003). However, the requirement for each ligand in liver regeneration is not the same. Liver regeneration after PH proceeds normally in *Tgfα* knock-out mice (Russell, Kaufmann, Sitaric, Luetteke, & Lee, 1996), whereas *amphiregulin*^{null} and *HB-EGF* knockout mice show suppression of proliferation in early liver regeneration after PH (Berasain et al., 2005; Mitchell et al., 2005). Consistent with the studies on EGFR ligands, elevated phosphorylation of EGFR is observed after PH (Stolz, Mars, Petersen, Kim, & Michalopoulos, 1999). Mice with liver-specific deletion of *Egfr* have an increased mortality rate, defects in G1 to S phase transition, and delayed liver regeneration after PH (Natarajan, Wagner, & Sibilia, 2007). A study using short hairpin RNA against EGFR to interfere with EGFR signaling also found suppression of hepatocyte proliferation after PH (Paranjpe et al., 2010).

In sum, liver regeneration involves multiple signaling pathways responsible for stress-response and growth promotion, some of which are homologous to pathways that play crucial roles in other regeneration models. In addition, though regenerative growth directly involves parenchymal cells, the same cell type that receives injury signals, it also requires the cooperation of other cell types.

3.4. Regeneration from dedifferentiated cells: Amphibian limb

Urodele amphibians, including axolotl and *Xenopus*, are a special group of vertebrates that can regenerate many of their body parts. Regeneration of limbs after amputation involves wound healing, blastema formation, cell proliferation, and redevelopment to form a new limb (Brockes & Kumar, 2008; Yokoyama, 2008). Formation and proliferation of a blastema involve a complex coordination among different cell types. After limb amputation, epidermal cells migrate to cover the wound, and then thicken to a multilayered apical epithelial cap (AEC). ECM is degraded, releasing dermal and muscle cells that migrate toward the AEC, dedifferentiate, and form a blastema beneath it (Yokoyama, 2008). Migration of fibroblasts requires transforming growth factor β (TGFβ) signaling (Levesque et al., 2007). Following amputation of an axolotl limb, weak expression of TGFβ can be detected 6 h after amputation, and strong upregulation is observed at 48 h, which lasts for days. Blocking TGFβ signaling in regenerating limbs using a specific inhibitor, SB431542, inhibited limb regeneration. When the inhibitor was added right after amputation, formation of a wound epidermis still occurred, but was slightly delayed. No fibroblast migration was observed after wound healing, and the blastema was absent.

The AEC is responsible for production of several essential signaling molecules that promote blastema cell proliferation. High expression of Wnt3a is observed at the inner layer of the AEC in regenerating *Xenopus* limbs (Yokoyama, Ogino, Stoick-Cooper, Grainger, & Moon, 2007). Blocking Wnt/β-catenin signaling in amputated limbs of *Xenopus* at stage 52 by expressing an inhibitor of the Wnt/β-catenin pathway, Dickkopf-1, blocked limb regeneration, but not normal limb development (Kawakami et al., 2006; Yokoyama et al., 2007). Conversely, activation of the Wnt/β-catenin pathway can induce regeneration in *Xenopus* at stage 53–54, when their regenerative potential is normally greatly diminished (Kawakami et al., 2006). The Blockage of Wnt/β-catenin signaling suppresses expression of Fibroblast growth factor-8 (FGF-8) in regenerating limbs (Yokoyama et al., 2007). FGFs are a group of mitogens that promote vertebrate limb growth during both development and regeneration. Several studies have found that FGF-8 and FGF-10 are highly expressed in the AEC and blastema of regenerating limbs, but not after amputation in nonregenerative limbs (Christensen, Weinstein, & Tassava, 2002; Han, An, & Kim, 2001; Yokoyama et al., 2000). Moreover, exogenous FGF-10 is sufficient to induce complete limb regeneration in *Xenopus* nonregenerative limb stumps (Yokoyama, Ide, & Tamura, 2001). Thus limb regeneration involves a clear dedifferentiation step and cooperation of multiple tissues and signals to promote regenerative growth.

4. CONTROL OF REGENERATIVE GROWTH BY ORGAN PATTERNING

The regulation of cell proliferation induced during epimorphic regeneration must be precisely controlled such that neither too little nor too much new tissue is produced. Studies of intercalary regeneration within insect limbs were particularly influential in promoting the hypothesis that growth is influenced by a gradient of positional information (Day & Lawrence, 2000; French, Bryant, & Bryant, 1976; Haynie & Bryant, 1976). According to this hypothesis, during developmental or regenerative growth, cell proliferation is stimulated by the juxtaposition of cells with different positional values, as occurs when a portion of a limb is excised. Conversely, growth is not stimulated when cells from similar positions are juxtaposed. These kinds of observations led to the suggestion that growth continues until a gradient of positional values within an organ becomes too flat to stimulate proliferation.

A molecular mechanism that could achieve this type of growth regulation was identified with the discovery of the Fat branch of the Hippo signaling pathway and its role in growth control (Reddy & Irvine, 2008; Staley & Irvine, 2012). Fat is an atypical cadherin protein that negatively regulates growth through activation of the Hippo pathway. Fat activity is controlled by two proteins expressed in gradients, its binding partner Dachsous (Ds), and a Golgi-localized kinase called Four-jointed (Fj) that modulates binding between Ds and fat. Juxtaposing cells that express different levels of Ds or Fj stimulates cell proliferation by inhibiting Fat activity, whereas uniformly expressing Ds or Fj increases Fat activity and inhibits cell proliferation (Rogulja, Rauskolb, & Irvine, 2008; Willecke, Hamaratoglu, Sansores-Garcia, Tao, & Halder, 2008). These kinds of studies have suggested a model in which growth is influenced by the steepness of the Ds and Fj gradients, analogous to the proposed influence of gradients of positional values on growth first suggested by intercalary regeneration experiments.

Intriguingly, Fj and Ds are expressed in proximal-distal gradients along each segment of developing insect legs and so are well positioned to be molecular manifestations of the positional values gradients suggested by regeneration studies (Bando et al., 2009; Clark et al., 1995; Villano & Katz,

1995). The idea that these genes could influence insect leg regeneration has been tested by combining studies of leg regeneration in crickets with techniques for knocking down gene expression using RNAi (Bando et al., 2009). Although precise manipulations of expression levels were not possible, these studies did reveal that components of the Fat and Hippo pathways are indeed required for normal leg regeneration.

5. SUPPRESSION OF REGENERATIVE GROWTH

As in developmental growth, growth inhibitors also exist in regenerative growth to prevent overgrowth. As regeneration proceeds, the proliferation rate decreases and redifferentiation occurs. This involves the downregulation of progrowth signals and upregulation of antigrowth signals. Growth inhibitors need to be downregulated during the initiation of regeneration, and then activated at the correct time to prevent tissue overgrowth and to allow differentiation to occur. Many studies on suppressors of regenerative growth have focused on the termination phase of mammalian liver regeneration and on comparison of nonregenerative versus regenerative tissues.

5.1. Termination of liver regeneration

Pathways identified as important inhibitors of liver regeneration during the termination phase include TGFβ signaling and integrin signaling, which mediates communication between the ECM and regenerating cells.

TGFβ signaling has multiple roles in regulating cell proliferation, differentiation, and migration in different contexts. Responses to TGFβ family signals depend on cell type, ligands, and receptors. The TGFβ family includes two main subfamilies, the TGFβ-Activin-Nodal subfamily and the Bone morphogenetic protein (BMP) subfamily (Massague, 2012). Some BMPs are required for limb regeneration (Beck, Christen, Barker, & Slack, 2006; Yu et al., 2010). The TGFβ subfamily is required for formation of a wound epithelium and blastema during amphibian limb regeneration because of its ability to regulate fibroblast migration (Levesque et al., 2007). However, TGFβ and Activin are generally considered as inhibitors of regenerative growth because of their role in suppressing cell proliferation. Deletion of type II TGFβ receptor (TGFBR2) in hepatocytes accelerated hepatocyte proliferation and increased liver mass/body weight ratio after PH, suggesting an inhibitory role of TGFβ in liver regeneration (Romero-Gallo et al., 2005). Another group examined hepatocyte-specific

Tgfbr2 knockout mice and found that the increase of hepatocyte proliferation in knockout mice was transient. This could be due to compensatory production of Activin A, another TGFβ family protein, in regenerating mutant livers. Blockage of Activin A signaling in *Tgfbr2*$^{-}$ mice increased hepatocyte proliferation during liver regeneration, suggesting that Activin A and TGFβ collaborate to inhibit hepatocyte proliferation (Oe et al., 2004). Elevation of TGFβ expression begins soon after PH (Michalopoulos, 2007; Pahlavan et al., 2006). However, as a suppressor of hepatocyte proliferation, TGFβ activity must be carefully limited during initiation of regeneration. Some inhibitors of TGFβ activity have been identified in regenerating livers, including IGF binding protein-1 and certain microRNAs (Leu, Crissey, & Taub, 2003; Yuan et al., 2011).

Integrins mediate communication between the ECM and attached cells. Integrin-linked kinase (ILK) is a kinase activated by binding to matrix-occupied integrins, and it transmits signals modulating multiple cellular processes. Loss of ILK disrupts cell-ECM communication. Liver-specific deletion of ILK in mice increased hepatocyte proliferation, resulting in a larger liver size (Gkretsi et al., 2008). Mice with liver-specific knockout of ILK are also defective in termination of proliferation after PH, and livers grow 58% larger than their original size, implicating ILK as an important factor promoting termination of liver regeneration (Apte et al., 2009).

Glypican 3 (GPC3), a glycosylphosphatidylinositol-anchored heparin sulfate proteoglycan, is another inhibitor of liver regeneration. GPC3 increases in cultured hepatocytes at late stages of HGF and EGF-stimulated growth, correlating with a decrease in proliferation rate. Moreover, knockdown of GPC3 in cultured hepatocytes increased their proliferation (Liu et al., 2009). Elevation of GPC3 expression was also observed during late stages of liver regeneration. Transgenic mice with hepatocyte-targeted overexpression of GPC3 developed normally, but exhibited a suppressed hepatocyte proliferation and lower liver weight during regeneration after PH (Liu et al., 2010). These results suggest a suppressive effect of GPC3 on liver regeneration, based on its effect on hepatocyte proliferation.

5.2. Nonregenerative tissues versus regenerative tissues

Many invertebrates, and some vertebrates such as urodele amphibians, can regenerate functional organs. However, mammals lose most of their regeneration capacity after birth. Instead, amputation or wounding in most mammalian organs such as skin, lung, or kidney only results in wound healing with scar formation, during which the injured tissue is replaced by fibroblasts and ECM secreted from fibroblasts (Gurtner, Werner, Barrandon, & Longaker, 2008). Even in the liver, which has a strong regeneration capacity, chronic injury can instead induce a scar-forming healing process, fibrosis. Fibrosis is a major cause of organ failure in many chronic diseases (Wynn, 2008). Determining what prevents mammalian organs from regenerating and ultimately inducing regeneration in mammals are major goals of regeneration studies. Mammals can achieve complete regeneration with no scar formation in response to damage during embryonic or fetal stages. Comparisons have been made between scar-forming wound healing in adult mammals and regeneration in fetal mammals or other vertebrates such as amphibians. Higher levels of inflammation, TGFβ activity, and ECM deposition occur in scar-forming wound healing than in regeneration (Gurtner et al., 2008), which suggests that these factors contribute to regeneration–suppression.

In regeneration-competent organs, ECM degradation is an early response after amputation or wounding, which allows release of differentiated cells and their migration to the wound site where they undergo dedifferentiation (Brockes & Kumar, 2008). ECM degradation is accomplished by MMPs. Upregulation of MMPs is observed at early stages of regeneration in many regeneration-competent organs, such as *Hydra*, *Drosophila* imaginal discs, amphibian limbs, and zebrafish fins (Bai et al., 2005; Carinato, Walter, & Henry, 2000; Shimizu et al., 2002; Yan et al., 2000; Yang, Gardiner, Carlson, Nugas, & Bryant, 1999). Conversely, abnormal ECM deposition and lack of MMP upregulation are observed in regeneration-deficient axolotl mutants and in *Xenopus* (Del Rio-Tsonis, Washabaugh, & Tsonis, 1992; Santosh et al., 2011). Collagen, a major component of the scar tissue, is an inhibitor of limb regeneration. In axolotl, nerve deviation at skin wounds in limb is enough to induce blastema formation, and this nerve-induced blastema is equivalent to an amputation-induced blastema (Endo et al., 2004; Satoh, Gardiner, Bryant, & Endo, 2007). The rate of blastema induction by nerves decreases as collagen accumulates. Moreover, placing acellular collagen sheets over deviated nerves blocks blastema formation, indicating the suppressive role of collagen accumulation in tissue regeneration (Satoh, Hirata, & Makanae, 2012).

The major source of collagen in scar-forming wound healing is activated fibroblasts, myofibroblasts, which are contractile cells causing wound contraction, a process seen in scar-forming wound healing but not in regeneration. Blocking wound contraction and fibroblast accumulation at wound sites by degradable scaffolds with molecules such as integrin ligands can promote regeneration in skin and nerves in adult mammals (Yannas, 2005, 2013). Activation of fibroblasts and secretion of collagen from myofibroblasts require TGFβ signaling (Wynn, 2008; Zeisberg & Kalluri, 2013). Blockage of TGFβ signaling can dissolve fibrotic tissues and restore normal regeneration. For example, treatment with TGFβ neutralizing antibodies blocked scar formation at adult rodent skin wounds and resulted in normal dermis (Shah, Foreman, & Ferguson, 1992, 1994). Activation of BMPs, which counteract TGFβ signaling, can promote regeneration and reverse fibrosis in the kidney (Sugimoto et al., 2012).

Thus, studies of both the termination phase of regeneration and nonregenerative tissues unveiled important roles for TGFβ and ECM in suppression of regenerative growth. This growth inhibition effect is normally required to prevent tissue overgrowth that might otherwise be caused by excessive growth-promoting signals induced during regeneration, but it is also a key factor resulting in regeneration deficiency in adult mammals.

6. CONCLUSION

Tissues under homeostasis balance progrowth signals with antigrowth signals. When injury occurs, this balance is rapidly disrupted, and through stress-response mechanisms, the progrowth signals become predominant. The duration and level of elevation of progrowth signals help determine the extent of regenerative growth. In regeneration-competent organs, progrowth signals are prolonged and augmented through collaboration of multiple pathways, including Wnt, BMP, EGFR, JAK-STAT, and Hippo, which allow sufficient parenchymal cells to be generated from blastema or stem cells to replace missing tissue. As regeneration progresses, progrowth signals gradually decline and antigrowth signals increase, leading to termination of growth. In nonregenerating organs, progrowth signals are also transiently activated after injury, triggering mild cell proliferation to accomplish wound healing. However, antigrowth signals, including TGFβ and collagen, rapidly increase, thus blocking regeneration. In some cases, regeneration can be induced in nonregenerative tissues by manipulating the balance between progrowth signals and antigrowth signals. Ongoing studies of how growth is controlled and organized in regenerating organs should facilitate further attempts to promote growth of normally nonregenerating organs.

ACKNOWLEDGMENTS

Research on regeneration in KDI's lab is supported by Human Frontiers Science Program Grant RGP0016/2010 and the Howard Hughes Medical Institute.

REFERENCES

Akerman, P., Cote, P., Yang, S. Q., McClain, C., Nelson, S., Bagby, G. J., et al. (1992). Antibodies to tumor necrosis factor-alpha inhibit liver regeneration after partial hepatectomy. *The American Journal of Physiology, 263*, G579–G585.

Apte, U., Gkretsi, V., Bowen, W. C., Mars, W. M., Luo, J. H., Donthamsetty, S., et al. (2009). Enhanced liver regeneration following changes induced by hepatocyte-specific genetic ablation of integrin-linked kinase. *Hepatology, 50*, 844–851.

Bai, S., Thummel, R., Godwin, A. R., Nagase, H., Itoh, Y., Li, L., et al. (2005). Matrix metalloproteinase expression and function during fin regeneration in zebrafish: Analysis of MT1-MMP, MMP2 and TIMP2. *Matrix Biology: Journal of the International Society for Matrix Biology, 24*, 247–260.

Bando, T., Mito, T., Maeda, Y., Nakamura, T., Ito, F., Watanabe, T., et al. (2009). Regulation ofleg size and shape by the Dachsous/Fat signalling pathway during regeneration. *Development, 136*, 2235–2245.

Beck, C. W., Christen, B., Barker, D., & Slack, J. M. (2006). Temporal requirement for bone morphogenetic proteins in regeneration of the tail and limb of Xenopus tadpoles. *Mechanisms of Development, 123*, 674–688.

Behrens, A., Sibilia, M., David, J. P., Mohle-Steinlein, U., Tronche, F., Schutz, G., et al. (2002). Impaired postnatal hepatocyte proliferation and liver regeneration in mice lacking c-jun in the liver. *The EMBO Journal, 21*, 1782–1790.

Berasain, C., Garcia-Trevijano, E. R., Castillo, J., Erroba, E., Lee, D. C., Prieto, J., et al. (2005). Amphiregulin: An early trigger of liver regeneration in mice. *Gastroenterology, 128*, 424–432.

Bergantinos, C., Corominas, M., & Serras, F. (2010). Cell death-induced regeneration in wing imaginal discs requires JNK signalling. *Development, 137*, 1169–1179.

Bergantinos, C., Vilana, X., Corominas, M., & Serras, F. (2010). Imaginal discs: Renaissance of a model for regenerative biology. *BioEssays: News and Reviews in Molecular, Cellular and Developmental Biology, 32*, 207–217.

Block, G. D., Locker, J., Bowen, W. C., Petersen, B. E., Katyal, S., Strom, S. C., et al. (1996). Population expansion, clonal growth, and specific differentiation patterns in primary cultures of hepatocytes induced by HGF/SF, EGF and TGF alpha in a chemically defined (HGM) medium. *The Journal of Cell Biology, 132*, 1133–1149.

Bogoyevitch, M. A., Ngoei, K. R., Zhao, T. T., Yeap, Y. Y., & Ng, D. C. (2010). c-Jun N-terminal kinase (JNK) signaling: Recent advances and challenges. *Biochimica et Biophysica Acta, 1804*, 463–475.

Bohm, F., Kohler, U. A., Speicher, T., & Werner, S. (2010). Regulation of liver regeneration by growth factors and cytokines. *EMBO Molecular Medicine, 2*, 294–305.

Bosch, M., Baguna, J., & Serras, F. (2008). Origin and proliferation of blastema cells during regeneration of Drosophila wing imaginal discs. *The International Journal of Developmental Biology, 52*, 1043–1050.

Bosch, M., Serras, F., Martin-Blanco, E., & Baguna, J. (2005). JNK signaling pathway required for wound healing in regenerating Drosophila wing imaginal discs. *Developmental Biology, 280*, 73–86.

Brockes, J. P., & Kumar, A. (2008). Comparative aspects of animal regeneration. *Annual Review of Cell and Developmental Biology, 24*, 525–549.

Campbell, L. J., & Crews, C. M. (2008). Wound epidermis formation and function in urodele amphibian limb regeneration. *Cellular and Molecular Life Sciences, 65*, 73–79.

Carinato, M. E., Walter, B. E., & Henry, J. J. (2000). Xenopus laevis gelatinase B (Xmmp-9): Development, regeneration, and wound healing. *Developmental Dynamics : An Official Publication of the American Association of Anatomists, 217*, 377–387.

Carlson, B. M. (2005). Some principles of regeneration in mammalian systems. *Anatomical Record. Part B: New Anatomist, 287*, 4–13.

Chen, F. (2012). JNK-induced apoptosis, compensatory growth, and cancer stem cells. *Cancer Research, 72*, 379–386.

Chera, S., Ghila, L., Dobretz, K., Wenger, Y., Bauer, C., Buzgariu, W., et al. (2009). Apoptotic cells provide an unexpected source of Wnt3 signaling to drive hydra head regeneration. *Developmental Cell, 17*, 279–289.

Chera, S., Ghila, L., Wenger, Y., & Galliot, B. (2011). Injury-induced activation of the MAPK/CREB pathway triggers apoptosis-induced compensatory proliferation in hydra head regeneration. *Development, Growth & Differentiation, 53*, 186–201.

Christensen, R. N., Weinstein, M., & Tassava, R. A. (2002). Expression of fibroblast growth factors 4, 8, and 10 in limbs, flanks, and blastemas of Ambystoma. *Developmental Dynamics: An Official Publication of the American Association of Anatomists, 223*, 193–203.

Clark, H. F., Brentrup, D., Schneitz, K., Bieber, A., Goodman, C., & Noll, M. (1995). Dachsous encodes a member of the cadherin superfamily that controls imaginal disc morphogenesis in Drosophila. *Genes & Development, 9*, 1530–1542.

Cordero, J. B., Stefanatos, R. K., Scopelliti, A., Vidal, M., & Sansom, O. J. (2012). Inducible progenitor-derived Wingless regulates adult midgut regeneration in Drosophila. *The EMBO Journal, 31*, 3901–3917.

Cressman, D. E., Greenbaum, L. E., DeAngelis, R. A., Ciliberto, G., Furth, E. E., Poli, V., et al. (1996). Liver failure and defective hepatocyte regeneration in interleukin-6-deficient mice. *Science, 274*, 1379–1383.

Das, M., Garlick, D. S., Greiner, D. L., & Davis, R. J. (2011). The role of JNK in the development of hepatocellular carcinoma. *Genes & Development, 25*, 634–645.

Day, S. J., & Lawrence, P. A. (2000). Measuring dimensions: The regulation of size and shape. *Development, 127*, 2977–2987.

Del Rio-Tsonis, K., Washabaugh, C. H., & Tsonis, P. A. (1992). The mutant axolotl short toes exhibits impaired limb regeneration and abnormal basement membrane formation. *Proceedings of the National Academy of Sciences of the United States of America, 89*, 5502–5506.

Endo, T., Bryant, S. V., & Gardiner, D. M. (2004). A stepwise model system for limb regeneration. *Developmental Biology, 270*, 135–145.

Fan, Y., & Bergmann, A. (2008a). Apoptosis-induced compensatory proliferation. The Cell is dead. Long live the Cell! *Trends in Cell Biology, 18*, 467–473.

Fan, Y., & Bergmann, A. (2008b). Distinct mechanisms of apoptosis-induced compensatory proliferation in proliferating and differentiating tissues in the Drosophila eye. *Developmental Cell, 14*, 399–410.

French, V., Bryant, P. J., & Bryant, S. V. (1976). Pattern regulation in epimorphic fields. *Science, 193*, 969–981.

Gkretsi, V., Apte, U., Mars, W. M., Bowen, W. C., Luo, J. H., Yang, Y., et al. (2008). Liver-specific ablation of integrin-linked kinase in mice results in abnormal histology, enhanced cell proliferation, and hepatomegaly. *Hepatology, 48*, 1932–1941.

Grusche, F. A., Degoutin, J. L., Richardson, H. E., & Harvey, K. F. (2011). The Salvador/Warts/Hippo pathway controls regenerative tissue growth in Drosophila melanogaster. *Developmental Biology, 350*, 255–266.

Gurtner, G. C., Werner, S., Barrandon, Y., & Longaker, M. T. (2008). Wound repair and regeneration. *Nature, 453*, 314–321.

Han, M. J., An, J. Y., & Kim, W. S. (2001). Expression patterns of Fgf-8 during development and limb regeneration of the axolotl. *Developmental Dynamics: An Official Publication of the American Association of Anatomists, 220*, 40–48.
Haynie, J. L., & Bryant, P. J. (1976). Intercalary regeneration in imaginal wing disk of Drosophila melanogaster. *Nature, 259*, 659–662.
Herrera, S. C., Martin, R., & Morata, G. (2013). Tissue homeostasis in the wing disc of drosophila melanogaster: Immediate response to massive damage during development. *PLoS Genetics, 9*, e1003446.
Huh, C. G., Factor, V. M., Sanchez, A., Uchida, K., Conner, E. A., & Thorgeirsson, S. S. (2004). Hepatocyte growth factor/c-met signaling pathway is required for efficient liver regeneration and repair. *Proceedings of the National Academy of Sciences of the United States of America, 101*, 4477–4482.
Huh, J. R., Guo, M., & Hay, B. A. (2004). Compensatory proliferation induced by cell death in the Drosophila wing disc requires activity of the apical cell death caspase Dronc in a nonapoptotic role. *Current Biology, 14*, 1262–1266.
Iimuro, Y., & Fujimoto, J. (2010). TLRs, NF-kappaB, JNK, and liver regeneration. *Gastroenterology Research and Practice, 2010*. Article ID: 598109. http://www.hindawi.com/journals/grp/2010/598109/cta/.
Jiang, H., & Edgar, B. A. (2011). Intestinal stem cells in the adult Drosophila midgut. *Experimental Cell Research, 317*, 2780–2788.
Jiang, H., Grenley, M. O., Bravo, M. J., Blumhagen, R. Z., & Edgar, B. A. (2011). EGFR/Ras/MAPK signaling mediates adult midgut epithelial homeostasis and regeneration in Drosophila. *Cell Stem Cell, 8*, 84–95.
Jiang, H., Patel, P. H., Kohlmaier, A., Grenley, M. O., McEwen, D. G., & Edgar, B. A. (2009). Cytokine/Jak/Stat signaling mediates regeneration and homeostasis in the Drosophila midgut. *Cell, 137*, 1343–1355.
Karpowicz, P., Perez, J., & Perrimon, N. (2010). The Hippo tumor suppressor pathway regulates intestinal stem cell regeneration. *Development, 137*, 4135–4145.
Kawakami, Y., Rodriguez Esteban, C., Raya, M., Kawakami, H., Marti, M., Dubova, I., et al. (2006). Wnt/beta-catenin signaling regulates vertebrate limb regeneration. *Genes & Development, 20*, 3232–3237.
King, R. S., & Newmark, P. A. (2012). The cell biology of regeneration. *The Journal of Cell Biology, 196*, 553–562.
Kiso, S., Kawata, S., Tamura, S., Inui, Y., Yoshida, Y., Sawai, Y., et al. (2003). Liver regeneration in heparin-binding EGF-like growth factor transgenic mice after partial hepatectomy. *Gastroenterology, 124*, 701–707.
Kondo, S., Senoo-Matsuda, N., Hiromi, Y., & Miura, M. (2006). DRONC coordinates cell death and compensatory proliferation. *Molecular and Cellular Biology, 26*, 7258–7268.
Leu, J. I., Crissey, M. A., & Taub, R. (2003). Massive hepatic apoptosis associated with TGF-beta1 activation after Fas ligand treatment of IGF binding protein-1-deficient mice. *The Journal of Clinical Investigation, 111*, 129–139.
Levesque, M., Gatien, S., Finnson, K., Desmeules, S., Villiard, E., Pilote, M., et al. (2007). Transforming growth factor: Beta signaling is essential for limb regeneration in axolotls. *PLoS One, 2*, e1227.
Li, F., Huang, Q., Chen, J., Peng, Y., Roop, D. R., Bedford, J. S., et al. (2010). Apoptotic cells activate the "phoenix rising" pathway to promote wound healing and tissue regeneration. *Science Signaling, 3*, ra13.
Li, W., Liang, X., Kellendonk, C., Poli, V., & Taub, R. (2002). STAT3 contributes to the mitogenic response of hepatocytes during liver regeneration. *The Journal of Biological Chemistry, 277*, 28411–28417.

Liu, B., Bell, A. W., Paranjpe, S., Bowen, W. C., Khillan, J. S., Luo, J. H., et al. (2010). Suppression of liver regeneration and hepatocyte proliferation in hepatocyte-targeted glypican 3 transgenic mice. *Hepatology, 52*, 1060–1067.

Liu, Y., Michalopoulos, G. K., & Zarnegar, R. (1994). Structural and functional characterization of the mouse hepatocyte growth factor gene promoter. *The Journal of Biological Chemistry, 269*, 4152–4160.

Liu, B., Paranjpe, S., Bowen, W. C., Bell, A. W., Luo, J. H., Yu, Y. P., et al. (2009). Investigation of the role of glypican 3 in liver regeneration and hepatocyte proliferation. *The American Journal of Pathology, 175*, 717–724.

Massague, J. (2012). TGFbeta signalling in context. *Nature Reviews. Molecular Cell Biology, 13*, 616–630.

Mattila, J., Omelyanchuk, L., Kyttala, S., Turunen, H., & Nokkala, S. (2005). Role of Jun N-terminal Kinase (JNK) signaling in the wound healing and regeneration of a Drosophila melanogaster wing imaginal disc. *The International Journal of Developmental Biology, 49*, 391–399.

Michalopoulos, G. K. (2007). Liver regeneration. *Journal of Cellular Physiology, 213*, 286–300.

Mitchell, C., Nivison, M., Jackson, L. F., Fox, R., Lee, D. C., Campbell, J. S., et al. (2005). Heparin-binding epidermal growth factor-like growth factor links hepatocyte priming with cell cycle progression during liver regeneration. *The Journal of Biological Chemistry, 280*, 2562–2568.

Morata, G., Shlevkov, E., & Perez-Garijo, A. (2011). Mitogenic signaling from apoptotic cells in Drosophila. *Development, Growth & Differentiation, 53*, 168–176.

Moreno, E., Yan, M., & Basler, K. (2002). Evolution of TNF signaling mechanisms: JNK-dependent apoptosis triggered by Eiger, the Drosophila homolog of the TNF superfamily. *Current Biology, 12*, 1263–1268.

Morgan, T. H. (1901). Regeneration and liability to injury. *Science, 14*, 235–248.

Natarajan, A., Wagner, B., & Sibilia, M. (2007). The EGF receptor is required for efficient liver regeneration. *Proceedings of the National Academy of Sciences of the United States of America, 104*, 17081–17086.

Oe, S., Lemmer, E. R., Conner, E. A., Factor, V. M., Leveen, P., Larsson, J., et al. (2004). Intact signaling by transforming growth factor beta is not required for termination of liver regeneration in mice. *Hepatology, 40*, 1098–1105.

Page-McCaw, A. (2008). Remodeling the model organism: Matrix metalloproteinase functions in invertebrates. *Seminars in Cell & Developmental Biology, 19*, 14–23.

Pahlavan, P. S., Feldmann, R. E., Jr., Zavos, C., & Kountouras, J. (2006). Prometheus' challenge: Molecular, cellular and systemic aspects of liver regeneration. *The Journal of Surgical Research, 134*, 238–251.

Paranjpe, S., Bowen, W. C., Tseng, G. C., Luo, J. H., Orr, A., & Michalopoulos, G. K. (2010). RNA interference against hepatic epidermal growth factor receptor has suppressive effects on liver regeneration in rats. *The American Journal of Pathology, 176*, 2669–2681.

Park, H. D., Ortmeyer, A. B., & Blankenbaker, D. P. (1970). Cell division during regeneration in Hydra. *Nature, 227*, 617–619.

Perez-Garijo, A., Martin, F. A., & Morata, G. (2004). Caspase inhibition during apoptosis causes abnormal signalling and developmental aberrations in Drosophila. *Development, 131*, 5591–5598.

Perez-Garijo, A., Shlevkov, E., & Morata, G. (2009). The role of Dpp and Wg in compensatory proliferation and in the formation of hyperplastic overgrowths caused by apoptotic cells in the Drosophila wing disc. *Development, 136*, 1169–1177.

Reddien, P. W. (2013). Specialized progenitors and regeneration. *Development, 140*, 951–957.

Reddien, P. W., & Sanchez Alvarado, A. (2004). Fundamentals of planarian regeneration. *Annual Review of Cell and Developmental Biology, 20*, 725–757.

Reddy, B. V., & Irvine, K. D. (2008). The Fat and Warts signaling pathways: New insights into their regulation, mechanism and conservation. *Development, 135*, 2827–2838.

Ren, F., Wang, B., Yue, T., Yun, E. Y., Ip, Y. T., & Jiang, J. (2010). Hippo signaling regulates Drosophila intestine stem cell proliferation through multiple pathways. *Proceedings of the National Academy of Sciences of the United States of America, 107*, 21064–21069.

Rogulja, D., Rauskolb, C., & Irvine, K. D. (2008). Morphogen control of wing growth through the Fat signaling pathway. *Developmental Cell, 15*, 309–321.

Romero-Gallo, J., Sozmen, E. G., Chytil, A., Russell, W. E., Whitehead, R., Parks, W. T., et al. (2005). Inactivation of TGF-beta signaling in hepatocytes results in an increased proliferative response after partial hepatectomy. *Oncogene, 24*, 3028–3041.

Russell, W. E., Kaufmann, W. K., Sitaric, S., Luetteke, N. C., & Lee, D. C. (1996). Liver regeneration and hepatocarcinogenesis in transforming growth factor-alpha-targeted mice. *Molecular Carcinogenesis, 15*, 183–189.

Ryoo, H. D., Gorenc, T., & Steller, H. (2004). Apoptotic cells can induce compensatory cell proliferation through the JNK and the Wingless signaling pathways. *Developmental Cell, 7*, 491–501.

Sabapathy, K., Hochedlinger, K., Nam, S. Y., Bauer, A., Karin, M., & Wagner, E. F. (2004). Distinct roles for JNK1 and JNK2 in regulating JNK activity and c-Jun-dependent cell proliferation. *Molecular Cell, 15*, 713–725.

Sakata, H., Takayama, H., Sharp, R., Rubin, J. S., Merlino, G., & LaRochelle, W. J. (1996). Hepatocyte growth factor/scatter factor overexpression induces growth, abnormal development, and tumor formation in transgenic mouse livers. *Cell Growth & Differentiation: The Molecular Biology Journal of the American Association for Cancer Research, 7*, 1513–1523.

Santosh, N., Windsor, L. J., Mahmoudi, B. S., Li, B., Zhang, W., Chernoff, E. A., et al. (2011). Matrix metalloproteinase expression during blastema formation in regeneration-competent versus regeneration-deficient amphibian limbs. *Developmental Dynamics: An Official Publication of the American Association of Anatomists, 240*, 1127–1141.

Satoh, A., Gardiner, D. M., Bryant, S. V., & Endo, T. (2007). Nerve-induced ectopic limb blastemas in the Axolotl are equivalent to amputation-induced blastemas. *Developmental Biology, 312*, 231–244.

Satoh, A., Hirata, A., & Makanae, A. (2012). Collagen reconstitution is inversely correlated with induction of limb regeneration in Ambystoma mexicanum. *Zoological Science, 29*, 191–197.

Schwabe, R. F., Bradham, C. A., Uehara, T., Hatano, E., Bennett, B. L., Schoonhoven, R., et al. (2003). c-Jun-N-terminal kinase drives cyclin D1 expression and proliferation during liver regeneration. *Hepatology, 37*, 824–832.

Seki, E., Tsutsui, H., Iimuro, Y., Naka, T., Son, G., Akira, S., et al. (2005). Contribution of Toll-like receptor/myeloid differentiation factor 88 signaling to murine liver regeneration. *Hepatology, 41*, 443–450.

Shah, M., Foreman, D. M., & Ferguson, M. W. (1992). Control of scarring in adult wounds by neutralising antibody to transforming growth factor beta. *Lancet, 339*, 213–214.

Shah, M., Foreman, D. M., & Ferguson, M. W. (1994). Neutralising antibody to TGF-beta 1,2 reduces cutaneous scarring in adult rodents. *Journal of Cell Science, 107*(Pt. 5), 1137–1157.

Shaw, R. L., Kohlmaier, A., Polesello, C., Veelken, C., Edgar, B. A., & Tapon, N. (2010). The Hippo pathway regulates intestinal stem cell proliferation during Drosophila adult midgut regeneration. *Development, 137*, 4147–4158.

Shimizu, H., Zhang, X., Zhang, J., Leontovich, A., Fei, K., Yan, L., et al. (2002). Epithelial morphogenesis in hydra requires de novo expression of extracellular matrix components and matrix metalloproteinases. *Development, 129*, 1521–1532.

Smith-Bolton, R. K., Worley, M. I., Kanda, H., & Hariharan, I. K. (2009). Regenerative growth in Drosophila imaginal discs is regulated by Wingless and Myc. *Developmental Cell, 16*, 797–809.

Staley, B. K., & Irvine, K. D. (2010). Warts and Yorkie mediate intestinal regeneration by influencing stem cell proliferation. *Current Biology, 20*, 1580–1587.

Staley, B. K., & Irvine, K. D. (2012). Hippo signaling in Drosophila: Recent advances and insights. *Developmental Dynamics: An Official Publication of the American Association of Anatomists, 241*, 3–15.

Stolz, D. B., Mars, W. M., Petersen, B. E., Kim, T. H., & Michalopoulos, G. K. (1999). Growth factor signal transduction immediately after two-thirds partial hepatectomy in the rat. *Cancer Research, 59*, 3954–3960.

Sugimoto, H., LeBleu, V. S., Bosukonda, D., Keck, P., Taduri, G., Bechtel, W., et al. (2012). Activin-like kinase 3 is important for kidney regeneration and reversal of fibrosis. *Nature Medicine, 18*, 396–404.

Sun, G., & Irvine, K. D. (2011). Regulation of Hippo signaling by Jun kinase signaling during compensatory cell proliferation and regeneration, and in neoplastic tumors. *Developmental Biology, 350*, 139–151.

Tanaka, E. M., & Reddien, P. W. (2011). The cellular basis for animal regeneration. *Developmental Cell, 21*, 172–185.

Villano, J. L., & Katz, F. N. (1995). Four-jointed is required for intermediate growth in the proximal-distal axis in Drosophila. *Development, 121*, 2767–2777.

Vinarsky, V., Atkinson, D. L., Stevenson, T. J., Keating, M. T., & Odelberg, S. J. (2005). Normal newt limb regeneration requires matrix metalloproteinase function. *Developmental Biology, 279*, 86–98.

Wang, P., & Hou, S. X. (2010). Regulation of intestinal stem cells in mammals and Drosophila. *Journal of Cellular Physiology, 222*, 33–37.

Westwick, J. K., Weitzel, C., Leffert, H. L., & Brenner, D. A. (1995). Activation of Jun kinase is an early event in hepatic regeneration. *The Journal of Clinical Investigation, 95*, 803–810.

Willecke, M., Hamaratoglu, F., Sansores-Garcia, L., Tao, C., & Halder, G. (2008). Boundaries of Dachsous Cadherin activity modulate the Hippo signaling pathway to induce cell proliferation. *Proceedings of the National Academy of Sciences of the United States of America, 105*, 14897–14902.

Wuestefeld, T., Klein, C., Streetz, K. L., Betz, U., Lauber, J., Buer, J., et al. (2003). Interleukin-6/glycoprotein 130-dependent pathways are protective during liver regeneration. *The Journal of Biological Chemistry, 278*, 11281–11288.

Wynn, T. A. (2008). Cellular and molecular mechanisms of fibrosis. *The Journal of Pathology, 214*, 199–210.

Yamada, Y., Kirillova, I., Peschon, J. J., & Fausto, N. (1997). Initiation of liver growth by tumor necrosis factor: deficient liver regeneration in mice lacking type I tumor necrosis factor receptor. *Proceedings of the National Academy of Sciences of the United States of America, 94*, 1441–1446.

Yan, L., Leontovich, A., Fei, K., & Sarras, M. P., Jr. (2000). Hydra metalloproteinase 1: A secreted astacin metalloproteinase whose apical axis expression is differentially regulated during head regeneration. *Developmental Biology, 219*, 115–128.

Yang, E. V., & Bryant, S. V. (1994). Developmental regulation of a matrix metalloproteinase during regeneration of axolotl appendages. *Developmental Biology, 166*, 696–703.

Yang, E. V., Gardiner, D. M., Carlson, M. R., Nugas, C. A., & Bryant, S. V. (1999). Expression of Mmp-9 and related matrix metalloproteinase genes during axolotl limb regeneration. *Developmental Dynamics: An Official Publication of the American Association of Anatomists, 216*, 2–9.

Yannas, I. V. (2005). Similarities and differences between induced organ regeneration in adults and early foetal regeneration. *Journal of the Royal Society, Interface/the Royal Society, 2*, 403–417.

Yannas, I. V. (2013). Emerging rules for inducing organ regeneration. *Biomaterials*, *34*, 321–330.
Yokoyama, H. (2008). Initiation of limb regeneration: The critical steps for regenerative capacity. *Development, Growth & Differentiation*, *50*, 13–22.
Yokoyama, H., Ide, H., & Tamura, K. (2001). FGF-10 stimulates limb regeneration ability in Xenopus laevis. *Developmental Biology*, *233*, 72–79.
Yokoyama, H., Ogino, H., Stoick-Cooper, C. L., Grainger, R. M., & Moon, R. T. (2007). Wnt/beta-catenin signaling has an essential role in the initiation of limb regeneration. *Developmental Biology*, *306*, 170–178.
Yokoyama, H., Yonei-Tamura, S., Endo, T., Izpisua Belmonte, J. C., Tamura, K., & Ide, H. (2000). Mesenchyme with fgf-10 expression is responsible for regenerative capacity in Xenopus limb buds. *Developmental Biology*, *219*, 18–29.
Yu, L., Han, M., Yan, M., Lee, E. C., Lee, J., & Muneoka, K. (2010). BMP signaling induces digit regeneration in neonatal mice. *Development*, *137*, 551–559.
Yuan, B., Dong, R., Shi, D., Zhou, Y., Zhao, Y., Miao, M., et al. (2011). Down-regulation of miR-23b may contribute to activation of the TGF-beta1/Smad3 signalling pathway during the termination stage of liver regeneration. *FEBS Letters*, *585*, 927–934.
Zeisberg, M., & Kalluri, R. (2013). Cellular mechanisms of tissue fibrosis. 1. Common and organ-specific mechanisms associated with tissue fibrosis. *American Journal of Physiology. Cell Physiology*, *304*, C216–C225.

CHAPTER FIVE

Cell Death: A Program to Regenerate

Sophie Vriz*,†, Silke Reiter‡, Brigitte Galliot‡,1
*Collège de France, Center for Interdisciplinary Research in Biology (CIRB), Paris, France
†University Paris-Diderot, Paris, France
‡Department of Genetics and Evolution, University of Geneva, Switzerland
[1]Corresponding author: e-mail address: brigitte.galliot@unige.ch

Contents

Abstract

Recent studies in *Drosophila*, *Hydra*, planarians, zebrafish, mice, indicate that cell death can open paths to regeneration in adult animals. Indeed injury can induce cell death, itself triggering regeneration following an immediate instructive mechanism, whereby the dying cells release signals that induce cellular responses over short and/or long-range distances. Cell death can also provoke a sustained derepressing response through the elimination of cells that suppress regeneration in homeostatic conditions. Whether common properties support what we name "regenerative cell death," is currently unclear. As key parameters, we review here the injury proapoptotic signals, the signals released by the dying cells, the cellular responses, and their respective timing. ROS appears as a common signal triggering cell death through MAPK and/or JNK pathway activation. But the modes of ROS production vary, from a brief pulse upon wounding, to repeated waves as observed in the zebrafish fin where ROS supports two peaks of cell

Current Topics in Developmental Biology, Volume 108
ISSN 0070-2153
http://dx.doi.org/10.1016/B978-0-12-391498-9.00002-4

death. Indeed regenerative cell death can be restricted to the injury phase, as in *Hydra*, *Drosophila*, or biphasic, immediate, and delayed, as in planarians and zebrafish. The dying cells release in a caspase-dependent manner a variety of signaling molecules, cytokines, growth factors, but also prostaglandins or ATP as recorded in *Drosophila*, *Hydra*, mice, and zebrafish, respectively. Interestingly, the ROS-producing cells often resist to cell death, implying a complex paracrine mode of signaling to launch regeneration, involving ROS-producing cells, ROS-sensing cells that release signaling molecules upon caspase activation, and effector cells that respond to these signals by proliferating, migrating, and/or differentiating.

ABBREVIATIONS

Dpp decapentaplegic (*Drosophila* growth factor that belongs to the BMP/TGF-β family)
DUOX dual oxidase
EC enterocytes
EGF epithelial growth factor
Hpa hours postamputation
JNK c-Jun N-terminal kinase
MAPK mitogen-activated protein kinase
NADPH oxidase nicotinamide adenine dinucleotide phosphate-oxidase
PCD programmed cell death
PG prostaglandin
ROS reactive oxygen species
Wg wingless (*Drosophila* growth factor that belongs to the Wnt family)

1. INTRODUCTION TO REGENERATIVE CELL DEATH

Regeneration of body parts, appendages, or internal organs needs the orchestration in time and space of several genetic programs, from wound healing to complete restoration of the missing part, with reestablishment of the initial function. This reactivation of developmental programs is easily triggered by injury in a wide variety of species, while in others, like most adult mammals, lesion/amputation leads to wound healing with scar formation (Brockes & Kumar, 2008; Sanchez Alvarado & Tsonis, 2006; Tanaka & Reddien, 2011; Wood & Eming, 2012). This observation suggests that injury activates distinct signaling cascades in these two contexts, either a signaling restricted to wound healing, or a signaling giving rise to wound healing and full restoration of tissues, organs, and appendages (Murawala, Tanaka, & Currie, 2012). If true, a detailed inventory of the injury-induced signals throughout the animal kingdom should provide valuable pieces of information to understand how developmental programs can be reactivated upon wounding, and thus contribute to the development of regenerative medicine.

Although regeneration is a widespread phenomenon among metazoans, a limited number of model systems actually provide the conditions for a parallel analysis of wound healing in the absence or presence of regenerative processes. Among them, we find the *Hydra* polyps and the planarian non-parasitic flatworms that can both regenerate any missing part of their body (Galliot & Ghila, 2010; King & Newmark, 2012), the adult holothuries (sea cucumbers) that regenerate their gut as well as their nervous system (Mashanov & Garcia-Arraras, 2011; San Miguel-Ruiz, Maldonado-Soto, & Garcia-Arraras, 2009), *Drosophila* that provides a genetic model for epithelial wound healing at embryonic, larval, and adult stages (Galko & Krasnow, 2004; Razzell, Wood, & Martin, 2011), but also regenerates its imaginal discs as larva (Bergantinos, Vilana, Corominas, & Serras, 2010; Bryant, 1971), and repair its midgut as adult (Amcheslavsky, Jiang, & Ip, 2009; Buchon, Broderick, Poidevin, Pradervand, & Lemaitre, 2009). Among vertebrates, the teleost fish (e.g., zebrafish) that regenerate their fins, heart, and brain (Choi & Poss, 2012; Kizil & Brand, 2011; Nakatani, Kawakami, & Kudo, 2007), the *Xenopus* tadpoles that regenerate their appendages (Slack, Lin, & Chen, 2008), and the adult urodeles (axolotl and newts) that perfectly heal their wounds while regenerating their appendages and a variety of organs (Brockes & Kumar, 2002; Murawala et al., 2012), also provide valuable models for this comparison. By contrast, adult mammals maintain their tissues through active self-renewal and can repair some tissues (e.g., skin, liver, bone, lung) but most of them repair their wounds with scars, and are unable to regenerate the 3D organization of their organs or appendages. Moreover, these tissue repair capacities decline upon aging (see in Sousounis, Baddour, & Tsonis, 2014).

At adulthood, size, morphology complexity, tissue complexity, and cellular behaviors largely vary between regenerative models. The body structure of *Hydra* polyps and planarians, although not similar, is relatively simple, comprising 12–15 differentiated cell types and large stocks of stem cells (Reddien & Sanchez Alvarado, 2004; Steele, 2002). Regeneration assays are usually performed on adult animals about 10 mm long; however, in both species, the adult body size dramatically varies according to the feeding rhythm, even though regeneration remains highly robust during starvation (Galliot & Ghila, 2010; Gonzalez-Estevez, Felix, Rodriguez-Esteban, & Aboobaker, 2012; Romero & Baguna, 1991). By contrast adult *Drosophila,* which exhibit a complex anatomy, do not renew much of their somatic tissues, mainly their gut that is used as a model of tissue repair (Buchon, Broderick, et al., 2013). At the other end of the spectrum, limbs of axolotls are several centimeters long, exhibit a complex 3D structure with tissues composed of a variety of differentiated cell types, which regenerate the missing structures from lineage-restricted progenitors (Tanaka & Reddien, 2011). Time provides a useful scale to measure these differences in complexity of regenerative processes, for example, head regeneration takes about 3 days in *Hydra,* a couple of weeks in planarians, whereas limb regeneration in urodeles takes about 2 months. However despite this heterogeneity, regeneration relies on a succession of common steps from injury to regenerative outgrowth, that is, wound healing, blastema formation, differentiation and growth, and the duration of the initial phases (wound healing, induction of the regenerative program) appear similar (Fig. 5.1).

In most regenerative contexts, progenitor cells proliferate to form a blastema, and two nonmutually exclusive mechanisms operate to produce this pool of progenitors, either by recruiting stem cells located in proximity or at distance of the lesion site, or by inducing the dedifferentiation of the cells located in the vicinity of the lesion (Brockes & Kumar, 2008; Kragl et al., 2009; Poss, 2010). In both situations, the tissues that provide progenitor cells are subjected to an important histolysis, a process affecting tissues subjacent to the wound over a 1–2 mm distance in vertebrates (Weiss & Rosenbaum, 1967). Although initially described in the notochord and cartilage of metamorphic toads by Carl Vogt in 1842, and subsequently repeatedly reported in developmental as well as nondevelopmental contexts (Clarke & Clarke, 1996), the role of programmed cell death (PCD) or apoptosis in developmental processes was then neglected and rediscovered 120 years later (Kerr, Wyllie, & Currie, 1972).

In regenerative studies cell death was considered as a side effect of injury and has not been taken into consideration until recently. In 1977, Haynie and Bryant reported that despite a 60% cell loss induced by irradiation, *Drosophila* wing imaginal discs were able to fully recover their size and shape through intensive cell proliferation (Haynie & Bryant, 1977). In parallel radiologists irradiating tumoral tissues (skin) of patients noted that dying cells induce proliferation of their neighboring cells and proposed the concept of "altruistic cell death" (Kondo, 1988). Thanks to a genetic strategy designed in *Drosophila* larval imaginal discs to produce "undead" cells, that is, cells where PCD is induced by activating the initiator caspases but simultaneously inhibited by expressing the baculovirus protein p35 that inhibits the effector caspase (Huh, Guo, & Hay, 2004; Perez-Garijo, Martin, & Morata, 2004; Ryoo, Gorenc, & Steller, 2004), cells

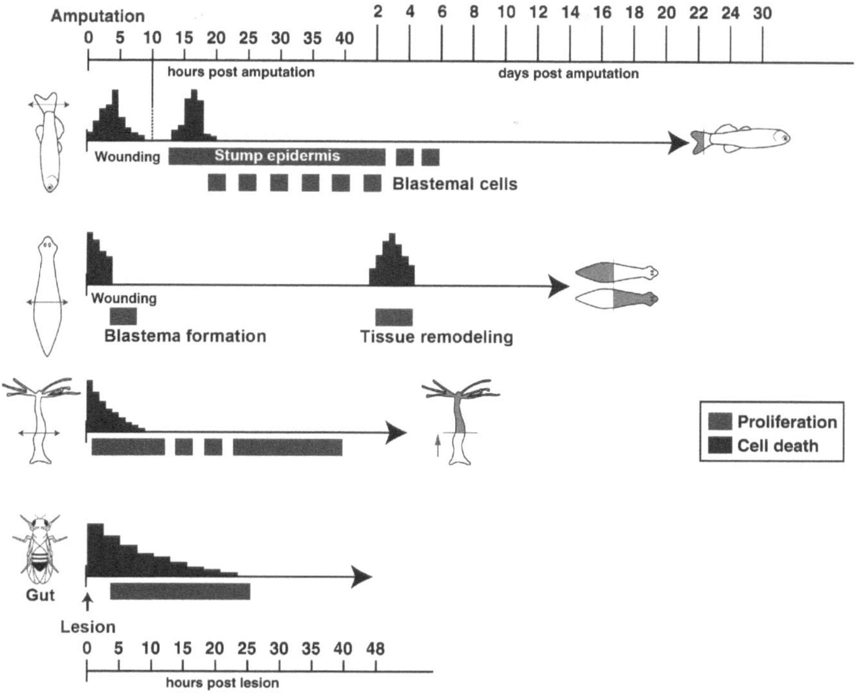

Figure 5.1 Injury-induced cell death (red) and cell proliferation (blue) events in *Danio rerio* regenerating its caudal fin, *S. mediterranean* regenerating its body, *Hydra vulgaris* regenerating its head after mid-gastric bisection, *Drosophila melanogaster* repairing its midgut. Fin regeneration in *Danio*, completed in 3 weeks, shows two peaks of apoptosis, the second one being required for the proliferation of the stump epidermis and the recruitment of blastema cells (Gauron et al., 2013). Body regeneration in *Schmidtea* (planarian), completed in 10–14 days, also exhibits two peaks of apoptosis, the second one associated with a round of proliferation, is involved in tissue remodeling (Pellettieri et al., 2010). Head regeneration in *Hydra*, completed in 3 days, displays one burst of apoptosis, directly promoting cell division (Chera et al., 2009). Infection with pathogenic bacteria leads to cell death of enterocytes in adult *Drosophila*, which stimulates the proliferation of intestine stem cells and the repair of the gut (Amcheslavsky et al., 2009; Buchon, Broderick, Poidevin, Pradervand, & Lemaitre, 2009). Time scales are indicated at the top and bottom; regenerated structures are colored yellow. (See color plate.)

remain indefinitely apoptotic. Hence the concept of apoptosis-induced proliferation, via c-Jun N-terminal kinase (JNK) activation and release of growth factors by the dying cells, was evidenced (see in Bergmann & Steller, 2010; Morata, Shlevkov, & Perez-Garijo, 2011; Worley, Setiawan, & Hariharan, 2012). However, it should be stressed that in both contexts, *Drosophila* imaginal discs and mammalian irradiated tissues, cell death is artificially induced.

The repair and/or regenerative contexts where injury spontaneously induces cell death are numerous, in vertebrates as well as in invertebrates. One can cite adult newts regenerating their retina (Kaneko, Matsumoto, & Hanyu, 1999) or their forelimb (Vlaskalin, Wong, & Tsilfidis, 2004), flatworms regenerating their body (Hwang, Kobayashi, Agata, Ikeo, & Gojobori, 2004; Pellettieri et al., 2010), *Xenopus* tadpoles regenerating their tail (Tseng, Adams, Qiu, Koustubhan, & Levin, 2007), *Hydra* regenerating their head (Chera et al., 2009), adult *Drosophila* repairing their midgut (Amcheslavsky et al., 2009; Buchon, Broderick, et al., 2009), mice regenerating their skin or their liver (Li et al., 2010b), sea cucumbers regenerating their gut (Mashanov, Zueva, Rojas-Catagena, & Garcia-Arraras, 2010) or their radial nerve cord (Mashanov, Zueva, & Garcia-Arraras, 2013), and adult zebrafish regenerating their fins (Gauron et al., 2013). When cell death appears instrumental for regeneration, we name the process "regenerative cell death." In several of these contexts (*Hydra, Drosophila* larvae, zebrafish, mice), it was shown that caspase activation is actually necessary for triggering the signaling that drives cellular remodeling and regeneration (Bergmann & Steller, 2010), and then one can talk of regenerative apoptosis. However, in contexts where caspase activation takes place, one observes a large heterogeneity in terms of cell death timing, signals inducing cell death, signals released by the dying cells, and cellular responses driven by these signals. To tentatively evidence some common rules between the various contexts where regenerative apoptosis operates, we propose here a comparative analysis of this process in vertebrate as well as invertebrate contexts, focusing on (i) the timing of injury-induced cell death, (ii) the injury signals that drive cell death, (iii) the caspase-dependent signaling turned on in the dying cells, and finally (iv) the cellular responses induced by the signals released by the dying cells.

2. WHAT TIMING FOR SPONTANEOUS CELL DEATH DURING REGENERATION?

2.1. Single immediate wave of cell death in *Hydra* regenerating its head

Since 1744 *Hydra* provides a simple model for regeneration studies, historically the first one (Galliot, 2012). *Hydra* is a tube shape freshwater cnidarian, about 10 mm long, with an apical pole, named head, centered on a unique opening, the mouth/anus, surrounded by a ring of tentacles, and a basal disc at the other extremity. *Hydra* consists of two cell layers named ectoderm and endoderm, separated by an acellular layer named mesoglea. The myoepithelial cells that form an epidermis in the ectoderm and a gastrodermis in the endoderm arise from two distinct stem cell populations. Moreover, the multipotent stem cells, located in the ectoderm, provide sensorymotor and ganglia neurons, stinging cells (nematocytes), and germ cells in the ectoderm, but also nerve cells and gland cells in the endoderm. After mid-gastric bisection, regeneration of the apical half (also named head) is observed in about 3 days, that of the basal half (also named foot) in less than 3 days (Fig. 5.1).

The analysis of the immediate cellular remodeling induced upon bisection showed a massive wave of cell death in head-regenerating tips, peaking after 1 h, while remaining limited in foot-regenerating tips (Chera et al., 2009). This asymmetric wave of apoptosis plays a key role in the initiation of head regeneration, as dying cells transiently release Wnt3, which activates the b-catenin pathway in the surrounding progenitors and promotes Wnt3 expression in the epithelial cells. Interestingly, the cells that undergo cell death belong to the interstitial lineage exclusively, either as progenitors or as differentiated cells (neurons, nematocytes, and gland cells). In fact, only a mild increase in the number of epithelial dying cells was noted in the vicinity of the wound, indicating that the epithelial cells in *Hydra* are extremely resistant to cell death, whatever the way used to induce cell death (Reiter, Crescenzi, Galliot, & Buzgariu, 2012). Inhibition of apoptosis leads to abortion of the head regeneration process in a large number of animals, whereas induction of apoptosis in the foot regenerating part leads to ectopic head formation (Chera et al., 2009). Heat-shock-induced cell death in non-regenerating *Hydra* does not lead to ectopic heads or ectopic regenerating structures. Therefore, high level of cell death in regenerating tips suffices to convert them to head regeneration, but cell death in the absence of wounding is not sufficient to trigger a regenerative program (Chera et al., 2009).

Hydra regeneration is a rapid process and it is possible that two peaks observed in planaria and zebrafish are superimposed or even merged in *Hydra*. There is actually a second phase of cell proliferation during head regeneration, starting about 24 h after amputation, involving all stem cell populations, epithelial and interstitial (Holstein et al., 1991; Miljkovic-Licina et al., 2007). This wave of cell proliferation is restricted to the presumptive head region, supporting the differentiation of the apical nervous system and the patterning of the regenerating head; it requires the expression of transcription factors as the paired-like homeobox gene *prdl-a* and the parahox gene *cnox-2* (Gauchat et al., 1998; Miljkovic-Licina et al., 2007). A wave of cell death preceding this second phase of cell proliferation was never recorded, but also was never thoroughly monitored.

2.2. Two distinct peaks of cell death in planarian regenerating its body

The freshwater planarian *Schmidtea mediterranea* is able to fully regenerate its body axis in about two weeks after gastric bisection. Regeneration does not involve a dedifferentiation process but the mobilization of well-identified stem cells, the neoblasts. In response to amputation the neoblasts, which are spread all over the body, increase their mitotic rate and migrate to the wound site where they stop proliferating and form a blastema. Then blastemal cells differentiate to replace the missing structures (Reddien & Sanchez Alvarado, 2004). Two distinct phases of cell death are observed during planaria regeneration, the first wave is strictly localized, taking place during the first 4 h in a 100-μm thick area on both sides of the bisection plane, whereas the second one, recorded 3 days later, is diffuse (Hwang et al., 2004; Pellettieri et al., 2010). The first peak is not specific to regeneration, rather linked to wound healing as also observed after a small incision. In contrast the second peak of cell death is specific to regeneration.

The widely spread second peak of cell death exhibits a magnitude that depends on the amputation level along the animal axis and on the regeneration type: The number of dying cells is very high in planarians regenerating their head or trunk, but close to the baseline level in animals regenerating their tail (Pellettieri et al., 2010). This second episode of cell death affects exclusively the differentiated cells, as neoblasts do not undergo cell death during regeneration. Hence this second apoptotic event suggests some systemic response to wounding, which allows the remodeling of preexisting structures (Pellettieri et al., 2010). Interestingly, these two peaks of cell death match the modulations of mitotic activity observed at 4–8 h postamputation (hpa) and 2–3 days postamputation, with a relative minimum at 24 hpa (Salo & Baguna, 1984; Wenemoser & Reddien, 2010). Hence in planarians, injury-induced cell death is tightly regulated in time and space, a regulation that correlates with the proliferative responses. However, the signaling function of the apoptotic cells remains unclear, particularly regarding the regulation of neoblast proliferation.

2.3. Two peaks of cell death in the adult zebrafish regenerating its caudal fin

All teleost fish regenerate organs and appendages at adulthood (Nakatani et al., 2007) and do so at a moderate speed, faster than amphibians, but slower than *Hydra*. The caudal fin regeneration is a popular model in which the amputated structure is replaced thanks to the formation of a blastema. This process involves the recruitment of progenitor cells through dedifferentiation of the stump mesenchymal cells (Nakatani et al., 2007). Blastema formation includes four successive steps, the two first ones restricted to the wound epidermis, the two last ones involving the formation and the function of the blastema (Gauron et al., 2013; Santos-Ruiz, Santamaria, Ruiz-Sanchez, & Becerra, 2002). During the first step (from 0 to 10 hpa), the lateral epidermis migrates to cover the wound in the absence of any cell proliferation and connective tissues form a clot. During this period, massive cell death is detected at the amputation plane, both apoptotic and non-apoptotic, but not specific to regeneration as also observed during skin wound healing (Gauron et al., 2013). Then, at the time of wound closure (10 hpa) the level of cell death gets transiently back to that observed in non-amputated fins. Soon after, a second burst of cell death, mainly apoptotic this time, arises in the stump epidermis, up to five ray segments away from the amputation plane (about 1 mm). This second burst of cell death is both specific and necessary to regeneration, as not observed during mere wound healing, and preventing blastema formation when inhibited (Gauron et al., 2013). Indeed cells from the same compartment, the stump epidermis, start to proliferate as a consequence of apoptosis (Gauron et al., 2013), an essential step as any treatment reducing epidermal proliferation impairs blastema formation (Dufourcq & Vriz, 2006; Jazwinska, Badakov, & Keating, 2007).

The third phase starts around 15 hpa with the recruitment of the prospective blastemal cells from the stump mesenchyme. These cells reenter the cell cycle and, while dividing, migrate toward the wound. This second proliferative event, which directly concerns the blastemal cells, is the criterion used to identify progenitor cells that have dedifferentiated and contribute to blastema formation. Even if in fish (Christen, Robles, Raya, Paramonov, & Izpisua Belmonte, 2010) and amphibians (Jhamb et al., 2011; Maki et al., 2009), these progenitor cells express stem cell markers as *klf4, pou5f1, sox2*, etc., an *in vivo* visualization of the recruitment process is currently missing. Finally at 36 hpa, the functional blastema is formed and a morphogenetic field is established (Poleo, Brown, Laforest, & Akimenko, 2001; Santos-Ruiz et al.,

2002), with complex signaling networks that further regulate proliferation and patterning (Stoick-Cooper, Moon, & Weidinger, 2007).

In summary, during zebrafish adult caudal fin regeneration, cell death occurs twice, first linked to the initial wound-healing process, and second as a burst strictly regulated in time and space in the wound epidermis, essential to blastema formation. Notably apoptosis does not occur in the mesenchyme and is no longer detected once the blastema is formed (Fig. 5.1). This implies that cells from the epidermis are the only cells to be sensitive to apoptosis-inducing signals, whereas the mesenchymal/blastemal cells are resistant to them, either actively expressing antiapoptotic factors and/or unable to undergo apoptosis. The role of injury-induced cell death is at least double, first directly triggering the proliferation of the stump epidermis, and later involved in the recruitment of blastemal cells to promote their entry into S-phase. The signaling pathways engaged by the apoptotic cells for this latter role are still elusive. As in *Drosophila* larvae regenerating their imaginal disc or adult *Drosophila* repairing their gut, apoptosis in regenerating zebrafish likely engage short- and long-distance responses.

2.4. Cell death in enterocytes supports gut repair in adult *Drosophila*

Adult *Drosophila* is able to maintain its midgut through self-renewal in homeostatic conditions, but also to repair it in response to injury (Buchon et al., 2013; Jiang & Edgar, 2011; Lee & Brey, 2013). Upon enteric bacterial infection, or after ingestion of cytotoxic agents, or upon expression of proapoptotic genes, the enterocytes leave the epithelium through delamination, then undergo cell death. The stressed enterocytes release signals, namely, epithelial growth factor (EGF) that promotes the rapid asymmetric division of the intestine stem cells, which self-renew and provide progenitors ready to differentiate (Amcheslavsky et al., 2009; Buchon, Broderick, et al., 2009; Jiang, Grenley, Bravo, Blumhagen, & Edgar, 2011). In case of oral infection with pathogenic bacteria, this massive wave of cell death that can affect up to 50% of the enterocytes, is induced as early as 30 min after exposure to bacteria and persists for hours, as long as bacterial infection is present (Buchon, Broderick, Kuraishi, & Lemaitre, 2010).

2.5. Sustained cell death in holothuria regenerating their gut or their nervous system

Echinoderms can regenerate external body parts (arms, spines) and internal organs (gut, visceral mass, nervous system) following self-induced (autonomy) or predator-related mutilations. In some species a piece of arm detached from the body can regenerate the whole body, which in the zoo of regenerative models put echinoderms on the podium of champions. After evisceration, gut regeneration relies on the remodeling of the remaining tissues. Cell death and cell proliferation have been documented in the see cucumber *Holothuria glaberrima* regenerating its gut (Mashanov & Garcia Arraras, 2014; Mashanov & Garcia-Arraras, 2011). Upon elimination of the intestine, cell death is observed in about 5% of the cells in the outer layer (mesothelium) during the first week, while surviving cells undergo dedifferentiation, cell cycle reentry, proliferation, and differentiation of a new luminal epithelium (Mashanov et al., 2010). The signaling function of these dying cells has not been reported yet; however, most mesothelial cells strongly express antiapoptotic genes as survivin and mortalin, suggesting that injury-induced cell death is tightly controlled. A recent RNAseq transcriptional profiling during intestinal regeneration in the sea cucumber *Apostichopus japonicas* did not identify changes related to cell death regulation except a downregulation of the effector of apoptosis gelsolin (Sun, Yang, Chen, Ma, & Lin, 2013). A similar cell death pattern was recorded in *H. glaberrima* regenerating its radial nerve cord (central nervous system), affecting about 4% of the cells located in the lesion area. This represents a 20 × increase in cell death during the first postinjury week, most of these cells being glial and nerve cells (Mashanov et al., 2013).

In conclusion, three main features emerge from this comparative analysis of the timing of regenerative cell death: (1) Cell death is observed in the vicinity of the wound, but it can also be recorded far from the injury site, implying a long-distance relay for the proapoptotic signals, (2) cell death is immediately detected after injury, but in some contexts, a second delayed wave of cell death is observed, with a pattern distinct from the first wave, (3) a restricted subset of cells are sensitive to the cell death inducing signals linked to injury, as only few cell types undergo cell death in an injured tissue.

3. INJURY SIGNALS TRIGGERING REGENERATIVE CELL DEATH

Several types of molecules have been characterized as immediate transcription-independent damage signals, including the Ca^{2+} ion and the nucleoside triphosphate ATP, which separately or together can activate nicotinamide adenine dinucleotide phosphate (NADPH) enzymes to produce the oxidizer hydrogen peroxide (H_2O_2) (Cordeiro & Jacinto, 2013). At high intracellular levels H_2O_2 acts as a potent cell death inducer through mitogen-activated protein kinase (MAPK) phosphatase inactivation and sustained JNK activation (Circu & Aw, 2010; Kamata et al., 2005). However, if the role of these damage signals in wound-healing processes is well established (for review, see Cordeiro & Jacinto, 2013), their role in regenerative processes is largely unknown. We review here recent data indicating that reactive oxygen species (ROS) signaling is involved in regenerative cell death. A direct link between ROS production and induction of cell death in regenerative contexts was first shown in adult *Drosophila* repairing their midgut (for review, see Buchon et al., 2013; Lee & Brey, 2013). Briefly, the pathogenic bacteria release in the intestinal lumen large amounts of uracil, a signal that directly activates the NADPH oxidase dual oxidase (DUOX) in enterocytes (ECs) leading to ROS production by these cells (Ha et al., 2009; Lee et al., 2013). ROS have a microbicidal function in homeostatic conditions, but high levels of ROS also elicit the delamination and death of damaged ECs as a response to pathogenic infection (Buchon, Broderick, Chakrabarti, & Lemaitre, 2009). These damaged ECs release cytokine-like signals that induce the proliferation of the neighboring intestine stem cells (Fig. 5.2). In the presence of nonlethal pathogenic bacteria, lower levels of uracil produce lower levels of ROS that lead to the replacement of the stressed enterocytes by quick differentiation of enteroblasts (Lee et al., 2013). Thus, ROS levels appear critical for directing distinct types of cellular response in the *Drosophila* midgut.

In zebrafish larvae, amputation of the caudal fin induces an immediate and transient wave of ROS production by the DUOX enzyme, and these ROS molecules act at several levels, attracting the leucocytes toward the wound and promoting the regrowth of sensory axons (Niethammer, Grabher, Look, & Mitchison, 2009; Rieger & Sagasti, 2011). In adult zebrafish regenerating their caudal fin, ROS are also rapidly detected at the wound but this time ROS production is sustained, reaching a peak value between 12 and 16 hpa. In this period, ROS signaling triggers two parallel events, JNK activation and a delayed burst of apoptosis observed between 15 and 18 hpa (Gauron et al., 2013). Both apoptosis and JNK activation are essential for blastema formation; however, it should be noted that here apoptosis does not seem to rely on JNK activation, as still present when JNK is pharmacologically inhibited.

In mammalian cells high levels of ROS inactivate MAPK phosphatases, and this inactivation leads to the sustained activity of the JNK and p38 MAP kinases, which activate caspases (Kamata et al., 2005; Ravindran, Gupta,

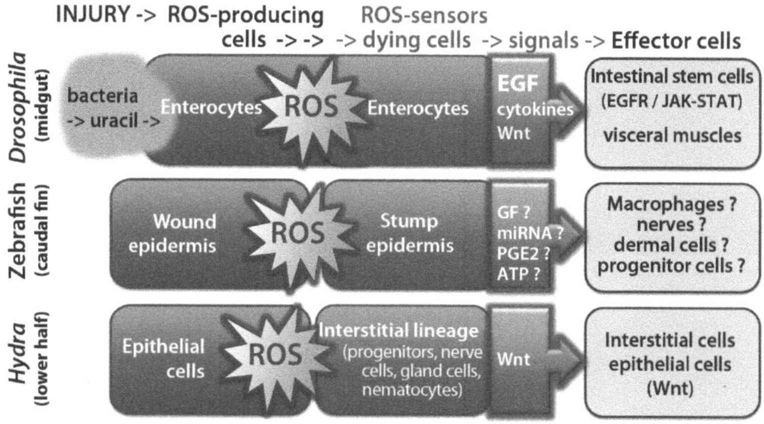

Figure 5.2 Paracrine signaling linked to regenerative cell death. Adult *Drosophila* repairing its midgut, adult zebrafish regenerating its caudal fin, or *Hydra* polyps regenerating their apical half produce reactive oxygen species (ROS) as an injury-induced signal that induces cell death and activate regeneration. Note that in *Drosophila,* the same cell type, enterocytes, but not necessarily the same cells, produce high levels of ROS and undergo cell death. In zebrafish, the delayed ROS production by the wound epidermis triggers cell death in the stump epidermis. In *Hydra* the epithelial cells at the wound edge that produce ROS resist to cell death. The signals that are transiently emitted by the apoptotic cells (red background) instruct the effector cells (gray background) involved in the regenerative processes (see text for details). (For interpretation of the references to color in this figure legend, the reader is referred to the online version of this chapter.)

Agrawal, Bala Bhaskar, & Lakshmana Rao, 2011). But oxidative stress also leads to cell death via activation of the upstream MAPK activator, the apoptosis signal-regulated kinase (ASK). In the absence of stress, thioredoxin, a small oxido-reductase enzyme, maintains ASK inactive and/or promotes its degradation, two processes that are no longer efficient in the presence of ROS (Finkel, 2011; Liu & Min, 2002; Ray, Huang, & Tsuji, 2012; Saitoh et al., 1998). In *Hydra*, injury-induced cell death in head-regenerating tips relies on the activation of the MAPK/CREB pathway (Chera, Ghila, Wenger, & Galliot, 2011; Kaloulis, Chera, Hassel, Gauchat, & Galliot, 2004), a process that appears to be ROS dependent as firstly ROS are produced within minutes upon bisection at the wound edges (Fig. 5.3) and secondly such high levels of ROS are required for injury-induced MAPK activation and injury-induced apoptosis (Reiter et al. in preparation).

In *Xenopus tropicalis* tadpole regenerating its tail and in the axolotl regenerating its limbs, proteomic and microarray analyses indicate that the NADPH metabolic pathway is induced shortly after amputation, suggesting

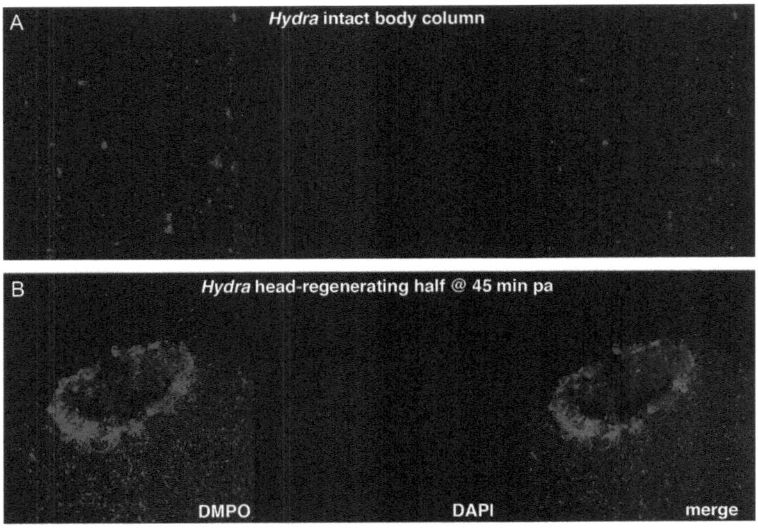

Figure 5.3 ROS signals in intact and bisected *Hydra*. ROS were detected by DMPO spin trap immunostaining in intact (A) or bisected (B) *Hydra* polyps (*sf-1* strain). In homeostatic conditions (A), few cells constitutively produce ROS along the body column. Upon bisection (B) ROS are produced locally at the wound edges, detected here after 45 min. (See color plate.)

that oxidative signaling participates in blastema formation (Knapp et al., 2013; Love et al., 2011; Rao et al., 2009). Indeed a recent study showed that amputation-induced high levels of ROS are necessary, not for wound healing, but for tail regeneration in *Xenopus laevis* tadpoles (Love et al., 2013). Interestingly, this sustained production of ROS leads to the activation of the Wnt pathway, which in turn promotes cell proliferation and axial growth. Regenerative cell death was not investigated in this study, although apoptosis was previously reported to be present and necessary within the first day following tail amputation in *Xenopus* tadpoles (Tseng et al., 2007). One possible scenario would speculate that the death of either ROS-producing or ROS-sensing cells provide signals that instruct the tail regenerative program. Alternatively, ROS signaling can directly activate Wnt signaling in the absence of any cell death via nucleoredoxin (Funato, Michiue, Asashima, & Miki, 2006).

All together these data indicate that ROS levels efficiently control induction of regenerative cell death. One critical question that arises from this scenario is the regulation of ROS sensing/ROS resistance, as ROS sensitivity appears quite variable among the distinct cell types present in a given wounded tissue. In the *Drosophila* midgut, the enterocytes that produce high levels of ROS might be distinct from the damaged enterocytes that sense ROS and delaminate; in *Hydra* regenerating its head and in the adult zebrafish regenerating its fin, the ROS-sensing cells that enter regenerative cell death are distinct from the ROS-producing cells (Fig. 5.2). Differential ROS sensing is well known in the immune system where ROS produced by monocytes induce apoptosis of Natural Killer cells but have no effect on lymphocytes (Hansson, Asea, Ersson, Hermodsson, & Hellstrand, 1996). Possible regulators of ROS sensitivity are the levels of O_2 consumption and metabolic rate.

During amphibian metamorphosis, the different cell types exhibit different sensitivities to thyroid hormones, namely to triiodothyronine (T3) that controls cell proliferation and apoptosis (Ishizuya-Oka, 2011; Puzianowska-Kuznicka, Pietrzak, Turowska, & Nauman, 2006). T3 levels are controlled by deiodinases that metabolize thyroid hormones, via deiodinase 2 (D2) for activation, and deiodinase 3 (D3) for inactivation (Gereben et al., 2008). This regulation is possibly at work in zebrafish regenerating fins as D3 activity was found enhanced during the early phase of regeneration and required for blastema formation (Bouzaffour, Rampon, Ramauge, Courtin, & Vriz, 2010; Kester et al., 2009). Further studies will tell whether the D2 and D3 expression levels correlate with cell sensitivity or cell resistance to ROS-induced apoptosis, and whether local modulations of deiodinase levels control ROS-induced apoptosis and subsequent proliferation.

4. CASPASE ACTIVATION, A VERSATILE TOOL TO REGULATE REGENERATION

To promote regenerative processes, cell death can elicit two distinct mechanisms, one active, instructive, through the caspase-dependent release of signals by the dying cells, and another passive, relying on derepression, through the injury-induced elimination of cells that in homeostatic conditions deliver inhibitory signals on "proregenerative" neighboring cells (Ryoo et al., 2004; Simon, Berg, & Kirkham, 2009). Experimental evidences accumulated over the past 10 years have shown that both mechanisms potentially coexist in the same organism, even in the same regenerative context where they can add their successive effects (see below). Indeed these two mechanisms cover different time windows, rather quick and transient for the instructive mechanism, vanishing when apoptotic bodies are metabolized, while the derepressing mechanism can persist over hours or days, as long as the cells that repress the regenerative potential of their neighbors are not replaced by new ones. Finally, we also briefly mention in this section some contexts where injury-induced cell death is toxic for regeneration.

4.1. Active, instructive mechanism of regenerative cell death

Developmental biologists using imaginal discs in *Drosophila* larvae and radiologists measuring the effects of irradiation in mammalian adult tissues have proved the validity of this instructive mechanism by characterizing in dying cells the production of caspase-dependent signals that modulate the behavior of neighboring cells or of even more distant cells (reviewed in Bergantinos, Vilana, Corominas, & Serras, 2010; Bergmann & Steller, 2010; Worley et al., 2012; Zimmerman, Huang, Li, Liu, & Li, 2013). The different signals released by the apoptotic cells after injury are depicted in Fig 5.4.

4.1.1 Growth factors: Wnts, Dpp, hedgehog, EGF

To decipher the role of induced cell death in *Drosophila* larval imaginal discs, several groups designed an ingenious genetic strategy whereby the initiator caspase Dronc and the p35 baculovirus protein that inhibits the effector caspases Drice/Dcp-1 were coactivated in well-defined regions of the wing disc, pushing thus the cells to enter apoptosis but simultaneously blocking the completion of the apoptotic program (Huh et al., 2004; Perez-Garijo et al., 2004; Ryoo et al., 2004; Wells, Yoshida, & Johnston, 2006). This strategy showed that "undead" cells, which produce mitogenic paracrine signals, namely the classical morphogens decapentaplegic (Dpp) and wingless (Wg), trigger proliferation of their neighbors thanks to the activation of Dronc, JNK, and p53 (for review, see Bergmann & Steller, 2010; Morata et al., 2011). However, these studies characterized apoptosis-induced proliferation in wing discs where epithelial cells are all proliferating, differentiating later during pupation. When apoptosis-induced proliferation was tested in eye imaginal discs, where proliferating and differentiating cells coexist but are separated by the morphogenetic furrow, a distinct mechanism was identified in the differentiating cells. In this tissue, cell proliferation relies on the activation of the effector caspases DrICE/Dcp-1 and on the release of Hedgehog (and not Dpp or Wg) that triggers cell cycle reentry of the undifferentiated cell-cycle arrested neighbors (Fan & Bergmann, 2008b). Hence different mechanisms can drive apoptosis-induced proliferation, and the choice between these mechanisms is influenced by the status of the tissue where apoptosis takes place, proliferating or differentiating (Fan & Bergmann, 2008a).

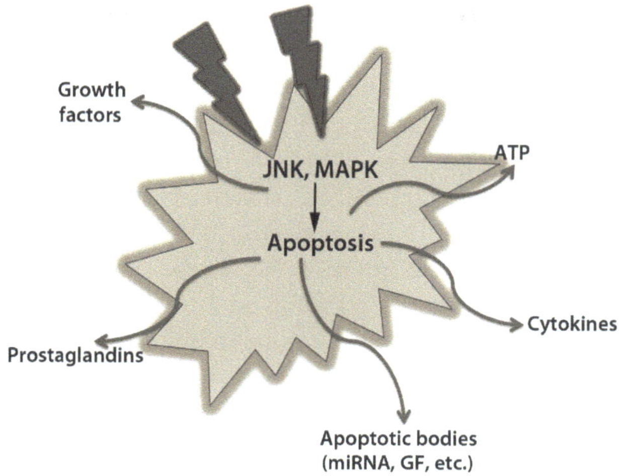

Figure 5.4 Active instructive mode of regenerative cell death. Scheme depicting a cell submitted to injury-induced signals (dark gray) activating the JNK and/or the MAPK pathways, which trigger cell death. This apoptotic cell potentially releases a variety of signals involved in tissue repair and/or regeneration: growth factors such as Wnt, BMPs, hedgehog, EGF, but also prostaglandins, apoptotic bodies that potentially contain miRNA and signaling molecules, cytokines, purines (ATP). Cellular outcomes of this paracrine signaling are diverse, including cell proliferation, cell migration, immune response, and neuronal activity. (For color version of this figure, the reader is referred to the online version of this chapter.)

To make life a little bit more complex, it also appeared that the "undead cell" strategy likely introduces some bias due to the persistent ectopic activation of JNK and the continuous delivery of growth factors. In fact in this experimental paradigm, apoptosis-induced proliferation can occur in imaginal discs mutated for Wg and Dpp, suggesting that these signals are actually dispensable (Morata et al., 2011). The Gal80 genetic system provides an alternative strategy, which allows the transient expression of proapoptotic genes by simply raising the temperature, triggering cell death for selected periods of time. Upon a 40-h cell death period, Wg was detected in the dying cells of the wing disc as expected, but was also found largely produced by the regenerating tissue where Wg stabilizes dMyc, leading to cell proliferation and regeneration (Smith-Bolton, Worley, Kanda, & Hariharan, 2009). A similar approach where cell death was induced for shorter periods of time, detected JNK activation in the healing tissue as well as two periods of cell proliferation, first immediate and transient in the injury area, and then delayed and sustained, affecting the growth of the whole disc (Bergantinos, Corominas, & Serras, 2010). In a third approach a massive and prolonged

(40–48 h) induction of cell death led to an immediate systemic response of the whole disc, which rapidly regenerates the damaged regions thanks to JNK signaling, allowing the apoptotic cells to induce proliferation over long distance, again in the absence of Wg or Dpp contribution (Herrera, Martin, & Morata, 2013).

All together these data point to JNK activation as necessary for the regeneration of the proliferating fly imaginal disc, but they also highlight some variability in the responses to injury-induced cell death depending on the extent and duration of cell death, with limited defects inducing a short-range signaling relying on Dpp and Wg signals on one side, and extended defects inducing a Dpp- and Wg-independent long-range signaling, on the other side. As recently stated "compensatory proliferation" is distinct from "apoptosis-induced proliferation": In the first case, cell proliferation (i) does not rely on the signals produced by the dying cells, (ii) engages the entire region/organ that senses tissue loss, with the proliferative cells migrating toward the damaged territory (Herrera et al., 2013). In the second case, cell proliferation is restricted to the damaged area and relies on the mitogenic signals released by the apoptotic cells (Mollereau et al., 2013). Compensatory proliferation, so far poorly understood, could actually respond to a passive, derepressing mechanism of regenerative cell death (see below). Although obtained in larval tissues artificially manipulated to trigger cell death, these results brought major conceptual advances on the various modes of regeneration and on the importance of the nonapoptotic functions of caspases.

In *Drosophila* larvae, normal apoptotic cells in imaginal discs also produce Wg whereas in adult *Drosophila*, stressed enterocytes produce EGF as main signal to trigger proliferation of intestine stem cells (see above), indicating that the damage-induced production of growth factors is not an experimental artifact. In *Hydra*, the apoptotic cells that are spontaneously produced in the head-regenerating tips upon mid-gastric bisection, also release growth factors. These dying cells release Wnt3, a signal that activates β-catenin signaling in the proliferating neighboring interstitial cells, and promote thus their rapid division (Chera et al., 2009). This early signaling is required for head regeneration, as head regeneration is inhibited by the presence of the caspase inhibitor z-vad-fmk in the medium, but rescued when the recombinant Wnt3 protein is added. Cell death also promotes the up-regulation of *Wnt3* expression is added in the surrounding epithelial cells, which is required for developing a sustained head organizing activity, responsible for apical patterning (Lengfeld et al., 2009). In bilaterians, Wnt signaling is required for blastema formation in the adult zebrafish regenerating its caudal fin (Stoick-Cooper et al., 2007), in the *Xenopus* tadpole regenerating its tail (Lin & Slack, 2008) or its limb (Yokoyama, Ogino, Stoick-Cooper, Grainger, & Moon, 2007), in the planarian regenerating the posterior part of its body (Gurley et al., 2010). However a link between injury-induced apoptosis and Wnt activation has not been established in these latter contexts.

4.1.2 Prostaglandins (PG9)E2

During apoptosis, caspases activate the Ca2+-independent phospholipase A2 (iPLA2) and trigger the production of prostaglandin (PG) E2, which in turn induces proliferation (Li, Huang, et al., 2010b). This PGE2-dependent induction of cell proliferation nearby the dying cells is essential for skin wound healing and for liver regeneration in mice (Li, Huang, et al., 2010b). It is worth noting that in the zebrafish hematopoietic stem cells PGE2 directly modulate the activity of the Wnt pathway. This direct PGE2/Wnt interaction, which is essential for liver and caudal fin regeneration in the adult zebrafish, is thus not restricted to the hematopoietic system and might be a shared property of regenerative tissues (Goessling et al., 2009). This induction of cell proliferation via apoptosis/PGE2 also exerts some unexpected effect on tumor cells. In anticancer chemotherapy and radiotherapy in particular, the major goal is to induce apoptosis of tumor cells. However, recent study demonstrated that dying cells stimulate tumor cell repopulation through a compensatory mechanism dependent on caspase and PGE2 signaling (Huang et al., 2011; Mao, Smith, Xie, & Wang, 2013). Hence apoptosis-induced PGE2 signaling seems to be involved in both regeneration and cancer progression.

4.1.3 Purinergic signaling via ATP

Apoptotic cells release ATP through the caspase-dependent activation of the pannexin-1 channel (Chekeni et al., 2010). The purinergic pathway is a widespread mode of signaling, present in both the plant and the animal kingdoms. Extracellular ATP can activate the P2X and P2Y receptors on adjacent cells or act via adenosine receptors or translocators once hydrolyzed by the ectonucleotidases CD39 and CD73 (also called nt5e). These two pathways are not mutually exclusive and are subject to considerable crosstalk (Stagg & Smyth, 2010). During regeneration, both extracellular ATP and adenosine are able to stimulate proliferation *in vivo* (Lin et al., 2007), with extracellular ATP mediating proliferation through MAPK activation (Tu et al., 2000). Extracellular purines (ATP or adenosine) can signal either directly to induce cell proliferation or indirectly by regulating the immune response to damage or nerve transmission.

ATP release also participates in the "find me signal," which attracts phagocytes to the lesion site (Elliott et al., 2009). Infiltration of macrophages is required for blastema formation in newts regenerating their limb regeneration (Godwin, Pinto, & Rosenthal, 2013) as well as in adult zebrafish regenerating their caudal fin and their heart (Huang et al., 2013; Li, Yan, Shi, Zhang, & Wen, 2012). Within the first 24 hpa these macrophages release proteases, growth factors, and cytokines of importance for tissue remodeling and stimulation of proliferation. Finally, ATP is an essential molecule in the communication between neuron and glia (Abbracchio, Burnstock, Verkhratsky, & Zimmermann, 2009), an essential function in systems where regeneration is nerve dependent (Geraudie & Singer, 1985; Kumar & Brockes, 2012; Singer, 1974).

4.1.4 Apoptotic bodies as signaling vehicles for miRNA

During apoptosis, blebbing of the cell membrane induces the formation of apoptotic bodies (also named microparticles) loaded with molecules. Upon calcium activation, these microparticles spread into the tissue (Hoang, Rampon, Freyssinet, Vriz, & Kerbiriou-Nabias, 2011), which allow them to carry biological information from apoptotic to nonapoptotic cells. Indeed apoptotic bodies that derive from endothelial cells contribute to the recruitment of progenitor cells thanks to the microRNA-126 (miR-126) they contain, which triggers the production of the chemokine CXCL12 in the recipient cells (Zernecke et al., 2009). A recent review focusing on microRNA transportation points to the efficacy of this mode of intercellular communication in various tissues (Vickers & Remaley, 2012). Interestingly, the CXCL12 (also named SDF1) chemokine not only recruits progenitors to protect from atherosclerosis, but also participates in epimorphic fin regeneration by interacting with the FGF pathway (Bouzaffour, Dufourcq, Lecaudey, Haas, & Vriz, 2009; Dufourcq & Vriz, 2006). These data indicate that the apoptotic bodies likely play an important role in the paracrine signaling supporting tissue repair and regeneration, specially as this mode of signaling can persist over long periods of time, longer than the signals directly released by the apoptotic cells.

4.2. Passive, derepressing mechanism of regenerative cell death

4.2.1 Neurotransmitters as inhibitors of tissue repair in urodeles

In the adult newt brain, the drug-induced elimination of dopaminergic neurons leads to the transient proliferation of ependymal progenitors and their subsequent differentiation into new dopaminergic neurons (Parish, Beljajeva, Arenas, & Simon, 2007). Further studies evidenced a direct negative paracrine control of neurogenesis by dopamine, as the ependymal progenitors, which express the dopamine receptor transiently resume cell proliferation in the absence of dopamine, but become quiescent upon dopaminergic signaling, when novel dopaminergic neurons differentiate (Berg, Kirkham, Wang, Frisen, & Simon, 2011). This discovery of a suppressive role of dopamine on the cycling activity of progenitors might be generalized to other neurotransmitters, each of them inhibiting a specific subset of stem cells (Berg, Belnoue, Song, & Simon, 2013). This mechanism does not seem to operate in mammals as evidenced by the development of the Parkinson disease upon degeneration of dopaminergic neurons. But understanding where lie the differences between mammals and urodeles would offer new ways to use cell death and manipulate neuronal regeneration.

4.2.2 The Hydra interstitial cells produce signals repressing regeneration

In *Hydra*, the epithelial cells from the ectodermal and endodermal layers suffice to drive head regeneration (Marcum & Campbell, 1978; Sugiyama & Fujisawa, 1978). However, in homeostatic conditions, the interstitial cells and their derivatives (nerve cells, stinging cells—nematocytes, nematoblasts—and gland cells) appear to repress the morphogenetic potential of these epithelial cells. This negative control was evidenced in the head-deficient *reg-16* strain, where the elimination of the interstitial cells could restore head regeneration (Sugiyama & Wanek, 1993). This "over-repression" of head regeneration by the interstitial cells in this strain was confirmed by transplantation experiments. Therefore, we speculate that in wild-type conditions, beside an instructive mode of regenerative cell death (i.e., the transient release of growth factors as Wnt3 by the dying cells), the elimination of the nerve cells, gland cells, nematoblasts, and interstitial progenitors in head-regenerating tips, which persists for about 30 h, contribute to head regeneration by derepressing locally the morphogenetic potential of the epithelial cells (Galliot, 2013). Candidate suppressors of regeneration in *Hydra* are neuropeptides that might replace cytokine signaling in cnidarians

(Bosch & Fujisawa, 2001), but also inhibitors of Wnt signaling that are expressed in gland cells along the body column (Guder et al., 2006).

4.3. Cell death as a toxic force preventing regeneration

In the chick embryo, neurogenesis of the central nervous system takes place until day 15 and regeneration of the neural tube is fully efficient until day 13, a process that does not rely on cell death but rather on the plasticity of neuronal progenitors (Whalley et al. 2009). However from day 13, that is, when angiogenesis takes place, injury induces massive apoptosis of the nervous system and the presence of dying cells prevents regeneration (Ferretti and Whalley, 2008). These data indicate that apoptosis can act as a brake rather than a promoter of regeneration. Here again it would be interesting to evaluate whether the negative role of cell death is strictly quantitative, i.e., the level of regeneration inhibition correlates with the level of cell death, and/or whether a different type of cell death is actually responsible for this negative role. In *H. glaberrima* regenerating its gut or its radial nerve cord (central nervous system), the cells involved in the regenerative process overexpress genes that appear to protect them from cell death, either mortalin and survivin in the mesothelium of the gut, or LTR retrotransposons in the glial cells (Mashanov, Zueva, & Garcia-Arraras, 2012; Mashanov et al., 2010). If confirmed, these results indicate that injury induces transcriptional regulations that restrict the level or the extent of cell death and thus ensure the conditions for an efficient regeneration.

5. A COMPLEX PARACRINE CONTROL OF REGENERATION THROUGH CELL DEATH

Immediate injury-induced cell death is a direct response to tissue injury, that is, observed at the wound leading edge independently of any regenerative process, and thus considered as part of the wound-healing process itself (Cordeiro & Jacinto, 2013). By contrast, when a later round of cell death is observed, it appears specific to regeneration as observed in planarians and zebrafish (Gauron et al., 2013; Pellettieri et al., 2010). Interestingly, in *Hydra* the immediate massive wave of cell death recorded in head-regenerating tips (Chera et al., 2009) does not belong to the wound-healing process as first it is not induced upon wounding, and second it is asymmetric, high on one side of the cut (head regenerating tips), but low on the other side (foot regenerating tips). In *Drosophila* where immunological, metabolic, and developmental studies led to a minutious dissection of the injury-induced signaling involved in midgut repair, the level of injury signals (uracil in case of pathogenic bacterial infection) directly modulates the level of ROS production, which when high leads to cell death and stem cell activation (Buchon et al., 2013; Lee et al., 2013).

Several questions arise from the studies discussed earlier. A first one concerns the injury-induced damages: are they qualitatively different when injury triggers wound healing only, or when injury triggers a regenerative response? Alternatively, are these two types of injury responses the result of the same signaling operating at different levels, for example, low levels triggering wound healing, and high levels a more complex signaling, necessary and sufficient to launch a regenerative response? The mechanisms at work in each context probably integrate qualitative and quantitative parameters as suggested by the recent model proposed by Lee et al. (2013) where a combination of intricate factors regulate gut homeostasis and gut repair in the adult *Drosophila*: the nature of the injury signal that varies with the pathogenicity of the bacterial strains, the duration of DUOX activation, and the resulting ROS levels.

A second important question concerns the specific features of cell death that drive regenerative processes. Although examples of compensatory proliferation induced by radiotherapy or chemotherapy are well established, it is generally assumed that dying cells do not systematically produce mitogenic factors that trigger proliferation of their neighbors. Therefore what are the properties of cell death that allow the delivery of signaling molecules by the dying cells, as such as the neighboring tissues change their behavior or even undergo reprogramming, thus promoting regeneration? A series of studies in *Drosophila* larvae have demonstrated the importance of the nonapoptotic functions of caspases in developing tissues (see above) and in cell fate specification (Hyman & Yuan, 2012; Kanuka et al., 2005; Kuranaga, 2012; Kuranaga & Miura, 2007). Therefore, activation of caspases and not cell death itself might be necessary to trigger a regenerative response. A recent study performed on human fibroblasts has shown that activation of caspases 3 and 8 in the absence of cell death is required for iPS reprogramming (Li et al., 2010a). As in *Drosophila*, the nonapoptotic functions of caspases might regulate the cellular plasticity required for tissue repair and regeneration.

A third type of questions relates to the various modes of injury-induced signaling in regenerating organisms. Injury-induced caspase activation can be local and immediate, but it can also be delayed or propagated to sites distant from injury (Gauron et al., 2013; Lee & Miura, 2014). In adult

Drosophila lesions of the cuticle are sensed far from the initial lesion as evidenced by the distant redox-dependent responses recorded in neuronal or intestinal tissues, both necessary for the survival of the whole animal (Nam, Jang, You, Lee, & Lee, 2012; Takeishi et al., 2013). Similar systemic wound response might be at work in the delayed and distant cell death responses as recorded in zebrafish, with ROS signaling at long distance to promote caspase activity in specific cell types (enterocytes in *Drosophila* and epidermal cells in zebrafish), while other cells are refractory to caspase induction. What are the mechanisms that propagate injury signals to distant sites? What are the molecular properties that maintain some cells resistant to cell death and others highly sensitive? Answers to these questions will open opportunities to manipulate injury-induced ROS signaling, caspase activation, and cell death in damaged tissues, organs, or appendages to possibly unlock and activate regenerative processes in mammalian contexts.

ACKNOWLEDGMENTS

The authors are grateful to France Maloumian for help with the iconography and to Nicolas Buchon for critical reading of this review. Work in the Galliot lab is supported by the Canton of Geneva, the Swiss National Science Foundation (SNF), the Human Frontier Science Program (HFSP), the NCCR "Frontiers in Genetics," and the Claraz donation. Work in the Vriz lab is supported by the University Paris Diderot and the PSL Research University.

REFERENCES

Abbracchio, M. P., Burnstock, G., Verkhratsky, A., & Zimmermann, H. (2009). Purinergic signalling in the nervous system: An overview. *Trends in Neurosciences*, *32*, 19–29.

Amcheslavsky, A., Jiang, J., & Ip, Y. T. (2009). Tissue damage-induced intestinal stem cell division in Drosophila. *Cell Stem Cell*, *4*, 49–61.

Berg, D. A., Belnoue, L., Song, H., & Simon, A. (2013). Neurotransmitter-mediated control of neurogenesis in the adult vertebrate brain. *Development*, *140*, 2548–2561.

Berg, D. A., Kirkham, M., Wang, H., Frisen, J., & Simon, A. (2011). Dopamine controls neurogenesis in the adult salamander midbrain in homeostasis and during regeneration of dopamine neurons. *Cell Stem Cell*, *8*, 426–433.

Bergantinos, C., Corominas, M., & Serras, F. (2010). Cell death-induced regeneration in wing imaginal discs requires JNK signalling. *Development*, *137*, 1169–1179.

Bergantinos, C., Vilana, X., Corominas, M., & Serras, F. (2010). Imaginal discs: Renaissance of a model for regenerative biology. *Bioessays*, *32*, 207–217.

Bergmann, A., & Steller, H. (2010). Apoptosis, stem cells, and tissue regeneration. *Science Signaling*, *3*, re8.

Bosch, T. C., & Fujisawa, T. (2001). Polyps, peptides and patterning. *BioEssays*, *23*, 420–427.

Bouzaffour, M., Dufourcq, P., Lecaudey, V., Haas, P., & Vriz, S. (2009). Fgf and Sdf-1 pathways interact during zebrafish fin regeneration. *PLoS One*, *4*, e5824.

Bouzaffour, M., Rampon, C., Ramauge, M., Courtin, F., & Vriz, S. (2010). Implication of type 3 deiodinase induction in zebrafish fin regeneration. *General and Comparative Endocrinology*, *168*, 88–94.

Brockes, J. P., & Kumar, A. (2002). Plasticity and reprogramming of differentiated cells in amphibian regeneration. *Nature Reviews. Molecular Cell Biology, 3*, 566–574.

Brockes, J. P., & Kumar, A. (2008). Comparative aspects of animal regeneration. *Annual Review of Cell and Developmental Biology, 24*, 525–549.

Bryant, P. J. (1971). Regeneration and duplication following operations in situ on the imaginal discs of Drosophila melanogaster. *Developmental Biology, 26*, 637–651.

Buchon, N., Broderick, N. A., Chakrabarti, S., & Lemaitre, B. (2009). Invasive and indigenous microbiota impact intestinal stem cell activity through multiple pathways in Drosophila. *Genes & Development, 23*, 2333–2344.

Buchon, N., Broderick, N. A., Kuraishi, T., & Lemaitre, B. (2010). Drosophila EGFR pathway coordinates stem cell proliferation and gut remodeling following infection. *BMC Biology, 8*, 152.

Buchon, N., Broderick, N. A., & Lemaitre, B. (2013). Gut homeostasis in a microbial world: Insights from Drosophila melanogaster. *Nature Reviews. Microbiology, 11*, 615–626.

Buchon, N., Broderick, N. A., Poidevin, M., Pradervand, S., & Lemaitre, B. (2009). Drosophila intestinal response to bacterial infection: Activation of host defense and stem cell proliferation. *Cell Host & Microbe, 5*, 200–211.

Chekeni, F. B., Elliott, M. R., Sandilos, J. K., Walk, S. F., Kinchen, J. M., Lazarowski, E. R., et al. (2010). Pannexin 1 channels mediate 'find-me' signal release and membrane permeability during apoptosis. *Nature, 467*, 863–867.

Chera, S., Ghila, L., Dobretz, K., Wenger, Y., Bauer, C., & Buzgariu, W. (2009). Apoptotic cells provide an unexpected source of Wnt3 signaling to drive hydra head regeneration. *Developmental Cell, 17*, 279–289.

Chera, S., Ghila, L., Wenger, Y., & Galliot, B. (2011). Injury-induced activation of the MAPK/CREB pathway triggers apoptosis-induced compensatory proliferation in hydra head regeneration. *Development, Growth & Differentiation, 53*, 186–201.

Choi, W. Y., & Poss, K. D. (2012). Cardiac regeneration. *Current Topics in Developmental Biology, 100*, 319–344.

Christen, B., Robles, V., Raya, M., Paramonov, I., & Izpisua Belmonte, J. C. (2010). Regeneration and reprogramming compared. *BMC Biology, 8*, 5.

Circu, M. L., & Aw, T. Y. (2010). Reactive oxygen species, cellular redox systems, and apoptosis. *Free Radical Biology & Medicine, 48*, 749–762.

Clarke, P. G., & Clarke, S. (1996). Nineteenth century research on naturally occurring cell death and related phenomena. *Anatomy and Embryology (Berlin), 193*, 81–99.

Cordeiro, J. V., & Jacinto, A. (2013). The role of transcription-independent damage signals in the initiation of epithelial wound healing. *Nature Reviews. Molecular Cell Biology, 14*, 249–262.

Dufourcq, P., & Vriz, S. (2006). The chemokine SDF-1 regulates blastema formation during zebrafish fin regeneration. *Development Genes and Evolution, 216*, 635–639.

Elliott, M. R., Chekeni, F. B., Trampont, P. C., Lazarowski, E. R., Kadl, A., Walk, S. F., et al. (2009). Nucleotides released by apoptotic cells act as a find-me signal to promote phagocytic clearance. *Nature, 461*, 282–286.

Fan, Y., & Bergmann, A. (2008a). Apoptosis-induced compensatory proliferation. The Cell is dead. Long live the Cell! *Trends in Cell Biology, 18*, 467–473.

Fan, Y., & Bergmann, A. (2008b). Distinct mechanisms of apoptosis-induced compensatory proliferation in proliferating and differentiating tissues in the Drosophila eye. *Developmental Cell, 14*, 399–410.

Ferretti, P., & Whalley, K. (2008). Successful neural regeneration in amniotes: the developing chick spinal cord. *Cellular and Molecular Life Sciences, 65*, 45–53.

Finkel, T. (2011). Signal transduction by reactive oxygen species. *The Journal of Cell Biology, 194*, 7–15.

Funato, Y., Michiue, T., Asashima, M., & Miki, H. (2006). The thioredoxin-related redox-regulating protein nucleoredoxin inhibits Wnt-beta-catenin signalling through dishevelled. *Nature Cell Biology*, *8*, 501–508.

Galko, M. J., & Krasnow, M. A. (2004). Cellular and genetic analysis of wound healing in Drosophila larvae. *PLoS Biology*, *2*, E239.

Galliot, B. (2012). Hydra, a fruitful model system for 270 years. *The International Journal of Developmental Biology*, *56*, 411–423.

Galliot, B. (2013). Injury-induced asymmetric cell death as a driving force for head regeneration in Hydra. *Development Genes and Evolution*, *223*, 39–52.

Galliot, B., & Ghila, L. (2010). Cell plasticity in homeostasis and regeneration. *Molecular Reproduction and Development*, *77*, 837–855.

Gauchat, D., Kreger, S., Holstein, T., & Galliot, B. (1998). prdl-a, a gene marker for *Hydra* apical differentiation related to triploblastic paired-like head-specific genes. *Development*, *125*, 1637–1645.

Gauron, C., Rampon, C., Bouzaffour, M., Ipendey, E., Teillon, J., Volovitch, M., et al. (2013). Sustained production of ROS triggers compensatory proliferation and is required for regeneration to proceed. *Scientific Reports*, *3*, 2084.

Geraudie, J., & Singer, M. (1985). Necessity of an adequate nerve supply for regeneration of the amputated pectoral fin in the teleost Fundulus. *The Journal of Experimental Zoology*, *234*, 367–374.

Gereben, B., Zavacki, A. M., Ribich, S., Kim, B. W., Huang, S. A., Simonides, W. S., et al. (2008). Cellular and molecular basis of deiodinase-regulated thyroid hormone signaling. *Endocrine Reviews*, *29*, 898–938.

Godwin, J. W., Pinto, A. R., & Rosenthal, N. A. (2013). Macrophages are required for adult salamander limb regeneration. *Proceedings of the National Academy of Sciences of the United States of America*, *110*, 9415–9420.

Goessling, W., North, T. E., Loewer, S., Lord, A. M., Lee, S., Stoick-Cooper, C. L., et al. (2009). Genetic interaction of PGE2 and Wnt signaling regulates developmental specification of stem cells and regeneration. *Cell*, *136*, 1136–1147.

Gonzalez-Estevez, C., Felix, D. A., Rodriguez-Esteban, G., & Aboobaker, A. A. (2012). Decreased neoblast progeny and increased cell death during starvation-induced planarian degrowth. *The International Journal of Developmental Biology*, *56*, 83–91.

Guder, C., Pinho, S., Nacak, T. G., Schmidt, H. A., Hobmayer, B., Niehrs, C., et al. (2006). An ancient Wnt-Dickkopf antagonism in Hydra. *Development*, *133*, 901–911.

Gurley, K. A., Elliott, S. A., Simakov, O., Schmidt, H. A., Holstein, T. W., & Sanchez Alvarado, A. (2010). Expression of secreted Wnt pathway components reveals unexpected complexity of the planarian amputation response. *Developmental Biology*, *347*, 24–39.

Ha, E. M., Lee, K. A., Seo, Y. Y., Kim, S. H., Lim, J. H., Oh, B. H., et al. (2009). Coordination of multiple dual oxidase-regulatory pathways in responses to commensal and infectious microbes in drosophila gut. *Nature Immunology*, *10*, 949–957.

Hansson, M., Asea, A., Ersson, U., Hermodsson, S., & Hellstrand, K. (1996). Induction of apoptosis in NK cells by monocyte-derived reactive oxygen metabolites. *The Journal of Immunology*, *156*, 42–47.

Haynie, J. L., & Bryant, P. J. (1977). The effects of X-rays on the proliferation dynamics of cells in the imaginal wing disc of Drosophila melanogaster. *Roux's Archives of Developmental Biology*, *183*, 85–100.

Herrera, S. C., Martin, R., & Morata, G. (2013). Tissue homeostasis in the wing disc of Drosophila melanogaster: Immediate response to massive damage during development. *PLoS Genetics*, *9*, e1003446.

Hoang, T. Q., Rampon, C., Freyssinet, J. M., Vriz, S., & Kerbiriou-Nabias, D. (2011). A method to assess the migration properties of cell-derived microparticles within a living tissue. *Biochimica et Biophysica Acta*, *1810*, 863–866.

Holstein, T. W., Hobmayer, E., & David, C. N. (1991). Pattern of epithelial cell cycling in hydra. *Developmental Biology*, *148*, 602–611.
Huang, Q., Li, F., Liu, X., Li, W., Shi, W., Liu, F. F., et al. (2011). Caspase 3-mediated stimulation of tumor cell repopulation during cancer radiotherapy. *Nature Medicine*, *17*, 860–866.
Huang, W. C., Yang, C. C., Chen, I. H., Liu, Y. M., Chang, S. J., & Chuang, Y. J. (2013). Treatment of glucocorticoids inhibited early immune responses and impaired cardiac repair in adult zebrafish. *PLoS One*, *8*, e66613.
Huh, J. R., Guo, M., & Hay, B. A. (2004). Compensatory proliferation induced by cell death in the Drosophila wing disc requires activity of the apical cell death caspase Dronc in a nonapoptotic role. *Current Biology*, *14*, 1262–1266.
Hwang, J. S., Kobayashi, C., Agata, K., Ikeo, K., & Gojobori, T. (2004). Detection of apoptosis during planarian regeneration by the expression of apoptosis-related genes and TUNEL assay. *Gene*, *333*, 15–25.
Hyman, B. T., & Yuan, J. (2012). Apoptotic and non-apoptotic roles of caspases in neuronal physiology and pathophysiology. *Nature Reviews. Neuroscience*, *13*, 395–406.
Ishizuya-Oka, A. (2011). Amphibian organ remodeling during metamorphosis: Insight into thyroid hormone-induced apoptosis. *Development, Growth & Differentiation*, *53*, 202–212.
Jazwinska, A., Badakov, R., & Keating, M. T. (2007). Activin-betaA signaling is required for zebrafish fin regeneration. *Current Biology*, *17*, 1390–1395.
Jhamb, D., Rao, N., Milner, D. J., Song, F., Cameron, J. A., Stocum, D. L., et al. (2011). Network based transcription factor analysis of regenerating axolotl limbs. *BMC Bioinformatics*, *12*, 80.
Jiang, H., & Edgar, B. A. (2011). Intestinal stem cells in the adult Drosophila midgut. *Experimental Cell Research*, *317*, 2780–2788.
Jiang, H., Grenley, M. O., Bravo, M. J., Blumhagen, R. Z., & Edgar, B. A. (2011). EGFR/Ras/MAPK signaling mediates adult midgut epithelial homeostasis and regeneration in Drosophila. *Cell Stem Cell*, *8*, 84–95.
Kaloulis, K., Chera, S., Hassel, M., Gauchat, D., & Galliot, B. (2004). Reactivation of developmental programs: The cAMP-response element-binding protein pathway is involved in hydra head regeneration. *Proceedings of the National Academy of Sciences of the United States of America*, *101*, 2363–2368.
Kamata, H., Honda, S., Maeda, S., Chang, L., Hirata, H., & Karin, M. (2005). Reactive oxygen species promote TNFalpha-induced death and sustained JNK activation by inhibiting MAP kinase phosphatases. *Cell*, *120*, 649–661.
Kaneko, Y., Matsumoto, G., & Hanyu, Y. (1999). The occurrence of apoptosis during retinal regeneration in adult newts. *Brain Research. Developmental Brain Research*, *117*, 225–228.
Kanuka, H., Kuranaga, E., Takemoto, K., Hiratou, T., Okano, H., & Miura, M. (2005). Drosophila caspase transduces Shaggy/GSK-3beta kinase activity in neural precursor development. *The EMBO Journal*, *24*, 3793–3806.
Kerr, J. F., Wyllie, A. H., & Currie, A. R. (1972). Apoptosis: A basic biological phenomenon with wide-ranging implications in tissue kinetics. *British Journal of Cancer*, *26*, 239–257.
Kester, M. H., Toussaint, M. J., Punt, C. A., Matondo, R., Aarnio, A. M., Darras, V. M., et al. (2009). Large induction of type III deiodinase expression after partial hepatectomy in the regenerating mouse and rat liver. *Endocrinology*, *150*, 540–545.
King, R. S., & Newmark, P. A. (2012). The cell biology of regeneration. *The Journal of Cell Biology*, *196*, 553–562.
Kizil, C., & Brand, M. (2011). Cerebroventricular microinjection (CVMI) into adult zebrafish brain is an efficient misexpression method for forebrain ventricular cells. *PLoS One*, *6*, e27395.
Knapp, D., Schulz, H., Rascon, C. A., Volkmer, M., Scholz, J., Nacu, E., et al. (2013). Comparative transcriptional profiling of the axolotl limb identifies a tripartite regeneration-specific gene program. *PLoS One*, *8*, e61352.

Kondo, S. (1988). Altruistic cell suicide in relation to radiation hormesis. *International Journal of Radiation Biology and Related Studies in Physics, Chemistry, and Medicine, 53*, 95–102.

Kragl, M., Knapp, D., Nacu, E., Khattak, S., Maden, M., Epperlein, H. H., et al. (2009). Cells keep a memory of their tissue origin during axolotl limb regeneration. *Nature, 460*, 60–65.

Kumar, A., & Brockes, J. P. (2012). Nerve dependence in tissue, organ, and appendage regeneration. *Trends in Neurosciences, 35*, 691–699.

Kuranaga, E. (2012). Beyond apoptosis: Caspase regulatory mechanisms and functions in vivo. *Genes to Cells, 17*, 83–97.

Kuranaga, E., & Miura, M. (2007). Nonapoptotic functions of caspases: Caspases as regulatory molecules for immunity and cell-fate determination. *Trends in Cell Biology, 17*, 135–144.

Lee, W. J., & Brey, P. T. (2013). How microbiomes influence metazoan development: Insights from history and drosophila modeling of gut-microbe interactions. *Annual Review of Cell and Developmental Biology, 29*, 571–592.

Lee, K. A., Kim, S. H., Kim, E. K., Ha, E. M., You, H., Kim, B., et al. (2013). Bacterial-derived uracil as a modulator of mucosal immunity and gut-microbe homeostasis in Drosophila. *Cell, 153*, 797–811.

Lee, W. J., & Miura, M. (2014). Mechanisms of systemic wound response in Drosophila. *Current Topics in Developmental Biology, 108*, in press.

Lengfeld, T., Watanabe, H., Simakov, O., Lindgens, D., Gee, L., Law, L., et al. (2009). Multiple Wnts are involved in Hydra organizer formation and regeneration. *Developmental Biology, 330*, 186–199.

Li, F., He, Z., Shen, J., Huang, Q., Li, W., Liu, X., et al. (2010). Apoptotic caspases regulate induction of iPSCs from human fibroblasts. *Cell Stem Cell, 7*, 508–520.

Li, F., Huang, Q., Chen, J., Peng, Y., Roop, D. R., Bedford, J. S., et al. (2010). Apoptotic cells activate the "phoenix rising" pathway to promote wound healing and tissue regeneration. *Science Signaling, 3*, ra13.

Li, L., Yan, B., Shi, Y. Q., Zhang, W. Q., & Wen, Z. L. (2012). Live imaging reveals differing roles of macrophages and neutrophils during zebrafish tail fin regeneration. *The Journal of Biological Chemistry, 287*, 25353–25360.

Lin, G., & Slack, J. M. (2008). Requirement for Wnt and FGF signaling in Xenopus tadpole tail regeneration. *Developmental Biology, 316*, 323–335.

Lin, J. H., Takano, T., Arcuino, G., Wang, X., Hu, F., Darzynkiewicz, Z., et al. (2007). Purinergic signaling regulates neural progenitor cell expansion and neurogenesis. *Developmental Biology, 302*, 356–366.

Liu, Y., & Min, W. (2002). Thioredoxin promotes ASK1 ubiquitination and degradation to inhibit ASK1-mediated apoptosis in a redox activity-independent manner. *Circulation Research, 90*, 1259–1266.

Love, N. R., Chen, Y., Bonev, B., Gilchrist, M. J., Fairclough, L., Lea, R., et al. (2011). Genome-wide analysis of gene expression during Xenopus tropicalis tadpole tail regeneration. *BMC Developmental Biology, 11*, 70.

Love, N. R., Chen, Y., Ishibashi, S., Kritsiligkou, P., Lea, R., Koh, Y., et al. (2013). Amputation-induced reactive oxygen species are required for successful Xenopus tadpole tail regeneration. *Nature Cell Biology, 15*, 222–228.

Maki, N., Suetsugu-Maki, R., Tarui, H., Agata, K., Del Rio-Tsonis, K., & Tsonis, P. A. (2009). Expression of stem cell pluripotency factors during regeneration in newts. *Developmental Dynamics, 238*, 1613–1616.

Mao, P., Smith, L., Xie, W., & Wang, M. (2013). Dying endothelial cells stimulate proliferation of malignant glioma cells via a caspase 3-mediated pathway. *Oncology Letters, 5*, 1615–1620.

Marcum, B. A., & Campbell, R. D. (1978). Development of Hydra lacking nerve and interstitial cells. *Journal of Cell Science, 29*, 17–33.

Mashanov, V., & Garcia Arraras, J. (2014). Post-embryonic organogenesis of the digestive tube: Why does it occur in worms and sea cucumbers but fails in humans? *Current Topics in Developmental Biology, 108*, in press.

Mashanov, V. S., & Garcia-Arraras, J. E. (2011). Gut regeneration in holothurians: A snapshot of recent developments. *The Biological Bulletin, 221*, 93–109.

Mashanov, V. S., Zueva, O. R., & Garcia-Arraras, J. E. (2012). Posttraumatic regeneration involves differential expression of long terminal repeat (LTR) retrotransposons. *Developmental Dynamics, 241*, 1625–1636.

Mashanov, V. S., Zueva, O. R., & Garcia-Arraras, J. E. (2013). Radial glial cells play a key role in echinoderm neural regeneration. *BMC Biology, 11*, 49.

Mashanov, V. S., Zueva, O. R., Rojas-Catagena, C., & Garcia-Arraras, J. E. (2010). Visceral regeneration in a sea cucumber involves extensive expression of survivin and mortalin homologs in the mesothelium. *BMC Developmental Biology, 10*, 117.

Miljkovic-Licina, M., Chera, S., Ghila, L., & Galliot, B. (2007). Head regeneration in wild-type hydra requires de novo neurogenesis. *Development, 134*, 1191–1201.

Mollereau, B., Perez-Garijo, A., Bergmann, A., Miura, M., Gerlitz, O., Ryoo, H. D., et al. (2013). Compensatory proliferation and apoptosis-induced proliferation: A need for clarification. *Cell Death and Differentiation, 20*, 181.

Morata, G., Shlevkov, E., & Perez-Garijo, A. (2011). Mitogenic signaling from apoptotic cells in Drosophila. *Development, Growth & Differentiation, 53*, 168–176.

Murawala, P., Tanaka, E. M., & Currie, J. D. (2012). Regeneration: The ultimate example of wound healing. *Seminars in Cell & Developmental Biology, 23*, 954–962.

Nakatani, Y., Kawakami, A., & Kudo, A. (2007). Cellular and molecular processes of regeneration, with special emphasis on fish fins. *Development, Growth & Differentiation, 49*, 145–154.

Nam, H. J., Jang, I. H., You, H., Lee, K. A., & Lee, W. J. (2012). Genetic evidence of a redox-dependent systemic wound response via Hayan protease-phenoloxidase system in Drosophila. *The EMBO Journal, 31*, 1253–1265.

Niethammer, P., Grabher, C., Look, A. T., & Mitchison, T. J. (2009). A tissue-scale gradient of hydrogen peroxide mediates rapid wound detection in zebrafish. *Nature, 459*, 996–999.

Parish, C. L., Beljajeva, A., Arenas, E., & Simon, A. (2007). Midbrain dopaminergic neurogenesis and behavioural recovery in a salamander lesion-induced regeneration model. *Development, 134*, 2881–2887.

Pellettieri, J., Fitzgerald, P., Watanabe, S., Mancuso, J., Green, D. R., & Sanchez Alvarado, A. (2010). Cell death and tissue remodeling in planarian regeneration. *Developmental Biology, 338*, 76–85.

Perez-Garijo, A., Martin, F. A., & Morata, G. (2004). Caspase inhibition during apoptosis causes abnormal signalling and developmental aberrations in Drosophila. *Development, 131*, 5591–5598.

Poleo, G., Brown, C. W., Laforest, L., & Akimenko, M. A. (2001). Cell proliferation and movement during early fin regeneration in zebrafish. *Developmental Dynamics, 221*, 380–390.

Poss, K. D. (2010). Advances in understanding tissue regenerative capacity and mechanisms in animals. *Nature Reviews. Genetics, 11*, 710–722.

Puzianowska-Kuznicka, M., Pietrzak, M., Turowska, O., & Nauman, A. (2006). Thyroid hormones and their receptors in the regulation of cell proliferation. *Acta Biochimica Polonica, 53*, 641–650.

Rao, N., Jhamb, D., Milner, D. J., Li, B., Song, F., Wang, M., et al. (2009). Proteomic analysis of blastema formation in regenerating axolotl limbs. *BMC Biology, 7*, 83.

Ravindran, J., Gupta, N., Agrawal, M., Bala Bhaskar, A. S., & Lakshmana Rao, P. V. (2011). Modulation of ROS/MAPK signaling pathways by okadaic acid leads to cell death via, mitochondrial mediated caspase-dependent mechanism. *Apoptosis, 16*, 145–161.

Ray, P. D., Huang, B. W., & Tsuji, Y. (2012). Reactive oxygen species (ROS) homeostasis and redox regulation in cellular signaling. *Cellular Signalling, 24*, 981–990.

Razzell, W., Wood, W., & Martin, P. (2011). Swatting flies: Modelling wound healing and inflammation in Drosophila. *Disease Models & Mechanisms, 4*, 569–574.

Reddien, P. W., & Sanchez Alvarado, A. (2004). Fundamentals of planarian regeneration. *Annual Review of Cell and Developmental Biology, 20*, 725–757.

Reiter, S., Crescenzi, M., Galliot, B., & Buzgariu, W. (2012). Hydra, a versatile model to study the homeostatic and developmental functions of cell death. *The International Journal of Developmental Biology, 56*, 593–604.

Rieger, S., & Sagasti, A. (2011). Hydrogen peroxide promotes injury-induced peripheral sensory axon regeneration in the zebrafish skin. *PLoS Biology, 9*, e1000621.

Romero, R., & Baguna, J. (1991). Quantitative cellular analysis of growth and reproduction in freshwater planarians (Turbellaria; Tricladida). I. A cellular description of the intact organism. *Invertebrate Reproduction and Development, 19*, 157–165.

Ryoo, H. D., Gorenc, T., & Steller, H. (2004). Apoptotic cells can induce compensatory cell proliferation through the JNK and the Wingless signaling pathways. *Developmental Cell, 7*, 491–501.

Saitoh, M., Nishitoh, H., Fujii, M., Takeda, K., Tobiume, K., Sawada, Y., et al. (1998). Mammalian thioredoxin is a direct inhibitor of apoptosis signal-regulating kinase (ASK) 1. *The EMBO Journal, 17*, 2596–2606.

Salo, E., & Baguna, J. (1984). Regeneration and pattern formation in planarians. I. The pattern of mitosis in anterior and posterior regeneration in Dugesia (G) tigrina, and a new proposal for blastema formation. *Journal of Embryology and Experimental Morphology, 83*, 63–80.

San Miguel-Ruiz, J. E., Maldonado-Soto, A. R., & Garcia-Arraras, J. E. (2009). Regeneration of the radial nerve cord in the sea cucumber Holothuria glaberrima. *BMC Developmental Biology, 9*, 3.

Sanchez Alvarado, A., & Tsonis, P. A. (2006). Bridging the regeneration gap: Genetic insights from diverse animal models. *Nature Reviews. Genetics, 7*, 873–884.

Santos-Ruiz, L., Santamaria, J. A., Ruiz-Sanchez, J., & Becerra, J. (2002). Cell proliferation during blastema formation in the regenerating teleost fin. *Developmental Dynamics, 223*, 262–272.

Simon, A., Berg, D., & Kirkham, M. (2009). Not lost in translation sensing the loss and filling the gap during regeneration. *Seminars in Cell & Developmental Biology, 20*, 691–696.

Singer, M. (1974). Trophic functions of the neuron. VI. Other trophic systems. Neurotrophic control of limb regeneration in the newt. *Annals of the New York Academy of Sciences, 228*, 308–322.

Slack, J. M., Lin, G., & Chen, Y. (2008). The Xenopus tadpole: A new model for regeneration research. *Cellular and Molecular Life Sciences, 65*, 54–63.

Smith-Bolton, R. K., Worley, M. I., Kanda, H., & Hariharan, I. K. (2009). Regenerative growth in Drosophila imaginal discs is regulated by Wingless and Myc. *Developmental Cell, 16*, 797–809.

Sousounis, K., Baddour, J. A., & Tsonis, P. A. (2014). Aging and regeneration in vertebrates. *Current Topics in Developmental Biology, 108*, in press.

Stagg, J., & Smyth, M. J. (2010). Extracellular adenosine triphosphate and adenosine in cancer. *Oncogene, 29*, 5346–5358.

Steele, R. E. (2002). Developmental signaling in Hydra: What does it take to build a "simple" animal? *Developmental Biology, 248*, 199–219.

Stoick-Cooper, C. L., Moon, R. T., & Weidinger, G. (2007). Advances in signaling in vertebrate regeneration as a prelude to regenerative medicine. *Genes & Development, 21*, 1292–1315.

Stoick-Cooper, C. L., Weidinger, G., Riehle, K. J., Hubbert, C., Major, M. B., Fausto, N., et al. (2007). Distinct Wnt signaling pathways have opposing roles in appendage regeneration. *Development, 134*, 479–489.

Sugiyama, T., & Fujisawa, T. (1978). Genetic analysis of developmental mechanisms in Hydra. II. Isolation and characterization of an interstitial cell-deficient strain. *Journal of Cell Science*, *29*, 35–52.

Sugiyama, T., & Wanek, N. (1993). Genetic analysis of developmental mechanisms in hydra. XXI. Enhancement of regeneration in a regeneration-deficient mutant strain by the elimination of the interstitial cell lineage. *Developmental Biology*, *160*, 64–72.

Sun, L., Yang, H., Chen, M., Ma, D., & Lin, C. (2013). RNA-Seq reveals dynamic changes of gene expression in key stages of intestine regeneration in the sea cucumber Apostichopus japonicas. *PLoS One*, *8*, e69441.

Takeishi, A., Kuranaga, E., Tonoki, A., Misaki, K., Yonemura, S., Kanuka, H., et al. (2013). Homeostatic epithelial renewal in the gut is required for dampening a fatal systemic wound response in Drosophila. *Cell Reports*, *3*, 919–930.

Tanaka, E. M., & Reddien, P. W. (2011). The cellular basis for animal regeneration. *Developmental Cell*, *21*, 172–185.

Tseng, A. S., Adams, D. S., Qiu, D., Koustubhan, P., & Levin, M. (2007). Apoptosis is required during early stages of tail regeneration in Xenopus laevis. *Developmental Biology*, *301*, 62–69.

Tu, M. T., Luo, S. F., Wang, C. C., Chien, C. S., Chiu, C. T., Lin, C. C., et al. (2000). P2Y(2) receptor-mediated proliferation of C(6) glioma cells via activation of Ras/Raf/MEK/MAPK pathway. *British Journal of Pharmacology*, *129*, 1481–1489.

Vickers, K. C., & Remaley, A. T. (2012). Lipid-based carriers of microRNAs and intercellular communication. *Current Opinion in Lipidology*, *23*, 91–97.

Vlaskalin, T., Wong, C. J., & Tsilfidis, C. (2004). Growth and apoptosis during larval forelimb development and adult forelimb regeneration in the newt (Notophthalmus viridescens). *Development Genes and Evolution*, *214*, 423–431.

Weiss, C., & Rosenbaum, R. M. (1967). Histochemical studies on cell death and histolysis during regeneration. I. Distribution of acid phosphomonoesterase activity in the normal, the regenerating and the resorbing forelimb of the larval spotted salamander, Amblystoma maculatum. *Journal of Morphology*, *122*, 203–229.

Wells, B. S., Yoshida, E., & Johnston, L. A. (2006). Compensatory proliferation in Drosophila imaginal discs requires Dronc-dependent p53 activity. *Current Biology*, *16*, 1606–1615.

Wenemoser, D., & Reddien, P. W. (2010). Planarian regeneration involves distinct stem cell responses to wounds and tissue absence. *Developmental Biology*, *344*, 979–991.

Whalley, K., Gogel, S., Lange, S., & Ferretti, P. (2009). Changes in progenitor populations and ongoing neurogenesis in the regenerating chick spinal cord. *Developmental Biology*, *332*, 234–245.

Wood, W., & Eming, S. A. (2012). Repair or regenerate—How can we tip the balance? *EMBO Reports*, *13*, 1040–1042.

Worley, M. I., Setiawan, L., & Hariharan, I. K. (2012). Regeneration and transdetermination in Drosophila imaginal discs. *Annual Review of Genetics*, *46*, 289–310.

Yokoyama, H., Ogino, H., Stoick-Cooper, C. L., Grainger, R. M., & Moon, R. T. (2007). Wnt/beta-catenin signaling has an essential role in the initiation of limb regeneration. *Developmental Biology*, *306*, 170–178.

Zernecke, A., Bidzhekov, K., Noels, H., Shagdarsuren, E., Gan, L., Denecke, B., et al. (2009). Delivery of microRNA-126 by apoptotic bodies induces CXCL12-dependent vascular protection. *Science Signaling*, *2*, ra81.

Zimmerman, M. A., Huang, Q., Li, F., Liu, X., & Li, C. Y. (2013). Cell death-stimulated cell proliferation: A tissue regeneration mechanism usurped by tumors during radiotherapy. *Seminars in Radiation Oncology*, *23*, 288–295.

CHAPTER SIX

Mechanisms of Systemic Wound Response in *Drosophila*

Won-Jae Lee[*,1], **Masayuki Miura**[†,1]
[*]School of Biological Science, Seoul National University and National Creative Research Initiative Center for Symbiosystem, Seoul, South Korea
[†]Department of Genetics, Graduate School of Pharmaceutical Sciences, The University of Tokyo, and Core Research for Evolutional Science and Technology, Japan Science and Technology Agency, Tokyo, Japan
[1]Corresponding authors: e-mail address: lwj@snu.ac.kr; miura@mol.f.u-tokyo.ac.jp

Contents

Abstract

In response to cellular and tissue losses caused by physical or chemical injuries, organisms must activate multiple wound repair systems at the cellular, tissue, and organismal levels. The systemic wound response (SWR) that occurs via interorgan communication between local wound sites and remote organs ensures that the host is protected efficiently in response to a local wound. The local wound response around the wound site is fairly well documented, but the molecular mechanisms that allow the host to launch SWR are poorly understood. Recent studies on the *Drosophila* adult model system have shown that the local wound response is not restricted to the wound site because it plays an essential role in generating signals transmitted to remote organs that subsequently achieve SWR. By exploiting the genetic methods available for investigating *Drosophila*, we are just beginning to understand the complex interorgan networks that operate during SWRs. This review discusses the basic processes involved with classical integumental wound responses and tissue regeneration, such as epithelial cell movement, hemocyte

Current Topics in Developmental Biology, Volume 108
ISSN 0070-2153
http://dx.doi.org/10.1016/B978-0-12-391498-9.00001-2

recruitment, apoptosis, melanization, and generation of reactive oxygen species, as well as the recently described intestinal epithelial cell renewal program that occurs in response to gut cell damages. Furthermore, we discuss how these local wound responses integrate with organ-to-organ communication to launch SWR. Genetic analysis of SWRs using the *Drosophila* model system will provide a unique opportunity to dissect the molecular mechanisms that control wound-induced organ-to-organ communication.

A Box with Definition

Integument: External layer that covers the insect surface. It is composed of a single layer of epithelia cells covered by extracellular cuticle.
Imaginal disc: Hollow sac of epithelial cells in larva that makes adult structures during metamorphosis.
Histoblast: Population of epithelial cells in larvae that can make adult epithelia during metamorphosis.

ABBREVIATIONS

BH3 bcl-2 homology domain
Chic chickadee
Ddc dopa decarboxylase
Diap1 *Drosophila* inhibitor of apoptosis protein 1
DPP decapentaplegic
ECR epithelial cell renewal
EB enteroblast
EE enteroendocrine
Grh grainy head
Hh hedgehog
ICE interleukin-1β-converting enzyme
IIS insulin/IGF signaling
ISC intestinal stem cell
LEC larval epidermal cells
LWR local wound response
MAPK mitogen-activated protein kinase
MMP-1 matrix metalloproteinase-1
Ple tyrosine hydrolase
PO phenoloxidase
Pvf PDGF- and VEGF-related factor
ROS reactive oxygen species
SH sulfhydryl
SOH sulfenic
SWR systemic wound response
Wg wingless
ZPA zone of polarizing activity

1. INTRODUCTION

Physical injury in the epithelial barrier of a multicellular organism provokes a diverse range of host responses, including local wound responses (LWRs) at the wound site and systemic wound responses (SWR) in the whole body (Belacortu & Paricio, 2011; Delessert, Wilson, Van Der Straeten, Dennis, & Dolferus, 2004; Foex, 1999; Lenz, Franklin, & Cheadle, 2007; Lowry & Calvano, 2008; Razzell, Wood, & Martin, 2011; Schilmiller & Howe, 2005). The LWR includes complex healing processes such as scab formation and reepithelization, which is required to maintain the integrity of epithelial tissue and limit excessive bleeding and inflammation. Because of the medical importance in humans, wound healing processes are subjected to intensive research using human cell cultures and mouse models. However, because wound response is confined to an *in vivo* condition, the molecular mechanism by which the host manages a local wound is only partially understood. Many important common aspects of wound response have been more clearly described in studies using genetic models of animals such as zebrafish and *Drosophila* (Galko & Krasnow, 2004; Mace, Pearson, & McGinnis, 2005; Niethammer, Grabher, Look, & Mitchison, 2009; Razzell et al., 2011). Using the advantages of elegant genetic tools, *Drosophila* has proven to be an ideal model to dissect molecular events governing wound response. The major advantages of *Drosophila* models in the research of wound response can be found in excellent recent reviews (Belacortu & Paricio, 2011; Razzell et al., 2011) and hence will not be treated herein. At present, genetic analyses in *Drosophila* using transgenic animals expressing wound-responsive reporter genes as well as genome-wide analysis of wound-responsive genes have been providing essential information on the basic features of LWR in the epithelia (Juarez, Patterson, Sandoval-Guillen, & McGinnis, 2011; Mace et al., 2005; Patterson et al., 2013; Razzell, Evans, Martin, & Wood, 2013). However, recent studies on *Drosophila* revealed that LWR is not restricted to the wound site; rather, an essential signal is sent out to remote organs to subsequently achieve an SWR (Nam, Jang, You, Lee, & Lee, 2012; Takeishi et al., 2013). This interorgan level of communication is likely a conserved and essential signature of wound responses in all multicellular organisms. In this review, we present conserved events of LWR, and discuss recent views regarding SWR in relationship with LWR based on observations from the *Drosophila* model of wound response.

2. BASIC PROCESSES INVOLVED IN LOCAL WOUND RESPONSES

Wound healing is a process that rapidly restores epithelial function after tissue damage. The immediate response that follows epidermal puncture wounding is scab formation, which is initiated by a clotting cascade. Reepithelialization is an essential process for wound repair (Wood et al., 2002). Diverse types of wounding methods, such as puncture wounding, UV irradiation, and physical or chemical damage, have been used on different tissues including the epidermis, the imaginal disc, and the intestine (Chatterjee & Ip, 2009; Razzell et al., 2011) (Fig. 6.1). Typically, a puncture wound experiment is performed by mechanically breaking the epidermal layer with a needle (nonsterile wound) or laser (sterile wound) on a stage 14 embryo or a 3rd instar larva, pupa, or adult (Razzell et al., 2011). In larvae, soon after the puncture, bleeding began and wound gaps closed by the formation of plugs within an hour. Next, melanized scabs form within a few hours through the activities of clotting and melanization cascades (Galko & Krasnow, 2004). In some experiments, pinching of the epidermis using dissecting forceps without breaking the cuticle is performed to exclude the clotting and infection processes from the epithelial wound response

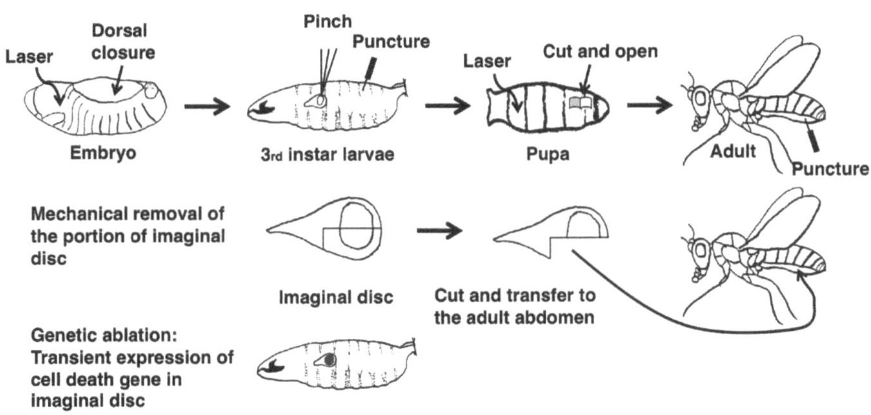

Figure 6.1 Wound models in *Drosophila*. Physical wounding methods in various stage of development are shown in upper part. In the middle, mechanical dissection of imaginal disc and its transfer to the adult abdomen for regeneration is shown. In the bottom, transient expression of cell death gene such as *reaper* or *eiger* under the tissue specific Gal4-driver induces cell death (marked as red circle in the imaginal disc) in the 3rd instar larvae. Disc regeneration occurs following conditional cell ablation. (For interpretation of the references to color in this figure legend, the reader is referred to the online version of this chapter.)

(Babcock et al., 2008; Diaz-Garcia & Baonza, 2013). However, this method does not cause melanization associated with scab formation due to the absence of cuticular disruption and bleeding (Pastor-Pareja, Wu, & Xu, 2008).

Using larval imaginal discs, the wound healing process during tissue regeneration has been intensely studied (Belacortu & Paricio, 2011). Mechanical cutting of the imaginal disc of the 3rd instar larvae (wing or leg discs) followed by grafting the imaginal disc pieces onto an adult abdomen is a classical experimental model for tissue regeneration (Hadron & Buck, 1962) (Fig. 6.1). The pinching method has also been applied for wounding the wing imaginal disc (Babcock et al., 2008; Diaz-Garcia & Baonza, 2013; Pastor-Pareja et al., 2008) (Fig. 6.1). Genetic ablation of developing tissue by expressing cell death gene can be also performed. However, continuous expression of cell death gene prevents wound healing or tissue regeneration process. Recently, genetic ablation of part of the imaginal disc by transiently expressing cell death-inducing genes, such as *Reaper*, *Hid*, or *Eiger*, in the 3rd instar larvae has been developed as another excellent model for tissue regeneration (Bergantinos, Corominas, & Serras, 2010; Herrera, Martin, & Morata, 2013; Smith-Bolton, Worley, Kanda, & Hariharan, 2009) (Fig. 6.1). The wound healing process of the imaginal disc in *Drosophila* larvae has been shown to be similar to that of epithelial wound healing. Here, we discuss some basic processes involved in LWR, which is essential initiation step for SWR.

2.1. Epithelial cell movement

The basic process of wound healing involves epithelial movement followed by attachment and fusion. Similar events are also observed during development, such as embryonic dorsal closure or pupal thorax and genital imaginal disc fusion. Dorsal closure is primarily regulated by JNK. JNK activation is also observed during larval or adult epithelial wound healing (Galko & Krasnow, 2004; Lesch, Jo, Wu, Fish, & Galko, 2010; Ramet, Lanot, Zachary, & Manfruelli, 2002) and larval imaginal disc wound healing (Bosch, Serras, Martin-Blanco, & Baguna, 2005). It has been demonstrated that genetic suppression of the JNK pathway impaired healing, highlighting the essential role of JNK in this process. Signaling by the mitogen-activated protein kinase (MAPK) cascade, which results in JNK activation, has been fairly well documented. This cascade consists of sequential activation of MAPKKKK Misshapen, MAPKKK Slipper, MAPKK Hemipterus, and MAPK/JNK Basket (Fig. 6.2). However, the upstream signal that initiates

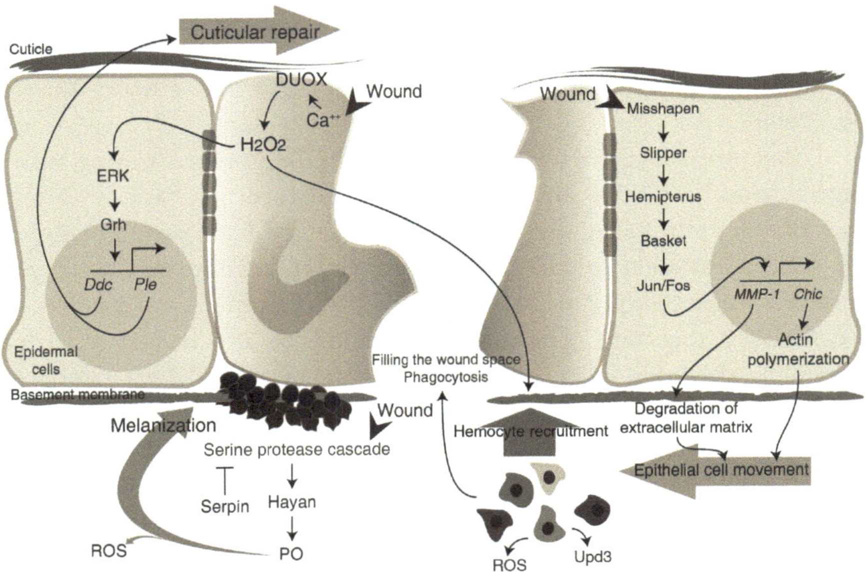

Figure 6.2 Local wound responses following integumental wound in *Drosophila*. Upon integumental wound, complex biological processes, such as cuticular repair, ROS generation, epithelial cell movement, hemocyte recruitment, and melanization, take place immediately around the wound site. See text for more details. (See color plate.)

the Basket-activating cascade remains unknown. One of the downstream targets of Basket is the transcription factor Jun/Fos, which modulates the expression of a diverse group of genes (Galko & Krasnow, 2004; Ramet et al., 2002). Matrix metalloproteinase-1 (MMP-1) is induced by Jun/Fos and it degrades extracellular matrix during wound healing (McClure, Sustar, & Schubiger, 2008; Srivastava, Pastor-Pareja, Igaki, Pagliarini, & Xu, 2007; Uhlirova & Bohmann, 2006), disc eversion in metamorphosis, disc regeneration, and tumor invasion in a *Drosophila* model (McClure et al., 2008; Srivastava, Pastor-Pareja, Igaki, Pagliarini, & Xu, 2007; Uhlirova & Bohmann, 2006). Other genes, such as Chickadee (Chic), a *Drosophila* profilin, regulate actin polymerization during epithelial movement (Jasper et al., 2001; Thomas, Rousset, & Noselli, 2009). The JNK pathway activation in epithelial cell movement during wound healing process is summarized in Fig. 6.2. In addition to the JNK pathway, grainy head (Grh)-mediated transcriptional regulation is known to regulate the expression of genes in response to wounding, such as *dopa decarboxylase* (*Ddc*) and *tyrosine hydrolase* (*Ple*), which are required for cuticular skeleton formation during wound healing (Mace et al., 2005) (Fig. 6.2). In this case, ERK acts as an upstream MAPK for Grh activation.

During the repair of epithelia, actin cables and filopodial and/or lamellipodial protrusions are observed at the wound edge. Expression of a dominant-negative form of Cdc42 prevented the sealing of the wound edge, indicating that a small Rho GTPase Cdc42 is required for filopodia formation (Wood et al., 2002). In larvae, when wounded epithelia are exposed to blood-borne ligand PDGF- and VEGF-related factors (Pvf), the epidermal Pvf receptor is activated and migration of epidermal cells is induced (Wu et al., 2009). Beyond the wounding margin, scattered syncytia and cell fusion events are also observed (Galko & Krasnow, 2004).

2.2. Hemocyte activation

Hemocytes are recruited to the damaged area to promote tissue repair and removal of dead cells (Babcock et al., 2008; Pastor-Pareja et al., 2008) (Fig. 6.2). Disruption of the basement membrane by wounding allows hemocytes an access to damaged epithelia and facilitates a systemic tissue damage response (Pastor-Pareja et al., 2008) (Fig. 6.2).

UV irradiation of the pupal retina leads to widespread apoptosis and tissue loss in the retina. In this condition, the ablation of hemocytes enhanced tissue damage, suggesting a protective functional role for hemocytes in preventing tissue damage. It has been shown that UV-damaged retina cells produce the PDGF-like growth factor Pvf-1 through the activity of the transcriptional regulator Schnurri. Pvf-1 in turn activates hemocytes by enhancing its phagocytic activity for removing dead cells, thereby ensuring the recovery and homeostasis of the damaged tissues (Kelsey, Luo, Bruckner, & Jasper, 2012).

When entire larvae were UV-irradiated, epidermal-specific DNA damage was induced. In response to such epidermal DNA damage, systemic responses are observed, including cytokine Upd3 expression in hemocytes, hemocyte expansion, and widespread melanization. In the early phase of the response to UV irradiation, systemic JAK/STAT signaling activation is observed in diverse tissues including in hemocytes, and the JAK/STAT activation state correlates with the hemocyte-mediated repression of systemic insulin/IGF signaling (IIS). In the late phase of the response, repression of IIS leads to FOXO-dependent induction of NF-κB/Relish in fat body, which is in turn associated with recovery of normal IIS levels, thereby maintaining systemic homeostasis. Therefore, cross-regulation between immune and metabolic signaling pathway across different tissues (hemocytes, fat body, and brain) regulates systemic responses to localized DNA damage (Karpac, Younger, & Jasper, 2011).

2.3. Apoptosis in wound and tissue damage responses

Tissue damage is often associated with caspase activation and cell death. Caspases are a family of cysteine proteases that are conserved in multicellular organisms and are essential executioners of apoptosis. In *Caenorhabditis elegans*, 131 cells undergo programmed cell death during development (Sulston & Horvitz, 1977), and the *Ced-3* gene was identified as an essential cell death-executing gene (Ellis & Horvitz, 1986). The first Ced-3 homolog discovered in mammals was interleukin-1β-converting enzyme, ICE, a cysteine protease responsible for the processing and secretion of pro-IL-1 (Cerretti et al., 1992; Miura, Zhu, Rotello, Hartwieg, & Yuan, 1993; Thornberry et al., 1992; Yuan, Shaham, Ledoux, Ellis, & Horvitz, 1993). ICE and Ced-3 are the founding members of the caspase family. Caspases cleave proteins at specific motifs, in which an aspartic acid residue is at the P1 position (Timmer & Salvesen, 2007). This unique cleavage site preference enables caspases to inactivate proteins or to modify protein functions by removing a portion of a target protein by site-specific cleavage. We discuss the general mode of caspase activation and the different roles of caspases in wound responses.

2.3.1 Caspase activation

Caspase activation mechanisms upon apoptotic stimulation have been intensively studied in mammals (Degterev, Boyce, & Yuan, 2003). Apoptotic caspases can be grouped into two categories: initiator caspases (such as caspase-8 or -9) or executioner caspases (such as caspase-3 or -7). Initiator caspases cleave the pro-form of an executioner caspase to convert it to its active form. In turn, the active form of an executioner caspase cleaves many target proteins in cells. Because of the strong enzymatic activity and relatively broad substrate specificity, hundreds of caspase-3 substrates have been identified and caspase-3 has been recognized as an apoptosis executioner (Dix, Simon, & Cravatt, 2008; Luthi & Martin, 2007).

Caspase activation is regulated through either the extrinsic or intrinsic signaling pathways. The extrinsic pathway is activated by death-inducing ligands, such as TNF-α and FasL. Fas or TNF receptor stimulation by ligand binding leads to the activation of an initiator caspase, caspase-8. Activated caspase-8 directly cleaves pro-caspase-3 to generate the active form of caspase-3, allowing the execution of apoptosis. The intrinsic pathway is widely used for most apoptotic caspase activation events in mammals. In this case, activated caspase-8 cleaves the Bcl-2 homology domain 3 (BH3)-only protein Bid to generate a truncated form of Bid (tBid) that activates the intrinsic pathway. Other BH3-only proteins are also induced upon apoptotic stimulation. The function of tBid and other BH3-only proteins is to promote the release of cytochrome *c* from mitochondria for stimulating cell death. Cytochrome *c* then forms an apoptosome complex by oligomerizing with Apaf-1 and pro-caspase-9. Apoptosome-mediated activation of caspase-9 results in the cleavage of pro-caspase-3 and the execution of apoptosis.

In *Drosophila*, mutations of the initiator caspase *Dronc* and the executioner caspase *Drice* prevent most embryonic apoptosis events (Chew et al., 2004; Daish, Mills, & Kumar, 2004; Kondo, Senoo-Matsuda, Hiromi, & Miura, 2006; Muro et al., 2006; Waldhuber, Emoto, & Petritsch, 2005; Xu, Li, Arcaro, Lackey, & Bergmann, 2005; Xu et al., 2006). The core machineries of the Apaf-1-mediated caspase-activating pathway have been conserved throughout evolution (Kuranaga & Miura, 2007). Although cytochrome *c* is not required for apoptososme formation in *Drosophila* (Shi, 2006; Yu, Wang, Acehan, Wang, & Akey, 2006), an *Apaf-1* (*dark/hac-1/dapaf-1*) mutant shows a similar phenotype as the *Dronc* and *Drice* mutants, suggesting that the Apaf-1/caspase-activating pathway is essential for apoptosis in *Drosophila* embryos (Kanuka et al., 1999; Rodriguez et al., 1999; Srivastava et al., 2007; Zhou, Song, Tittel, & Steller, 1999). In living cells, *Drosophila* inhibitor of apoptosis protein 1 (Diap1), an E3 ubiquitin ligase, inhibits Dronc. Apoptotic stimulation induces the Diap1-antagonists (i.e., Reaper, Hid, and Grim) that bind to Diap1 and induces self-ubiquitination, thus inducing the degradation of Diap1 through the ubiquitin–proteasome pathway (Gyrd-Hansen & Meier, 2010). Free Dronc, when forms the apoptosome holoenzyme with Apaf-1, is converted to its active form. Active form of Dronc then cleaves and activates the effector caspases Drice and death caspase-1 (Dcp-1 is another effector caspase), resulting in apoptosis execution.

An extrinsic pathway has also been identified in *Drosophila*. Eiger, a TNF-α homolog, binds to its receptor Wengen, resulting in the activation of JNK and promotion of caspase-dependent cell death (Igaki et al., 2002; Kanda, Igaki, Kanuka, Yagi, & Miura, 2002; Kauppila et al., 2003; Moreno, Yan, & Basler, 2002). Eiger-induced cell death can be significantly suppressed by the caspase inhibitor p35 in wing imaginal disc cells, but not in eye imaginal disc cells (Kanda, Igaki, Okano, & Miura, 2011; Moreno, Basler, & Morata, 2002). Therefore, the caspase dependency of Eiger-induced cell death probably varies between different tissues.

2.3.2 Caspase in cell proliferation

The *Drosophila* larval imaginal disc has a striking regenerative capacity. Haynie and Bryant reported that although 40–60% of cells in the larval wing disc are killed or inhibited during proliferation by ionizing radiation, the emerging flies present with a normal size and shape (Haynie & Bryant, 1977), a phenomenon called "compensatory proliferation." Ablation of the cells in the posterior compartment by expressing Ricin, a toxin of plant origin that inhibits protein synthesis in *Drosophila* wing imaginal discs, resulted in the stimulation of cell proliferation in the same area. This suggests that the mitogenic signal is produced in the ablated area to stimulate cell proliferation (Milan, Campuzano, & Garcia-Bellido, 1997). In this case, mitogenic signals could be produced by the dying cell itself or by immune cells recruited to the damaged area, or by neighboring cells activated following tissue injury.

The rapid elimination of the dying cells makes it difficult to examine the contribution of apoptotic signaling to the generation of mitogenic signals. The genetic manipulation of cell death by the expression of the caspase inhibitor p35 was performed to examine the role of caspases in tissue repair. In *Drosophila* and in mammals, p35 inhibits the effector caspases (such as Drice and Dcp-1 in *Drosophila*, and caspase-3 and -7 in mammals), but not the initiator caspase (such as Dronc in *Drosophila* and caspase-9 in mammals) (Hawkins et al., 2000; Meier, Silke, Leevers, & Evan, 2000). Expression of *p35* together with upstream apoptotic stimulation (e.g., irradiation or expression of caspase-activating genes such as *Reaper* or *Hid*) can generate apoptotic "undead" cells, in which the initiator caspase Dronc is activated while the executioner caspases are inhibited. The generation of undead cells in this pathway caused hyperplastic growth of the wing disc (Huh, Guo, & Hay, 2004; Perez-Garijo, Martin, & Morata, 2004; Ryoo, Gorenc, & Steller, 2004). In this case, the undead cells ectopically express wingless (Wg) and decapentaplegic (Dpp) to promote cell proliferation (Huh et al., 2004; Kondo et al., 2006; Perez-Garijo et al., 2004; Perez-Garijo, Martin, Struhl, & Morata, 2005; Ryoo et al., 2004; Wells, Yoshida, & Johnston, 2006). In this experimental system, JNK and p53 are activated downstream of Dronc, and these pathways seem to be involved in the induction of mitogen expression and tissue growth (Kondo et al., 2006; Ryoo et al., 2004; Wells & Johnston, 2012; Wells et al., 2006) (Fig. 6.3). This hypothesis is corroborated by the finding that the activation of JNK signaling independent of apoptosis can induce proliferation and tissue growth (Bergantinos et al., 2010; Perez-Garijo et al., 2005;

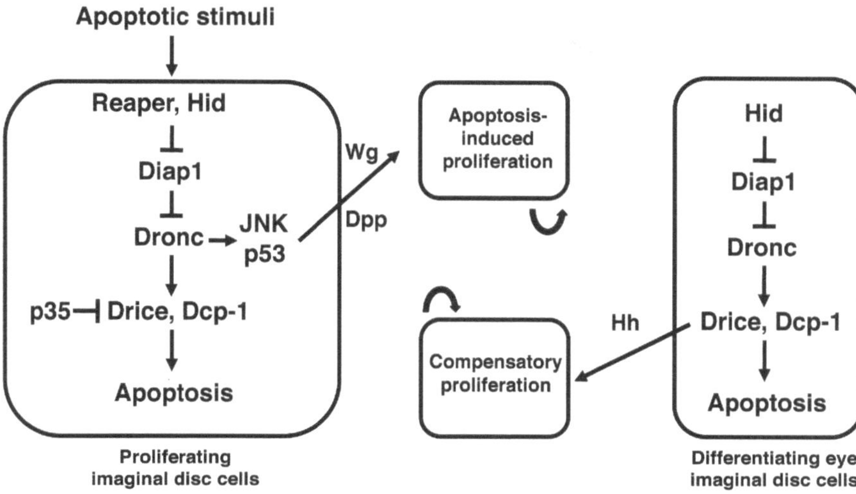

Figure 6.3 Genetic cascade of cell death and apoptosis-induced proliferation in *Drosophila*. Expression of IAP antagonist, Reaper, and/or Hid is induced upon apoptotic stimulation. Initiator caspase Dronc is activated, then executioner caspase Drice and Dcp1 is activated to induce apoptosis. When undead cells were generated by expressing executioner caspase inhibitor p35, JNK and p53 are activated downstream of Dronc, and morphogen Wg and Dpp expression is induced to promote proliferation of neighboring cells in proliferating imaginal disc cells. When cell death was induced in photoreceptor neurons, Hh is induced downstream of executioner caspases to promote proliferation of postmitotic but uncommitted neighboring cells in eye imaginal disc. See text for more details.

Perez-Garijo, Shlevkov, & Morata, 2009; Smith-Bolton et al., 2009; Suissa, Ziv, Dinur, Arama, & Gerlitz, 2011; Warner & Longmore, 2010). JNK and p53 signaling pathways act both upstream and downstream of the apoptotic pathway, resulting in a feedback amplification loop for apoptosis signaling under stress conditions (Shlevkov & Morata, 2012; Warner & Longmore, 2010; Wells et al., 2006), thereby promoting compensatory proliferation under conditions of cellular stress. Both *Drosophila* and human p53 can stimulate JNK activity, where p53 physically binds to diphosphorylated JNK and potentiates JNK activation by preventing its rapid dephosphorylation by the JNK phosphatases (Gowda, Zhou, Chadwell, & McEwen, 2012). Thus, the p53/JNK complex may work together, downstream of Dronc, although the mechanisms of Dronc-mediated activation of p53/JNK are currently unknown. In case of the *Reaper*-induced cell death, expression of *Reaper* leads to the degradation of Diap1 (that normally facilitates degradation of DTRAF1 through the ubiquitin–proteasome system), thereby stabilizing DTRAF1 and activating the JNK pathway (Kuranaga et al., 2002).

Although, the ectopic activation of Wg and Dpp is observed in apoptotic cells, Dpp/Wg are not required for compensatory proliferation after radiation-induced apoptosis in the wing disc (Perez-Garijo et al., 2009). However, ectopic Dpp/Wg signaling is required for the hyperplastic overgrowth of the wing disc when apoptotic undead cells are artificially generated. Apart from the role of Dpp/Wg in apoptosis-induced cell proliferation in the proliferating wing disc (as a majority of these cells do not differentiate and extensively proliferate during the larval stage), another mechanism was discovered in the differentiating eye tissue. When Hid was expressed in the differentiating field of the *Drosophila* pupal retina to induce apoptosis, dying postmitotic photoreceptor cells induced the expression of hedgehog (Hh). Hh then triggered the cell cycle reentry of neighboring cells that are postmitotic but still uncommitted. Unlike the apoptosis-induced proliferation in the wing imaginal disc, where the initiator caspase Dronc but not the effector caspases are involved, Hid-induced proliferation is prevented by coexpressing *p35*; thus, the effector caspases Drice and Dcp-1 are required for apoptosis-induced proliferation in the eye discs (Fan & Bergmann, 2008) (Fig. 6.3).

2.3.3 Apoptosis and regeneration

Since caspase-activated cells can generate mitogens, cell death may also have a potential to promote proliferation of blastema during regeneration. Recently, caspases and the apoptotic process have been discovered to play active roles during regeneration. Some animals such as Hydra, a classical animal model for regeneration, show striking regenerative capacity. Trembley first showed that when Hydra was cut into half, each part regenerates into a complete body (Trembley, 1744). Galliot and her colleagues showed that massive apoptosis was induced at the head-regenerating tip after amputation at the level of the midgut in Hydra. If effector caspase activity is inhibited by administration of the inhibitor, both expression of the morphogen Wnt3 from dying cells and regeneration were impaired (Chera et al., 2009; Galliot & Chera, 2010). When ectopic apoptosis was induced, the ectopic head was generated in the apoptotic area through activation of the Wnt3/β-catenin pathway. This experiment clearly showed that apoptosis is both necessary and sufficient for head regeneration. The requirement of effector caspases for regeneration is also observed in the regeneration of the *Xenopus* tadpole tail following amputation (Tseng, Adams, Qiu, Koustubhan, & Levin, 2007).

Classical experiments by Hadorn and colleagues showed the striking potential of *Drosophila* imaginal disc regeneration (Hadron & Buck, 1962). When the mature imaginal disc was fragmented and transplanted into the adult abdomen, a disc fragment could regenerate the rest of the disc with the pattern regulatory rules (intercalary regeneration) (Haynie & Bryant, 1976). Such tissue-ablation experiments required microsurgical skills and were technically difficult to perform. To overcome this difficulty, Hariharan and his colleagues set up an experiment in which a large portion of the wing imaginal disc was genetically ablated; subsequently, the authors observed complete regeneration of the wing following tissue ablation (Smith-Bolton et al., 2009). In this experiment, apoptosis was transiently induced in a portion of the wing imaginal disc by expression of the Diap1 antagonist *Reaper* or TNF-α/*Eiger*, such that dying cells were extruded from the epithelium. As observed during compensatory proliferation or apoptosis-induced proliferation, Wg is expressed in the damaged tissue (Smith-Bolton et al., 2009). However, unlike the apoptosis-induced proliferation observed in the condition of the apoptotic undead cell (i.e., *Reaper* and caspase inhibitor *p35* are coexpressed in the wing imaginal disc), Wg is predominantly expressed in the proliferating cells, but not in the dying cells during tissue regeneration (Smith-Bolton et al., 2009). In this case, the JNK and the Hippo pathways are required for regeneration (Bergantinos et al., 2010; Grusche, Degoutin, Richardson, & Harvey, 2011; Herrera et al., 2013; Sun & Irvine, 2011), while the requirement of Wg is still under debate (Herrera et al., 2013; Smith-Bolton et al., 2009).

Cellular behavior of the regenerating wing imaginal disc was also reported. At an early time point following cell ablation (by expressing the Diap1 antagonist Hid), localized cell ablation induced cell proliferation in the damaged area and also in other regions of the wing imaginal disc, indicating the systemic proliferative response in the wing imaginal disc. However, at later time points, the proliferation area is restricted to the damaged region. Another interesting observation is that the dying cells are basally extruded, which could affect the basement membrane architecture. Because JNK is activated in dying cells, *MMP-1* expression could be induced by JNK resulting in basement membrane degradation. This in turn may facilitate the hemocytes access to the dying cells and a subsequent hemocyte-mediated systemic response. The cells outside of damaged area migrate into the damaged area and the immigrant cells acquire the character of the repopulated area (Herrera et al., 2013).

2.3.4 The apoptotic cell as a signaling center for tissue remodeling

Massive cell death can be observed after wounding or after the induction of apoptosis. As mentioned above, in Hydra, ectopic cell death induced head regeneration, indicating that the cell death area has organizer activity (Chera et al., 2009). Morphogens, which are secreted from a localized source, have been identified as critical secreted molecules involved in inductive tissue interaction and pattern formation. Tissues and cells vary in their response to morphogens based on a concentration gradient. For example, signaling centers such as the zone of polarizing activity (ZPA) of the chick developing limb bud (Sanz-Ezquerro & Tickle, 2000), floor plate of the developing chick spinal cord (Homma, Yaginuma, & Oppenheim, 1994), and developing *Xenopus* notochord (Offner, Duval, Jamrich, & Durand, 2005) secrete the Shh morphogen. These areas interestingly undergo massive cell death, which suggests that apoptosis regulates the number of morphogen-expressing cells, thereby controlling the concentration pool of morphogens (Offner et al., 2005; Sanz-Ezquerro & Tickle, 2000). In addition to this passive role, apoptosis has been speculated to be actively involved in the expression and secretion of morphogens at these signaling centers as described below.

Massive cell death is also observed in tissue remodeling during metamorphosis of *Drosophila*. The adult abdominal epidermis is known to be generated from histoblasts. During the pupal stage, the polyploid larval epidermal cells (LECs) are replaced by histoblasts. In this process, coordination between LEC apoptosis and histoblast proliferation is observed. It has been shown that caspase activation in LECs is induced at the LEC/histoblast boundary in which the transition from S/G2 cell cycle of the boundary histoblasts is essential to induce nonautonomous LEC apoptosis. This replacement boundary, formed locally by communication between the proliferating and dying cells, is shown to contribute to a spatiotemporal pattern of tissue remodeling (Nakajima, Kuranaga, Sugimura, Miyawaki, & Miura, 2011). DPP is secreted by the boundary LECs and regulates the spread of histoblasts and their invasiveness (Ninov et al., 2010), thereby highlighting the importance of morphogens secreted by the dying cells for tissue remodeling.

2.4. Melanization

Because melanin formation at wound sites is one of the earliest and most visible events following epithelial wounding, it has been long believed that the host system in melanin formation plays an important role in LWR. The phenoloxidase (PO) enzyme is responsible for melanin formation, and PO activity can be found in both circulating hemolymph and the cuticular matrix (Ashida, 1990; Cerenius, Lee, & Soderhall, 2008; Kanost, Jiang, & Yu, 2004). PO catalyzes the oxygenation of monophenols to *o*-diphenols and the oxidation of *o*-diphenols to the corresponding *o*-quinones. Melanins are heteropolymers derived from *o*-quinones (Ashida, 1990; Cerenius et al., 2008; Kanost et al., 2004). Because PO is a highly reactive enzyme and its by-products have deleterious effects on the host, PO is normally present in its enzymatically inactive precursor form, Pro-PO (Ashida & Ohnishi, 1967). Pioneering studies in insects have shown that the CLIP-domain-containing serine protease cleaves the NH_2-terminal part of Pro-PO by limited proteolysis, thereby converting Pro-PO into PO (Ashida, 1990; Nam, Jang, Asano, & Lee, 2008). This Pro-PO-cleaving serine protease is itself activated by sequential activation of multiple upstream serine proteases via limited proteolysis (Fig. 6.2). The PO-activating serine protease cascade, like the vertebrate complement activation system, is initiated following recognition of microbial cells during infection or epithelial wound.

Genetic analysis of the PO-activating cascade in *Drosophila* has revealed the complexity of the enzyme cascade. The PO-activating cascade consists of serine proteases and their serine protease inhibitor (serpin) molecules that act as negative regulators of the cascade (Kanost, 1999). At present, three serine proteases (*MP1*, *MP2/PAE1*, and *Hayan*) and three serpins (*spn27A*, *spn28D*, and *spn77Ba*) are known to act as upstream components for PO activation (De Gregorio et al., 2002; Ligoxygakis et al., 2002; Nam et al., 2012; Scherfer et al., 2008; Tang, Kambris, Lemaitre, & Hashimoto, 2008). Analysis of animals with different serpins mutated or knocked down has revealed that *spn27A*, *spn28D*, and *spn77Ba* are involved mainly in hemolymph, the epithelial wound site, and the trachea, respectively (De Gregorio et al., 2002; Scherfer et al., 2008; Tang et al., 2008). Interestingly, *Hayan* mutation can suppress all of the spontaneous melanization phenotypes induced by lack of three different serpins (i.e., serpin knockdown or mutant animals), indicating that *Hayan* is operating in different tissues as a downstream component of different serpins in the PO-activating cascade (Nam et al., 2012). Using genetic models lacking or reduced in PO activity (such as a *Bc* mutant animal lacking circulating PO activity, and knockdown or mutant animals for upstream serine proteases) has begun to reveal the biological functions of PO (Ligoxygakis et al., 2002; Nam et al., 2012; Rizki, Rizki, & Bellotti, 1985). PO-dependent melanin formation at the wound is known to contribute to scab formation at the epithelial wound site, providing structural stability (Galko & Krasnow, 2004). Hemolymph PO activity also contributes to antimicrobial activities (Ayres & Schneider, 2008), antiparasitic activity by encapsulation of parasite eggs (Rizki, Rizki, & Carton, 1990), and SWR (Nam et al., 2012).

2.5. Generation of reactive oxygen species

The involvement of reactive oxygen species (ROS) has long been observed in the wound response of organisms (Moreira, Stramer, Evans, Wood, & Martin, 2010; Niethammer et al., 2009; Orozco-Cardenas, Narvaez-Vasquez, & Ryan, 2001; Roy, Khanna, Nallu, Hunt, & Sen, 2006; Schafer & Werner, 2008; Woo et al., 2010) and is used as immediate response (Cordeiro & Jacinto, 2013). ROS such as H_2O_2 are considered candidate molecules for the communication of wound signals because of their rapid diffusivity and ability to activate various intracellular signaling pathways. It is known that wounds can induce DUOX-dependent ROS (such as hydrogen peroxide) generation, which acts as an attractant, recruiting inflammatory cells in *Drosophila* embryos and zebrafish larvae (Moreira et al., 2010; Niethammer et al., 2009) (Fig. 6.2). As DUOX knockdown leads to incomplete wound repair in zebrafish larvae, DUOX-dependent ROS are required for the recruitment of immune cells and tissue repair (Niethammer et al., 2009). In *Drosophila*, DUOX was originally shown to play an antiseptic function in adult gut epithelia by producing microbicidal ROS (Ha, Oh, Bae, & Lee, 2005). In *Drosophila* embryo, wounding rapidly induces calcium mobilization through unknown mechanisms, which induces DUOX enzymatic activity via the calcium-responsive EF hands of DUOX (Razzell et al., 2013). As calcium mobilization is sufficient to induce DUOX-dependent H_2O_2 generation, wound-induced calcium mobilization acts as an upstream event of DUOX activation (Razzell et al., 2013) (Fig. 6.2). The DUOX-dependent H_2O_2 generation is further shown to be required for hemocyte recruitment at the wound site and for the activation of wound-responsive genes, such as *Ddc* and *Ple*, in an ERK and Grh-dependent manner (Fig. 6.2). Interestingly, H_2O_2 treatment without disrupting the membrane is sufficient to restore hemocyte recruitment and induce complete activation of *Ddc* and *Ple* expression; thus, DUOX-dependent H_2O_2 acts as a central molecule controlling wound response in the *Drosophila* embryo (Juarez et al., 2011). How does H_2O_2 regulate complex signaling pathways? A well-known mechanism of H_2O_2-sensitive signaling modulation is the oxidative inactivation of redox-sensitive proteins harboring low pK_a cysteine residues (Rhee, 2006). In this case, local H_2O_2 oxidizes the sulfhydryl group (–SH) of cysteine to sulfenic (–SOH), thereby modulating the target protein activity. For example, phosphoinositide 3-kinase signaling is modulated by redox-sensitive inactivation of PTEN, a tumor suppressor with a PtdIns(3,4,5)P_3 phosphatase activity, which in turn regulates diverse cellular activities such as cell proliferation and growth (Connor et al., 2005). In fact, the Src-family kinase in the immune cells of the zebrafish larvae is activated in response to H_2O_2, acting as a redox sensor (Yoo, Starnes, Deng, & Huttenlocher, 2011). Therefore, it is likely that the physiological concentration of H_2O_2 acts as a critical second messenger modulating downstream events of wound response. However, there is currently a poor understanding of the signaling components that act as redox sensors and the subsequent downstream events. In the *Drosophila* embryo wound model, it has been shown that trypsin-like protease activity acts as a downstream event of DUOX-dependent ROS generation, which is required for the induction of wound-responsive genes (Patterson et al., 2013). Although the exact relationship between ROS and protease activation remains to be elucidated, all of these observations highlight the importance of redox signaling in the initial and downstream events of wound response.

2.6. Intestinal epithelial cell renewal program in response to gut cell damage

Aged or injured cells must be replenished in adult organs or tissues to preserve tissue homeostasis. Adult stem cells in these tissues play a part in tissue regeneration by activating their self-renewing and differentiating properties. One of the best studied adult stem cells is the intestinal stem cell (ISC) in the *Drosophila* adult midgut (Micchelli & Perrimon, 2006; Ohlstein & Spradling, 2006, 2007). Under normal conditions, mature enterocytes (ECs) are replaced continuously every week to maintain gut cell homeostasis (Micchelli & Perrimon, 2006). This phenomenon is called gut cell renewal that is ensured by intestinal epithelial cell renewal (ECR) program. ISC has been identified as a source of replenishment for cells of various types in the gut epithelia. Upon stimulation, each ISC produces one daughter cell that retains the fate of its parent cell, and one postmitotic enteroblast that in turn differentiates into either an EC or an enteroendocrine cell (Micchelli & Perrimon, 2006; Ohlstein & Spradling, 2006, 2007).

Recent germ-free animal studies in *Drosophila* showed significant reductions in homeostatic gut cell regeneration compared to that of conventional animals harboring normal microflora (i.e., gut microbiota) (Buchon, Broderick, Poidevin, Pradervand, & Lemaitre, 2009; Shin et al., 2011). This observation suggested that gut microbiota influence "genetically programmed" host epithelial cell regeneration. Genetic studies have demonstrated that multiple signaling pathways, such as Wg (Lin, Xu, & Xi, 2008), Notch (Kapuria, Karpac, Biteau, Hwangbo, & Jasper, 2012; Ohlstein & Spradling, 2007), Hippo (Karpowicz, Perez, & Perrimon, 2010; Shaw et al., 2010), JNK (Biteau, Hochmuth, & Jasper, 2008), JAK-STAT (Buchon, Broderick, Poidevin, Pradervand, & Lemaitre, 2009; Cronin et al., 2009; Jiang et al., 2009), growth factor (Buchon, Broderick, Kuraishi, & Lemaitre, 2010; Jiang & Edgar, 2009; Xu et al., 2011), MAP kinase (Jiang, Grenley, Bravo, Blumhagen, & Edgar, 2011), and redox signaling (Hochmuth, Biteau, Bohmann, & Jasper, 2011; Lee, 2009), are deeply involved in ISC proliferation and differentiation.

Of note is the fact that gut cell renewal rates are not constant, but are highly dynamic depending on local situations. For example, the adult midgut is often subjected to varying degrees of tissue damage, as it is continuously exposed to xenobiotics and microorganisms derived from the environment. In this situation, the host accelerates gut cell renewal to compensate for the epithelial damage. For instance, naturally occurring *Drosophila* pathogens, such as *Pseudomonas entomophila* (life-threatening pathogen) and *Erwinia carotovora* (opportunistic pathogens), have been shown to induce stem cell proliferation and differentiation by activating JAK-STAT pathways (Buchon, Broderick, Chakrabarti, & Lemaitre, 2009; Buchon, Broderick, Poidevin, Pradervand, & Lemaitre, 2009). Furthermore, chemically induced colitis has been shown to enhance gut cell turnover rates (Amcheslavsky, Jiang, & Ip, 2009; Buchon, Broderick, Chakrabarti, & Lemaitre, 2009). Therefore, gut cell regeneration rates may be proportional to the extent of tissue damage. Some important questions remain to be clarified. For example, how do pathogens induce host cell damage such as gut cell apoptosis? And, how do damaged host cells signal to ISC to initiate the regeneration process? Future research must address these questions.

3. ORGANISMAL LEVEL OF SYSTEMIC WOUND RESPONSE IN *DROSOPHILA*

Tissue damage response is not only restricted to the damaged local tissue but also observed in tissues distant from the damaged site. This SWR via inter-tissue communication is an essential biological phenomenon in multicellular organisms, which ensures a whole-body level of host protection against local wounds in adult. Here we will review distinct SWR mechanisms in two remote tissues, neurons, and intestine, following integumental wounding in adult *Drosophila*.

3.1. Role of caspase in SWR

The hypomorphic allele of the caspase activator *dark* gene (*dpf-1*^{K1}) is a viable mutant in which caspase activation and apoptosis are impaired in the embryo (Kanuka et al., 1999). When adult *dpf-1*^{K1} flies were subjected to puncture wounds by pricking the cuticle with a microinjection needle (i.e., an integumental wound), the fly was highly sensitive to the wound and its viability subsequently and gradually decreased. However, wild-type flies normally survive under the same experimental conditions. Because LWR, such as scab formation, was similar between wild-type and *dpf-1*^{K1} flies, it was speculated that lethality resulted from the lack of SWR. Interestingly, caspase activation is observed in midgut ECs, distant from the primary wound site, in wild-type flies within 30 min of being wounded. Caspase-activated ECs were dying and apically extruded into the gut lumen. Inhibition of caspase activity in ECs of wild-type flies showed similar integumental wound-induced lethality as in the case of *dpf-1*^{K1} flies (Takeishi et al., 2013). Importantly, integumental wound-induced caspase activation is suppressed by overexpression of antioxidant enzymes in ECs, indicating that ROS in ECs are required for caspase activation. These observations also suggest that intestinal ROS generation and subsequent caspase activation in ECs are important components of SWR that occur even when a wound site is distant from the gut.

As discussed earlier, ECR program in the gut is ensured by ISC proliferation and differentiation for EC regeneration in gut epithelia. ECR has been extensively studied in the context of local EC damage, and ROS is established as one of the important factors affecting ECR (Hochmuth et al., 2011). In addition to ROS, locally damaged ECs secrete Upd3 to induce activation of the JAK/STAT signaling pathway, which is required for ECR program (Buchon et al., 2010; Jiang & Edgar, 2009; Jiang et al., 2009). Interestingly, in addition to direct EC damage, integumental wounds were found to be able to induce Upd3 expression in ECs within 24 h. These observations indicate that two essential factors (i.e., ROS and Upd3) are upregulated after integumental wounding and stimulate ISC proliferation and differentiation, which supports the concept of inter-tissue communication between integumental and gut epithelia.

Strikingly, integumental wounds induce ISC proliferation in wild-type flies, whereas wound-induced ISC proliferation is severely impaired in *dpf-1*^{K1} flies. Considering that EC cell death followed by ISC proliferation is the basis for ECR program, it is possible that integumental wounds somehow induce redox-dependent caspase activation in ECs, leading to EC death and ISC proliferation. Interestingly, when ISC proliferation was inhibited by expressing PTEN, integumental wound-induced caspase activation in ECs is also inhibited. In either case (i.e., inhibition of ISC proliferation or absence of caspase activation), integumental wounds could lead to increased lethality. Therefore, it is possible that integumental wounds induce SWR in the intestine where ECR program is initiated by reciprocally activating interactions between EC cell death and ISC proliferation/differentiation. This is required for dampening the lethal response following integumental wounds (Fig. 6.4).

An important question that needs clarification is why *dpf-1*^{K1} flies died following integumental wounds. Although the exact cause of lethality is presently unclear, an interesting observation was made. Hemolymph from wounded *dpf-1*^{K1} flies had a toxic effect when transferred to wild-type flies, suggesting that some toxic factors are generated in ECs following integumental wounding, and that caspase activity in ECs suppresses (or removes) these lethal toxic factors from being released into the hemolymph after wounding (Takeishi et al., 2013) (Fig. 6.4). To clarify the role of caspase and ECs in SWR, further research is needed to identify the lethal toxic factors and how they exert their effects.

3.2. Role of prophenoloxidase in SWR

The link between PO and ROS in SWR was first suspected following the observation that *Bc* or *Hayan* flies (lacking hemolymph PO activity) are susceptible to integumental wounding, whereas wild-type flies remain intact (Nam et al., 2012). Interestingly, integumental wound-induced lethality in *Bc* or *Hayan* flies can be largely rescued by enhancing the systemic H_2O_2 level (by reducing the antioxidant enzyme activity) (Nam et al., 2012). Because wound-induced host lethality is also observed in transgenic animals overexpressing H_2O_2-removing antioxidant enzymes, it is possible that some H_2O_2 is systemically produced and that it is required for host survival during SWR (Nam et al., 2012).

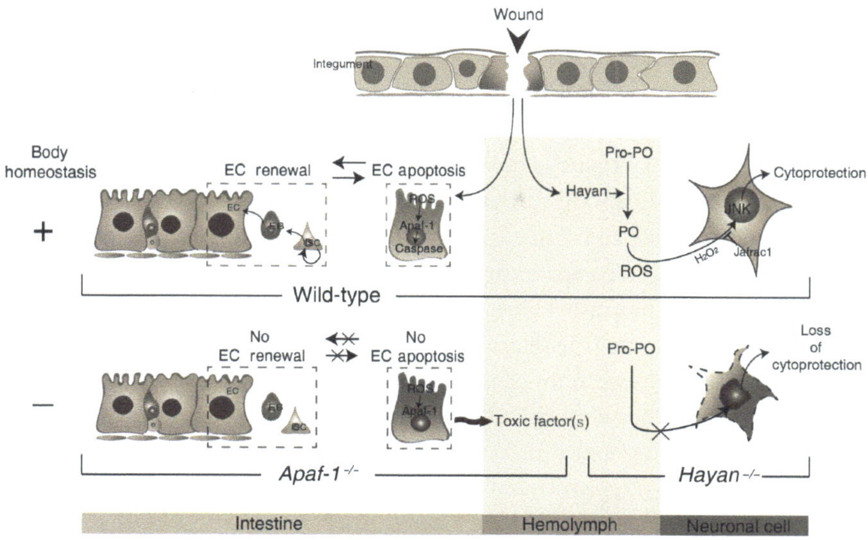

Figure 6.4 Systemic wound responses following integumental wound in *Drosophila*. Following integumental wounding, two remote tissues distant from an initial wound site, that is, neuronal cells and intestinal epithelial cells, received local wound information. In the intestinal cells, caspase activation in ECs (EC apoptosis) and subsequent ISC proliferation/differentiation (EC renewal) are required for efficient host protection. Note the reciprocally activating interactions between "EC apoptosis" and "EC renewal." *Drosophila* mutant for the caspase activator, *Apaf-1*$^{-/-}$, was defective in gut-cell turnover (i.e., absence of EC apoptosis leading to inhibition of EC renewal) and showed increased wound-induced lethality. In this mutant animal, a lethal toxic factor was released into the hemolymph following integumental wounding. Following integumental wounding, Hayan-dependent conversion of Pro-PO to active PO is one of the earliest events to occur in hemolymph. Subsequently, active PO induces hemolymph ROS that directly induce JNK-dependent cytoprotections in neurons. H_2O_2-removing enzyme, jafrac1, seems to be involved in the redox balance in these cells. *Hayan* mutant animals, *Hayan*$^{-/-}$, are defective in hemolymph PO activation and neuronal JNK activation, which results in high susceptibility to integumental wound. See text for more details. ISC, intestinal stem cells; EB, enteroblast; EE, enteroendocrine; Pro-PO, prophenoloxidase; PO, phenoloxidase. (See color plate.)

An important question concerns the source of ROS in SWR. Given that PO can produce ROS *in vitro*, and that melanin precursors are known to be potential sources of ROS (Komarov, Slepneva, Glupov, & Khramtsov, 2005; Nappi & Vass, 1993; Slepneva, Glupov, Sergeeva, & Khramtsov, 1999), hemolymph PO likely generates systemic wound-induced extracellular ROS *in vivo*. It has been shown that the production of local H_2O_2 at the wound site is assured by the action of the DUOX enzyme in *Drosophila* embryos and zebrafish larvae (Moreira et al., 2010;

Niethammer et al., 2009). It is accordingly conceivable that epithelial wounding induces the activation of two different ROS-generating enzymes, DUOX for local ROS and PO for systemic ROS.

How do PO-induced ROS promote host survival in SWR? It was known that neuronal tissue is highly susceptible to stress. For example, protection of neuronal cells by ectopic expression of cytoprotective gene is sufficient to confer host resistance to stress (Wang, Bohmann, & Jasper, 2003). Because cytoprotective gene expression is known to be mainly controlled by JNK activity and because H_2O_2 can act as a diffusible second messenger capable of affecting the JNK pathway in different tissues, it has been hypothesized that PO-induced hemolymph ROS directly induces JNK activity in neurons. Indeed, the wound-induced lethality in *Hayan* flies can be largely rescued by reducing neuronal antioxidant capacity (i.e., by introducing knockdown of H_2O_2-removing enzyme, jafrac1) or enhancing neuronal JNK activity (Nam et al., 2012). These observations support the idea of intertissue communication wherein integumental wounds induce a redox-dependent JNK activation in neurons through hemolymph PO activity (Fig. 6.4). Following integumental wounding, PO activation is one of the earliest events to occur, which is required for melanization at the wound site as well as antimicrobial and antiparasitic activity. Considering that infection inevitably accompanies integumental wounds, the role of PO in SWR, together with its previously identified immunological roles, would provide a considerable survival advantage to the host.

4. CONCLUSION

The wound response that occurs following epithelial injury has been studied extensively as a local tissue repair process. Generation of ROS, cytokines, growth factors, and morphogen secretions, as well as epithelial cell movement and hemocyte recruitment, are readily observed in damaged tissues. Similar events are also observed during dorsal closure and imaginal disc regeneration. It is increasingly evident that these local responses play an instructive role in the development of SWRs by facilitating interorgan communication. A recent genetic analysis in *Drosophila* demonstrated that two remote tissues distant from an initial integumental wound site, that is, neuronal cells and intestinal epithelial cells, received local wound information and launched SWR to provide organismal level protection. Both SWRs are interestingly redox-dependent because hemolymph ROS and intestinal ROS are required for SWR activation that occurs in the neurons and the intestine, respectively. In the case of intestine, organism level of protection is ensured by ROS-modulated and caspase-dependent gut ECR that inhibits the release of lethal toxic factor(s). However, in the case of neuronal cells, the exact molecular mechanism by which PO-dependent ROS protect host organism remains to be elucidated.

These observations in *Drosophila* SWR reinforce the hypothesis that wound responses are under redox control at the wound site in various animal taxa, ranging from *Drosophila* to mammals. These observations also suggest that the essential role of ROS can be extended to the neurons and gut following integumental wounding. However, many crucial questions remain unanswered. For example, why is SWR necessary? How does the LWR transmit a signaling cascade that initiates SWR? What are the genes or molecules responsible for interorgan communication? What are the exact biological processes that occur in the organs distant from the damaged site? Unfortunately, we cannot provide clear answers to these questions. Considering the complexities of organ-to-organ communication, the simple organization of internal tissues and the availability of genetic tools in *Drosophila*, as well as the capacity to perform unbiased genetic screening, *Drosophila* is one of the most suitable model organisms for dissecting the molecular mechanisms underlying SWR.

ACKNOWLEDGMENTS

This study was supported by the National Creative Research Initiative Program from National Research Foundation of Korea (Grant No. 2006-0050687 to W.-J.L), and by the Japanese Ministry of Education, Science, Sports, Culture, and Technology (to M. M.).

REFERENCES

Amcheslavsky, A., Jiang, J., & Ip, Y. T. (2009). Tissue damage-induced intestinal stem cell division in Drosophila. *Cell Stem Cell*, *4*(1), 49–61.

Ashida, M. (1990). The prophenoloxidase cascade in insect immunity. *Research in Immunology*, *141*(9), 908–910.

Ashida, M., & Ohnishi, E. (1967). Activation of pre-phenol oxidase in hemolymph of the silkworm, Bombyx mori. *Archives of Biochemistry and Biophysics*, *122*(2), 411–416.

Ayres, J. S., & Schneider, D. S. (2008). A signaling protease required for melanization in Drosophila affects resistance and tolerance of infections. *PLoS Biology*, *6*(12), 2764–2773.

Babcock, D. T., Brock, A. R., Fish, G. S., Wang, Y., Perrin, L., Krasnow, M. A., et al. (2008). Circulating blood cells function as a surveillance system for damaged tissue in Drosophila larvae. *Proceedings of the National Academy of Sciences of the United States of America*, *105*(29), 10017–10022.

Belacortu, Y., & Paricio, N. (2011). Drosophila as a model of wound healing and tissue regeneration in vertebrates. *Developmental Dynamics*, *240*(11), 2379–2404.

Bergantinos, C., Corominas, M., & Serras, F. (2010). Cell death-induced regeneration in wing imaginal discs requires JNK signalling. *Development*, *137*(7), 1169–1179.

Biteau, B., Hochmuth, C. E., & Jasper, H. (2008). JNK activity in somatic stem cells causes loss of tissue homeostasis in the aging Drosophila gut. *Cell Stem Cell*, *3*(4), 442–455.
Bosch, M., Serras, F., Martin-Blanco, E., & Baguna, J. (2005). JNK signaling pathway required for wound healing in regenerating Drosophila wing imaginal discs. *Developmental Biology*, *280*(1), 73–86.
Buchon, N., Broderick, N. A., Chakrabarti, S., & Lemaitre, B. (2009). Invasive and indigenous microbiota impact intestinal stem cell activity through multiple pathways in Drosophila. *Genes & Development*, *23*(19), 2333–2344.
Buchon, N., Broderick, N. A., Kuraishi, T., & Lemaitre, B. (2010). Drosophila EGFR pathway coordinates stem cell proliferation and gut remodeling following infection. *BMC Biology*, *8*, 152.
Buchon, N., Broderick, N. A., Poidevin, M., Pradervand, S., & Lemaitre, B. (2009). Drosophila intestinal response to bacterial infection: Activation of host defense and stem cell proliferation. *Cell Host & Microbe*, *5*(2), 200–211.
Cerenius, L., Lee, B. L., & Soderhall, K. (2008). The proPO-system: Pros and cons for its role in invertebrate immunity. *Trends in Immunology*, *29*(6), 263–271.
Cerretti, D. P., Kozlosky, C. J., Mosley, B., Nelson, N., Van Ness, K., Greenstreet, T. A., et al. (1992). Molecular cloning of the interleukin-1 beta converting enzyme. *Science*, *256*(5053), 97–100.
Chatterjee, M., & Ip, Y. T. (2009). Pathogenic stimulation of intestinal stem cell response in Drosophila. *Journal of Cellular Physiology*, *220*(3), 664–671.
Chera, S., Ghila, L., Dobretz, K., Wenger, Y., Bauer, C., Buzgariu, W., et al. (2009). Apoptotic cells provide an unexpected source of Wnt3 signaling to drive hydra head regeneration. *Developmental Cell*, *17*(2), 279–289.
Chew, S. K., Akdemir, F., Chen, P., Lu, W. J., Mills, K., Daish, T., et al. (2004). The apical caspase dronc governs programmed and unprogrammed cell death in Drosophila. *Developmental Cell*, 7(6), 897–907.
Connor, K. M., Subbaram, S., Regan, K. J., Nelson, K. K., Mazurkiewicz, J. E., Bartholomew, P. J., et al. (2005). Mitochondrial H2O2 regulates the angiogenic phenotype via PTEN oxidation. *The Journal of Biological Chemistry*, *280*(17), 16916–16924.
Cordeiro, J. V., & Jacinto, A. (2013). The role of transcription-independent damage signals in the initiation of epithelial wound healing. *Nature Reviews. Molecular Cell Biology*, *14*(4), 249–262.
Cronin, S. J., Nehme, N. T., Limmer, S., Liegeois, S., Pospisilik, J. A., Schramek, D., et al. (2009). Genome-wide RNAi screen identifies genes involved in intestinal pathogenic bacterial infection. *Science*, *325*(5938), 340–343.
Daish, T. J., Mills, K., & Kumar, S. (2004). Drosophila caspase DRONC is required for specific developmental cell death pathways and stress-induced apoptosis. *Developmental Cell*, 7(6), 909–915.
De Gregorio, E., Han, S. J., Lee, W. J., Baek, M. J., Osaki, T., Kawabata, S., et al. (2002). An immune-responsive Serpin regulates the melanization cascade in Drosophila. *Developmental Cell*, *3*(4), 581–592.
Degterev, A., Boyce, M., & Yuan, J. (2003). A decade of caspases. *Oncogene*, *22*(53), 8543–8567.
Delessert, C., Wilson, I. W., Van Der Straeten, D., Dennis, E. S., & Dolferus, R. (2004). Spatial and temporal analysis of the local response to wounding in Arabidopsis leaves. *Plant Molecular Biology*, *55*(2), 165–181.
Diaz-Garcia, S., & Baonza, A. (2013). Pattern reorganization occurs independently of cell division during Drosophila wing disc regeneration in situ. *Proceedings of the National Academy of Sciences of the United States of America*, *110*(32), 13032–13037.
Dix, M. M., Simon, G. M., & Cravatt, B. F. (2008). Global mapping of the topography and magnitude of proteolytic events in apoptosis. *Cell*, *134*(4), 679–691.

Ellis, H. M., & Horvitz, H. R. (1986). Genetic control of programmed cell death in the nematode C. elegans. *Cell, 44*(6), 817–829.

Fan, Y., & Bergmann, A. (2008). Distinct mechanisms of apoptosis-induced compensatory proliferation in proliferating and differentiating tissues in the Drosophila eye. *Developmental Cell, 14*(3), 399–410.

Foex, B. A. (1999). Systemic responses to trauma. *British Medical Bulletin, 55*(4), 726–743.

Galko, M. J., & Krasnow, M. A. (2004). Cellular and genetic analysis of wound healing in Drosophila larvae. *PLoS Biology, 2*(8), E239.

Galliot, B., & Chera, S. (2010). The Hydra model: Disclosing an apoptosis-driven generator of Wnt-based regeneration. *Trends in Cell Biology, 20*(9), 514–523.

Gowda, P. S., Zhou, F., Chadwell, L. V., & McEwen, D. G. (2012). p53 binding prevents phosphatase-mediated inactivation of diphosphorylated c-Jun N-terminal kinase. *The Journal of Biological Chemistry, 287*(21), 17554–17567.

Grusche, F. A., Degoutin, J. L., Richardson, H. E., & Harvey, K. F. (2011). The Salvador/Warts/Hippo pathway controls regenerative tissue growth in Drosophila melanogaster. *Developmental Biology, 350*(2), 255–266.

Gyrd-Hansen, M., & Meier, P. (2010). IAPs: from caspase inhibitors to modulators of NF-kappaB, inflammation and cancer. *Nature Reviews. Cancer, 10*(8), 561–574.

Ha, E. M., Oh, C. T., Bae, Y. S., & Lee, W. J. (2005). A direct role for dual oxidase in Drosophila gut immunity. *Science, 310*(5749), 847–850.

Hadron, E., & Buck, D. (1962). On the differentiation of transplanted wing imaginal disc fragments of Drosophila melanogaster. *Revue Suisse de Zoologie, 69*, 302–310.

Hawkins, C. J., Yoo, S. J., Peterson, E. P., Wang, S. L., Vernooy, S. Y., & Hay, B. A. (2000). The Drosophila caspase DRONC cleaves following glutamate or aspartate and is regulated by DIAP1, HID, and GRIM. *The Journal of Biological Chemistry, 275*(35), 27084–27093.

Haynie, J. L., & Bryant, P. J. (1976). Intercalary regeneration in imaginal wing disk of Drosophila melanogaster. *Nature, 259*(5545), 659–662.

Haynie, J. L., & Bryant, P. J. (1977). The effects of X-rays on the proliferation dynamics in the wing disc of Drosophila melanogaster. *Development Genes and Evolution, 183*, 85–100.

Herrera, S. C., Martin, R., & Morata, G. (2013). Tissue homeostasis in the wing disc of Drosophila melanogaster: Immediate response to massive damage during development. *PLoS Genetics, 9*(4), e1003446.

Hochmuth, C. E., Biteau, B., Bohmann, D., & Jasper, H. (2011). Redox regulation by Keap1 and Nrf2 controls intestinal stem cell proliferation in Drosophila. *Cell Stem Cell, 8*(2), 188–199.

Homma, S., Yaginuma, H., & Oppenheim, R. W. (1994). Programmed cell death during the earliest stages of spinal cord development in the chick embryo: A possible means of early phenotypic selection. *The Journal of Comparative Neurology, 345*(3), 377–395.

Huh, J. R., Guo, M., & Hay, B. A. (2004). Compensatory proliferation induced by cell death in the Drosophila wing disc requires activity of the apical cell death caspase Dronc in a nonapoptotic role. *Current Biology, 14*(14), 1262–1266.

Huh, J. R., Vernooy, S. Y., Yu, H., Yan, N., Shi, Y., Guo, M., et al. (2004). Multiple apoptotic caspase cascades are required in nonapoptotic roles for Drosophila spermatid individualization. *PLoS Biology, 2*(1), E15.

Igaki, T., Kanda, H., Yamamoto-Goto, Y., Kanuka, H., Kuranaga, E., Aigaki, T., et al. (2002). Eiger, a TNF superfamily ligand that triggers the Drosophila JNK pathway. *The EMBO Journal, 21*(12), 3009–3018.

Jasper, H., Benes, V., Schwager, C., Sauer, S., Clauder-Munster, S., Ansorge, W., et al. (2001). The genomic response of the Drosophila embryo to JNK signaling. *Developmental Cell, 1*(4), 579–586.

Jiang, H., & Edgar, B. A. (2009). EGFR signaling regulates the proliferation of Drosophila adult midgut progenitors. *Development*, *136*(3), 483–493.

Jiang, H., Grenley, M. O., Bravo, M. J., Blumhagen, R. Z., & Edgar, B. A. (2011). EGFR/Ras/MAPK signaling mediates adult midgut epithelial homeostasis and regeneration in Drosophila. *Cell Stem Cell*, *8*(1), 84–95.

Jiang, H., Patel, P. H., Kohlmaier, A., Grenley, M. O., McEwen, D. G., & Edgar, B. A. (2009). Cytokine/Jak/Stat signaling mediates regeneration and homeostasis in the Drosophila midgut. *Cell*, *137*(7), 1343–1355.

Juarez, M. T., Patterson, R. A., Sandoval-Guillen, E., & McGinnis, W. (2011). Duox, Flotillin-2, and Src42A are required to activate or delimit the spread of the transcriptional response to epidermal wounds in Drosophila. *PLoS Genetics*, 7(12), e1002424.

Kanda, H., Igaki, T., Kanuka, H., Yagi, T., & Miura, M. (2002). Wengen, a member of the Drosophila tumor necrosis factor receptor superfamily, is required for Eiger signaling. *The Journal of Biological Chemistry*, *277*(32), 28372–28375.

Kanda, H., Igaki, T., Okano, H., & Miura, M. (2011). Conserved metabolic energy production pathways govern Eiger/TNF-induced nonapoptotic cell death. *Proceedings of the National Academy of Sciences of the United States of America*, *108*(47), 18977–18982.

Kanost, M. R. (1999). Serine proteinase inhibitors in arthropod immunity. *Developmental and Comparative Immunology*, *23*(4–5), 291–301.

Kanost, M. R., Jiang, H., & Yu, X. Q. (2004). Innate immune responses of a lepidopteran insect, Manduca sexta. *Immunological Reviews*, *198*, 97–105.

Kanuka, H., Sawamoto, K., Inohara, N., Matsuno, K., Okano, H., & Miura, M. (1999). Control of the cell death pathway by Dapaf-1, a Drosophila Apaf-1/CED-4-related caspase activator. *Molecular Cell*, *4*(5), 757–769.

Kapuria, S., Karpac, J., Biteau, B., Hwangbo, D., & Jasper, H. (2012). Notch-mediated suppression of TSC2 expression regulates cell differentiation in the Drosophila intestinal stem cell lineage. *PLoS Genetics*, *8*(11), e1003045.

Karpac, J., Younger, A., & Jasper, H. (2011). Dynamic coordination of innate immune signaling and insulin signaling regulates systemic responses to localized DNA damage. *Developmental Cell*, *20*(6), 841–854.

Karpowicz, P., Perez, J., & Perrimon, N. (2010). The Hippo tumor suppressor pathway regulates intestinal stem cell regeneration. *Development*, *137*(24), 4135–4145.

Kauppila, S., Maaty, W. S., Chen, P., Tomar, R. S., Eby, M. T., Chapo, J., et al. (2003). Eiger and its receptor, Wengen, comprise a TNF-like system in Drosophila. *Oncogene*, *22*(31), 4860–4867.

Kelsey, E. M., Luo, X., Bruckner, K., & Jasper, H. (2012). Schnurri regulates hemocyte function to promote tissue recovery after DNA damage. *Journal of Cell Science*, *125*(Pt. 6), 1393–1400.

Komarov, D. A., Slepneva, I. A., Glupov, V. V., & Khramtsov, V. V. (2005). Superoxide and hydrogen peroxide formation during enzymatic oxidation of DOPA by phenoloxidase. *Free Radical Research*, *39*(8), 853–858.

Kondo, S., Senoo-Matsuda, N., Hiromi, Y., & Miura, M. (2006). DRONC coordinates cell death and compensatory proliferation. *Molecular and Cellular Biology*, *26*(19), 7258–7268.

Kuranaga, E., Kanuka, H., Igaki, T., Sawamoto, K., Ichijo, H., Okano, H., et al. (2002). Reaper-mediated inhibition of DIAP1-induced DTRAF1 degradation results in activation of JNK in Drosophila. *Nature Cell Biology*, *4*(9), 705–710.

Kuranaga, E., & Miura, M. (2007). Nonapoptotic functions of caspases: Caspases as regulatory molecules for immunity and cell-fate determination. *Trends in Cell Biology*, *17*(3), 135–144.

Lee, W. J. (2009). Bacterial-modulated host immunity and stem cell activation for gut homeostasis. *Genes & Development*, *23*(19), 2260–2265.

Lenz, A., Franklin, G. A., & Cheadle, W. G. (2007). Systemic inflammation after trauma. *Injury, 38*(12), 1336–1345.
Lesch, C., Jo, J., Wu, Y., Fish, G. S., & Galko, M. J. (2010). A targeted UAS-RNAi screen in Drosophila larvae identifies wound closure genes regulating distinct cellular processes. *Genetics, 186*(3), 943–957.
Ligoxygakis, P., Pelte, N., Ji, C., Leclerc, V., Duvic, B., Belvin, M., et al. (2002). A serpin mutant links Toll activation to melanization in the host defence of Drosophila. *The EMBO Journal, 21*(23), 6330–6337.
Lin, G., Xu, N., & Xi, R. (2008). Paracrine Wingless signalling controls self-renewal of Drosophila intestinal stem cells. *Nature, 455*(7216), 1119–1123.
Lowry, S. F., & Calvano, S. E. (2008). Challenges for modeling and interpreting the complex biology of severe injury and inflammation. *Journal of Leukocyte Biology, 83*(3), 553–557.
Luthi, A. U., & Martin, S. J. (2007). The CASBAH: A searchable database of caspase substrates. *Cell Death and Differentiation, 14*(4), 641–650.
Mace, K. A., Pearson, J. C., & McGinnis, W. (2005). An epidermal barrier wound repair pathway in Drosophila is mediated by grainy head. *Science, 308*(5720), 381–385.
McClure, K. D., Sustar, A., & Schubiger, G. (2008). Three genes control the timing, the site and the size of blastema formation in Drosophila. *Developmental Biology, 319*(1), 68–77.
Meier, P., Silke, J., Leevers, S. J., & Evan, G. I. (2000). The Drosophila caspase DRONC is regulated by DIAP1. *The EMBO Journal, 19*(4), 598–611.
Micchelli, C. A., & Perrimon, N. (2006). Evidence that stem cells reside in the adult Drosophila midgut epithelium. *Nature, 439*(7075), 475–479.
Milan, M., Campuzano, S., & Garcia-Bellido, A. (1997). Developmental parameters of cell death in the wing disc of Drosophila. *Proceedings of the National Academy of Sciences of the United States of America, 94*(11), 5691–5696.
Miura, M., Zhu, H., Rotello, R., Hartwieg, E. A., & Yuan, J. (1993). Induction of apoptosis in fibroblasts by IL-1 beta-converting enzyme, a mammalian homolog of the C. elegans cell death gene ced-3. *Cell, 75*(4), 653–660.
Moreira, S., Stramer, B., Evans, I., Wood, W., & Martin, P. (2010). Prioritization of competing damage and developmental signals by migrating macrophages in the Drosophila embryo. *Current Biology, 20*(5), 464–470.
Moreno, E., Basler, K., & Morata, G. (2002). Cells compete for decapentaplegic survival factor to prevent apoptosis in Drosophila wing development. *Nature, 416*(6882), 755–759.
Moreno, E., Yan, M., & Basler, K. (2002). Evolution of TNF signaling mechanisms: JNK-dependent apoptosis triggered by Eiger, the Drosophila homolog of the TNF superfamily. *Current Biology, 12*(14), 1263–1268.
Muro, I., Berry, D. L., Huh, J. R., Chen, C. H., Huang, H., Yoo, S. J., et al. (2006). The Drosophila caspase Ice is important for many apoptotic cell deaths and for spermatid individualization, a nonapoptotic process. *Development, 133*(17), 3305–3315.
Nakajima, Y., Kuranaga, E., Sugimura, K., Miyawaki, A., & Miura, M. (2011). Non-autonomous apoptosis is triggered by local cell cycle progression during epithelial replacement in Drosophila. *Molecular and Cellular Biology, 31*(12), 2499–2512.
Nam, H. J., Jang, I. H., Asano, T., & Lee, W. J. (2008). Involvement of pro-phenoloxidase 3 in lamellocyte-mediated spontaneous melanization in Drosophila. *Molecules and Cells, 26*(6), 606–610.
Nam, H. J., Jang, I. H., You, H., Lee, K. A., & Lee, W. J. (2012). Genetic evidence of a redox-dependent systemic wound response via Hayan protease-phenoloxidase system in Drosophila. *The EMBO Journal, 31*(5), 1253–1265.
Nappi, A. J., & Vass, E. (1993). Melanogenesis and the generation of cytotoxic molecules during insect cellular immune reactions. *Pigment Cell Research, 6*(3), 117–126.

Niethammer, P., Grabher, C., Look, A. T., & Mitchison, T. J. (2009). A tissue-scale gradient of hydrogen peroxide mediates rapid wound detection in zebrafish. *Nature, 459*(7249), 996–999.
Ninov, N., Menezes-Cabral, S., Prat-Rojo, C., Manjon, C., Weiss, A., Pyrowolakis, G., et al. (2010). Dpp signaling directs cell motility and invasiveness during epithelial morphogenesis. *Current Biology, 20*(6), 513–520.
Offner, N., Duval, N., Jamrich, M., & Durand, B. (2005). The pro-apoptotic activity of a vertebrate Bar-like homeobox gene plays a key role in patterning the Xenopus neural plate by limiting the number of chordin- and shh-expressing cells. *Development, 132*(8), 1807–1818.
Ohlstein, B., & Spradling, A. (2006). The adult Drosophila posterior midgut is maintained by pluripotent stem cells. *Nature, 439*(7075), 470–474.
Ohlstein, B., & Spradling, A. (2007). Multipotent Drosophila intestinal stem cells specify daughter cell fates by differential notch signaling. *Science, 315*(5814), 988–992.
Orozco-Cardenas, M. L., Narvaez-Vasquez, J., & Ryan, C. A. (2001). Hydrogen peroxide acts as a second messenger for the induction of defense genes in tomato plants in response to wounding, systemin, and methyl jasmonate. *Plant Cell, 13*(1), 179–191.
Pastor-Pareja, J. C., Wu, M., & Xu, T. (2008). An innate immune response of blood cells to tumors and tissue damage in Drosophila. *Disease Models & Mechanisms, 1*(2–3), 144–154, discussion 153.
Patterson, R. A., Juarez, M. T., Hermann, A., Sasik, R., Hardiman, G., & McGinnis, W. (2013). Serine proteolytic pathway activation reveals an expanded ensemble of wound response genes in Drosophila. *PLoS One, 8*(4), e61773.
Perez-Garijo, A., Martin, F. A., & Morata, G. (2004). Caspase inhibition during apoptosis causes abnormal signalling and developmental aberrations in Drosophila. *Development, 131*(22), 5591–5598.
Perez-Garijo, A., Martin, F. A., Struhl, G., & Morata, G. (2005). Dpp signaling and the induction of neoplastic tumors by caspase-inhibited apoptotic cells in Drosophila. *Proceedings of the National Academy of Sciences of the United States of America, 102*(49), 17664–17669.
Perez-Garijo, A., Shlevkov, E., & Morata, G. (2009). The role of Dpp and Wg in compensatory proliferation and in the formation of hyperplastic overgrowths caused by apoptotic cells in the Drosophila wing disc. *Development, 136*(7), 1169–1177.
Ramet, M., Lanot, R., Zachary, D., & Manfruelli, P. (2002). JNK signaling pathway is required for efficient wound healing in Drosophila. *Developmental Biology, 241*(1), 145–156.
Razzell, W., Evans, I. R., Martin, P., & Wood, W. (2013). Calcium flashes orchestrate the wound inflammatory response through DUOX activation and hydrogen peroxide release. *Current Biology, 23*(5), 424–429.
Razzell, W., Wood, W., & Martin, P. (2011). Swatting flies: Modelling wound healing and inflammation in Drosophila. *Disease Models & Mechanisms, 4*(5), 569–574.
Rhee, S. G. (2006). Cell signaling. H2O2, a necessary evil for cell signaling. *Science, 312*(5782), 1882–1883.
Rizki, T. M., Rizki, R. M., & Bellotti, R. A. (1985). Genetics of a Drosophila phenoloxidase. *Molecular & General Genetics, 201*(1), 7–13.
Rizki, T. M., Rizki, R. M., & Carton, Y. (1990). Leptopilina heterotoma and L. boulardi: Strategies to avoid cellular defense responses of Drosophila melanogaster. *Experimental Parasitology, 70*(4), 466–475.
Rodriguez, A., Oliver, H., Zou, H., Chen, P., Wang, X., & Abrams, J. M. (1999). Dark is a Drosophila homologue of Apaf-1/CED-4 and functions in an evolutionarily conserved death pathway. *Nature Cell Biology, 1*(5), 272–279.

Roy, S., Khanna, S., Nallu, K., Hunt, T. K., & Sen, C. K. (2006). Dermal wound healing is subject to redox control. *Molecular Therapy*, *13*(1), 211–220.

Ryoo, H. D., Gorenc, T., & Steller, H. (2004). Apoptotic cells can induce compensatory cell proliferation through the JNK and the Wingless signaling pathways. *Developmental Cell*, 7(4), 491–501.

Sanz-Ezquerro, J. J., & Tickle, C. (2000). Autoregulation of Shh expression and Shh induction of cell death suggest a mechanism for modulating polarising activity during chick limb development. *Development*, *127*(22), 4811–4823.

Schafer, M., & Werner, S. (2008). Oxidative stress in normal and impaired wound repair. *Pharmacological Research: The Official Journal of the Italian Pharmacological Society*, *58*(2), 165–171.

Scherfer, C., Tang, H., Kambris, Z., Lhocine, N., Hashimoto, C., & Lemaitre, B. (2008). Drosophila Serpin-28D regulates hemolymph phenoloxidase activity and adult pigmentation. *Developmental Biology*, *323*(2), 189–196.

Schilmiller, A. L., & Howe, G. A. (2005). Systemic signaling in the wound response. *Current Opinion in Plant Biology*, *8*(4), 369–377.

Shaw, R. L., Kohlmaier, A., Polesello, C., Veelken, C., Edgar, B. A., & Tapon, N. (2010). The Hippo pathway regulates intestinal stem cell proliferation during Drosophila adult midgut regeneration. *Development*, *137*(24), 4147–4158.

Shi, Y. (2006). Mechanical aspects of apoptosome assembly. *Current Opinion in Cell Biology*, *18*(6), 677–684.

Shin, S. C., Kim, S. H., You, H., Kim, B., Kim, A. C., Lee, K. A., et al. (2011). Drosophila microbiome modulates host developmental and metabolic homeostasis via insulin signaling. *Science*, *334*(6056), 670–674.

Shlevkov, E., & Morata, G. (2012). A dp53/JNK-dependant feedback amplification loop is essential for the apoptotic response to stress in Drosophila. *Cell Death and Differentiation*, *19*(3), 451–460.

Slepneva, I. A., Glupov, V. V., Sergeeva, S. V., & Khramtsov, V. V. (1999). EPR detection of reactive oxygen species in hemolymph of Galleria mellonella and Dendrolimus superans sibiricus (Lepidoptera) larvae. *Biochemical and Biophysical Research Communications*, *264*(1), 212–215.

Smith-Bolton, R. K., Worley, M. I., Kanda, H., & Hariharan, I. K. (2009). Regenerative growth in Drosophila imaginal discs is regulated by Wingless and Myc. *Developmental Cell*, *16*(6), 797–809.

Srivastava, A., Pastor-Pareja, J. C., Igaki, T., Pagliarini, R., & Xu, T. (2007). Basement membrane remodeling is essential for Drosophila disc eversion and tumor invasion. *Proceedings of the National Academy of Sciences of the United States of America*, *104*(8), 2721–2726.

Srivastava, M., Scherr, H., Lackey, M., Xu, D., Chen, Z., Lu, J., et al. (2007). ARK, the Apaf-1 related killer in Drosophila, requires diverse domains for its apoptotic activity. *Cell Death and Differentiation*, *14*(1), 92–102.

Suissa, Y., Ziv, O., Dinur, T., Arama, E., & Gerlitz, O. (2011). The NAB-Brk signal bifurcates at JNK to independently induce apoptosis and compensatory proliferation. *The Journal of Biological Chemistry*, *286*(17), 15556–15564.

Sulston, J. E., & Horvitz, H. R. (1977). Post-embryonic cell lineages of nematode, Caenorhabditis elegans. *Developmental Biology*, *56*(1), 110–156.

Sun, G., & Irvine, K. D. (2011). Regulation of Hippo signaling by Jun kinase signaling during compensatory cell proliferation and regeneration, and in neoplastic tumors. *Developmental Biology*, *350*(1), 139–151.

Takeishi, A., Kuranaga, E., Tonoki, A., Misaki, K., Yonemura, S., Kanuka, H., et al. (2013). Homeostatic epithelial renewal in the gut is required for dampening a fatal systemic wound response in Drosophila. *Cell Reports*, *3*(3), 919–930.

Tang, H., Kambris, Z., Lemaitre, B., & Hashimoto, C. (2008). A serpin that regulates immune melanization in the respiratory system of Drosophila. *Developmental Cell, 15*(4), 617–626.

Thomas, C., Rousset, R., & Noselli, S. (2009). JNK signalling influences intracellular trafficking during Drosophila morphogenesis through regulation of the novel target gene Rab30. *Developmental Biology, 331*(2), 250–260.

Thornberry, N. A., Bull, H. G., Calaycay, J. R., Chapman, K. T., Howard, A. D., Kostura, M. J., et al. (1992). A novel heterodimeric cysteine protease is required for interleukin-1 beta processing in monocytes. *Nature, 356*(6372), 768–774.

Timmer, J. C., & Salvesen, G. S. (2007). Caspase substrates. *Cell Death and Differentiation, 14*(1), 66–72.

Trembley, A. (1744). Meoires pour sevir al'hisoire d'un genre de polyps d'eao dounce. *JuH Verbeek, Leiden.*

Tseng, A. S., Adams, D. S., Qiu, D., Koustubhan, P., & Levin, M. (2007). Apoptosis is required during early stages of tail regeneration in Xenopus laevis. *Developmental Biology, 301*(1), 62–69.

Uhlirova, M., & Bohmann, D. (2006). JNK- and Fos-regulated Mmp1 expression cooperates with Ras to induce invasive tumors in Drosophila. *The EMBO Journal, 25*(22), 5294–5304.

Waldhuber, M., Emoto, K., & Petritsch, C. (2005). The Drosophila caspase DRONC is required for metamorphosis and cell death in response to irradiation and developmental signals. *Mechanisms of Development, 122*(7–8), 914–927.

Wang, M. C., Bohmann, D., & Jasper, H. (2003). JNK signaling confers tolerance to oxidative stress and extends lifespan in Drosophila. *Developmental Cell, 5*(5), 811–816.

Warner, S. J., & Longmore, G. D. (2010). Context-dependent compensatory proliferation in epithelial homeostasis and tumorigenesis. *Cell Cycle, 9*(20), 4037–4038.

Wells, B. S., & Johnston, L. A. (2012). Maintenance of imaginal disc plasticity and regenerative potential in Drosophila by p53. *Developmental Biology, 361*(2), 263–276.

Wells, B. S., Yoshida, E., & Johnston, L. A. (2006). Compensatory proliferation in Drosophila imaginal discs requires Dronc-dependent p53 activity. *Current Biology, 16*(16), 1606–1615.

Woo, H. A., Yim, S. H., Shin, D. H., Kang, D., Yu, D. Y., & Rhee, S. G. (2010). Inactivation of peroxiredoxin I by phosphorylation allows localized H(2)O(2) accumulation for cell signaling. *Cell, 140*(4), 517–528.

Wood, W., Jacinto, A., Grose, R., Woolner, S., Gale, J., Wilson, C., et al. (2002). Wound healing recapitulates morphogenesis in Drosophila embryos. *Nature Cell Biology, 4*(11), 907–912.

Wu, Y., Brock, A. R., Wang, Y., Fujitani, K., Ueda, R., & Galko, M. J. (2009). A blood-borne PDGF/VEGF-like ligand initiates wound-induced epidermal cell migration in Drosophila larvae. *Current Biology, 19*(17), 1473–1477.

Xu, D., Li, Y., Arcaro, M., Lackey, M., & Bergmann, A. (2005). The CARD-carrying caspase Dronc is essential for most, but not all, developmental cell death in Drosophila. *Development, 132*(9), 2125–2134.

Xu, N., Wang, S. Q., Tan, D., Gao, Y., Lin, G., & Xi, R. (2011). EGFR, Wingless and JAK/STAT signaling cooperatively maintain Drosophila intestinal stem cells. *Developmental Biology, 354*(1), 31–43.

Xu, D., Wang, Y., Willecke, R., Chen, Z., Ding, T., & Bergmann, A. (2006). The effector caspases drICE and dcp-1 have partially overlapping functions in the apoptotic pathway in Drosophila. *Cell Death and Differentiation, 13*(10), 1697–1706.

Yoo, S. K., Starnes, T. W., Deng, Q., & Huttenlocher, A. (2011). Lyn is a redox sensor that mediates leukocyte wound attraction in vivo. *Nature, 480*(7375), 109–112.

Yu, X., Wang, L., Acehan, D., Wang, X., & Akey, C. W. (2006). Three-dimensional structure of a double apoptosome formed by the Drosophila Apaf-1 related killer. *Journal of Molecular Biology*, *355*(3), 577–589.

Yuan, J., Shaham, S., Ledoux, S., Ellis, H. M., & Horvitz, H. R. (1993). The C. elegans cell death gene ced-3 encodes a protein similar to mammalian interleukin-1 beta-converting enzyme. *Cell*, *75*(4), 641–652.

Zhou, L., Song, Z., Tittel, J., & Steller, H. (1999). HAC-1, a Drosophila homolog of APAF-1 and CED-4 functions in developmental and radiation-induced apoptosis. *Molecular Cell*, *4*(5), 745–755.

CHAPTER SEVEN

Postembryonic Organogenesis of the Digestive Tube: Why Does It Occur in Worms and Sea Cucumbers but Fail in Humans?

Vladimir S. Mashanov, Olga Zueva, José E. García-Arrarás[1]
Department of Biology, University of Puerto Rico, Rio Piedras Campus, Rio Piedras, Puerto Rico, USA
[1]Corresponding author: e-mail address: jegarcia@hpcf.upr.edu

Contents

Abstract

We provide an integrative view of mechanisms that enable regeneration of the digestive tube in various animal models, including vertebrates, tunicates, echinoderms, insects, and flatworms. Two main strategies of regeneration of the endodermal luminal (mucosal) epithelium have evolved in metazoans. One of them involves proliferation of resident epithelial cells, while the other relies on recruitment of cells from extramucosal sources. In any of these two scenarios, either pluri-/multipotent stem cells or specialized differentiated cells can be used as the starting material. Posttraumatic visceral

Current Topics in Developmental Biology, Volume 108
ISSN 0070-2153
http://dx.doi.org/10.1016/B978-0-12-391498-9.00006-1

regeneration shares some common mechanisms with normal embryonic development as well as with organ homeostatic maintenance, but there are signaling pathways and/or cellular pools that are specific to the regenerative phenomena. Comparative analysis of the literature suggests that mammals share with spontaneously regenerating animals many of the regeneration-related adaptations and are able to efficiently repair components of their digestive tube at the level of individual tissues, but fail to do so at the whole-organ scale. We review what might cause this failure in the context of the current state of knowledge about various regenerative models.

If the latent capacity for regeneration persists [. . .], then the restoration of its overt expression should be possible if the mechanisms of its inhibition could be discovered and eventually rendered ineffectual

Richard Goss (1992)

1. INTRODUCTION

Comparative analysis of the mechanisms involved in gut regeneration is of considerable interest for two main reasons. First, the digestive tube as a whole is arguably one of the few organs whose homology throughout all major clades of the Eumetazoa raises few doubts. Therefore, it is interesting to see how mechanisms of tissue homeostasis and posttraumatic regeneration of the same structure have been evolving in animal phylogeny. Evolution of regenerative mechanisms is still one of the most obscure issues in developmental biology. There is no clear understanding in the field of why the regenerative potential varies so greatly from species to species. The more widely accepted explanation, advocated by Richard Goss (1992), assumes that regeneration is an intrinsic property of multicellular organisms, which has already been in place in all major ancestral groups and that could be lost in evolution, but not gained again. The second hypothesis is that regeneration has evolved independently de novo in some taxa as a result of secondary gains (Bely & Nyberg, 2010) and thus can depend on some locally evolved taxon-specific features, which have no direct counterparts in other organisms (Garza-Garcia, Driscoll, & Brockes, 2010). Studying regenerative mechanisms of the same conserved organ, such as the digestive tube, across animal phylogeny may therefore turn out very informative for understanding fundamental issues in evolutionary developmental biology.

The second reason attracting attention to comparative studies of gut regeneration is that digestive diseases are a serious burden on human society worldwide. For example, in the US alone, digestive diseases are diagnosed in about 35% of the population and account for ~10% of all deaths (Everhart & Ruhl, 2009). A better understanding of how structural and functional properties of the digestive tube are restored in spontaneously regenerating animals will eventually yield the necessary information to direct the design of new clinical approaches to cure bowel diseases in human patients.

The digestive tube shows stereotypical organization across phylogeny. As the name implies, it is generally a tubular structure with the central lumen communicating with the external environment through terminal openings. This inner cavity is lined with a single layer of endodermally derived luminal (mucosal) epithelium, whose functions and unique position within the body place it under a range of adverse conditions. First, since the gut lumen is technically an extension of the external environment, the luminal epithelium must act as an effective protective barrier to prevent pathogen invasion. Second, the very function of food digestion makes the gut lumen an extremely hostile environment for living cells. This implies that the luminal epithelium must be protected from chemical, enzymatic, and mechanical damage. On the other hand, this tissue cannot develop into a multilayered impermeable barrier, like the epidermis, because it has to perform its vitally important function of absorption of nutrients, ions, and water from the gut lumen. This multitude of functional requirements poses paradoxical constraints: the luminal epithelium must be thin and capable of efficient absorption, and, at the same time, it must be able to withstand a variety of harsh and potentially damaging conditions. Thus, across animal phylogeny, we see that the gut mucosal epithelium is a rapidly regenerating tissue with an extensive capacity to constantly replace its old and damaged cells.

Here, we provide an overview of what is known about normal physiological cell renewal and posttraumatic regeneration in the digestive system of various animal models with an attempt to develop an integrative view of the different strategies that have evolved to repair visceral damage and discuss what can be learned from comparative studies to improve the limited regenerative capacity in the human bowel.

2. MECHANISMS OF DIGESTIVE TRACT REGENERATION IN DIFFERENT ANIMAL MODELS

2.1. Mammals

2.1.1 Physiological self-renewal of the luminal epithelium

Physiological self-renewal of the intestinal epithelium has been extensively studied in the adult mammalian small bowel. In this organ, the mucosal epithelium is organized into a series of numerous projections, or villi, and depressions, called crypts of Lieberkühn (Fig. 7.1A). Not only does this organization increase the absorptive surface of the epithelium, but it also demarcates the differentiated (villi) and proliferative domains (crypts) within the tissue. Intestinal stem cells are localized to a well-defined niche at the base of the crypts, four to six cells per niche, interspersed between Paneth cells (Barker et al., 2007; Crosnier, Stamataki, & Lewis, 2006). Each crypt gives rise to ~300 new cells every day. The stem cells produce extensively dividing transit amplifying cells, whose progeny are displaced toward the crypt–villus junction where they differentiate into specialized cells (Figs. 7.1A and 7.2), such as enterocytes, goblet cells, and enteroendocrine cells, which then continue to gradually migrate toward the apex of the villus. After 2–7 days of performing their respective functions, the worn-out differentiated cells are shed into the bowel lumen (Barker et al., 2007; Crosnier et al., 2006). It has been recently shown that the intestinal crypt also harbors a second population of progenitor cells, which are relatively quiescent (slow-cycling) and, in normal physiological settings, give rise to only Paneth and enteroendocrine cells. In the regenerating intestinal mucosa, however, these cells acquire a much broader developmental potential, enter an extensively proliferative state, and give rise to all types of specialized epithelial cells (Buczacki et al., 2013). The separation of the differentiated and proliferative domains is a conserved feature of the intestinal luminal epithelium in vertebrates. In some species, such as zebrafish, crypts never develop, but stem cells are still maintained in the restricted zones of intervillus pockets (Crosnier et al., 2006).

2.1.2 Molecular control of the stem cell niche

The maintenance of the stem cell niche and segregation of the proliferative and differentiated domains (crypt and villi, respectively) is controlled by an interplay between Wnt, Notch, Hedgehog, and Bmp signaling. The canonical Wnt pathway is essential for maintenance of the proliferative activity in the crypt and, thus, for physiological cell turnover in the luminal epithelium of the small intestine (Crosnier et al., 2006). The crucial role of Wnt allowed the use of perturbations of its signaling to identify specific genetic markers of intestinal stem cells among the Wnt target genes. One such gene, Lgr5 (leucine-rich repeat-containing G-protein coupled receptor 5), has proved to be of particular value, as its expression allows easy identification of the stem cells with high specificity (Barker et al., 2007). Other markers expressed in the intestinal stem cells include Bmi1, but its expression domain

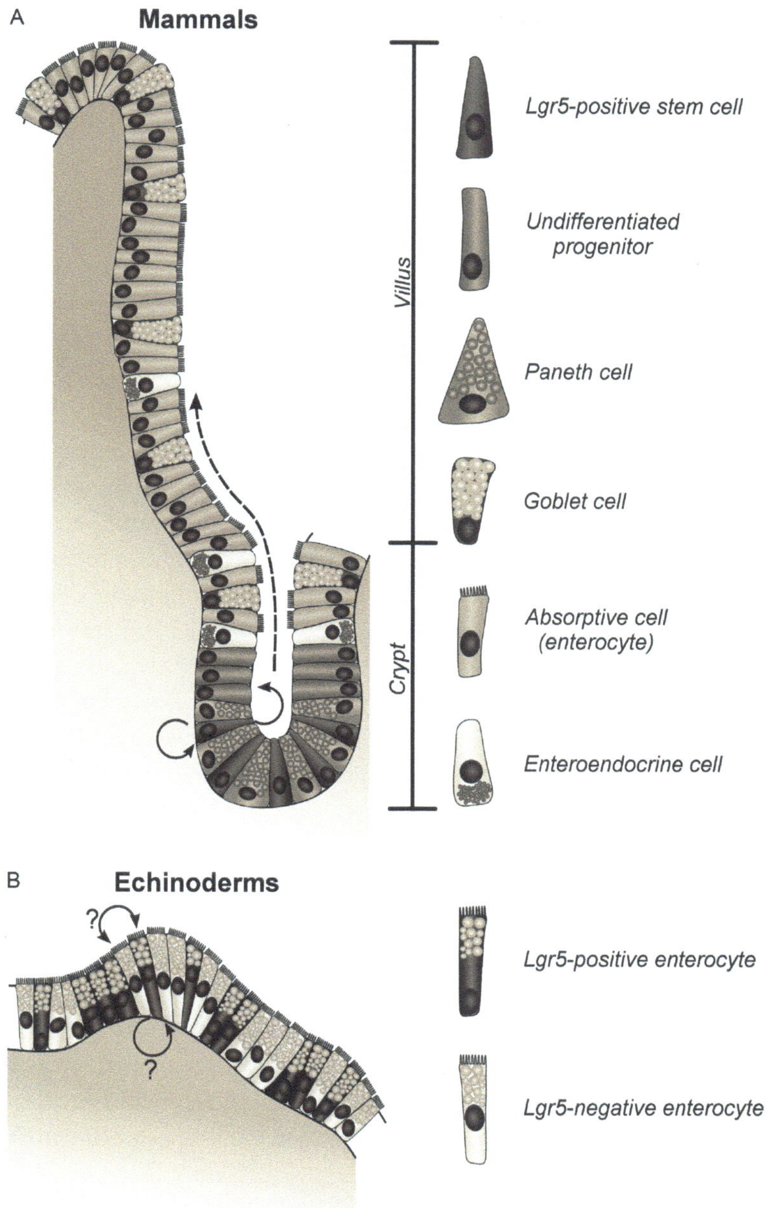

Figure 7.1 Diagram of organization of the mucosal epithelium in the mammalian small intestine (A) and in the digestive tube of a sea cucumber (B) highlighting the difference in cell renewal mechanisms. In mammals, Lgr5-positive intestinal stem cells are localized at the bottom of the crypt and both self-renew and produce rapidly dividing transit-amplifying progenitors (arrows), which gradually move apically (dashed arrow) and give

(Continued)

is broader as it also includes committed progenitors of specialized cells (Doupé & Jones, 2013; van Es et al., 2012).

2.1.3 Posttraumatic regeneration of the intestinal epithelium

Posttraumatic regenerative response of the mammalian digestive tube epithelium has been probed in a variety of injury paradigms and experimental settings, including surgical resection, gamma irradiation, transgenic ablation of stem cells, and exposure of the mucosal epithelium to burns and toxic chemicals. Although generally considered poor regenerators, mammals turned out to be capable of robust regeneration of their luminal epithelium after various types of injury. A number of studies have shown that there is far more developmental plasticity in the mucosal epithelium than was previously thought. For example, chemically burnt intestinal mucosa is readily healed, provided that subepithelial structures (the basal lamina and lamina propia) remain intact. The damaged necrotic cells are sloughed off into the gut lumen, while the subsequent reepithelialization proceeds very quickly. The surviving nongoblet epithelial cells surrounding the wound develop lamellipodia and migrate across the denuded basal lamina of the injured area. After just 2 h, 80–90% of the damaged mucosal surface is repaired (Feil et al., 1989). This early response is then followed by increased proliferation to compensate for the cell loss (Sturm & Dignass, 2008). Most importantly, the resident Lgr5^{+} intestinal stem cells are not the exclusive source of new cells in the regenerating intestinal epithelium (Fig. 7.2). *After selective ablation of the stem cells, either by making them susceptible to diphtheria toxin or by radiation-induced damage*, at least some of the committed transit-amplifying progenitors of both the absorptive and secretory lineages are capable of reverting back (dedifferentiating) to the Lgr5^{+} phenotype and thus replenishing the lost stem cell pool (Doupé & Jones, 2013; Tian et al., 2011; van Es et al., 2012). Notably, even differentiated cells are able to revert to the stem cell phenotype. For example, Paneth cells, the postmitotic differentiated cell component of the intestinal stem cell niche, have been recently shown to dedifferentiate, stop expressing their cell-type-specific markers,

Figure 7.1—Cont'd rise to the specialized cells. Paneth cells are the only differentiated cell type that remains in the crypt and constitute a component of the stem cell niche. In the digestive epithelium of echinoderms, unlike in mammals, Lgr5-positive cells are not spatially segregated into distinct zones, but are instead interspersed among Lgr5-negative enterocytes. Lineage relationships still remain to be determined in this model (arrows with question marks). (See color plate.)

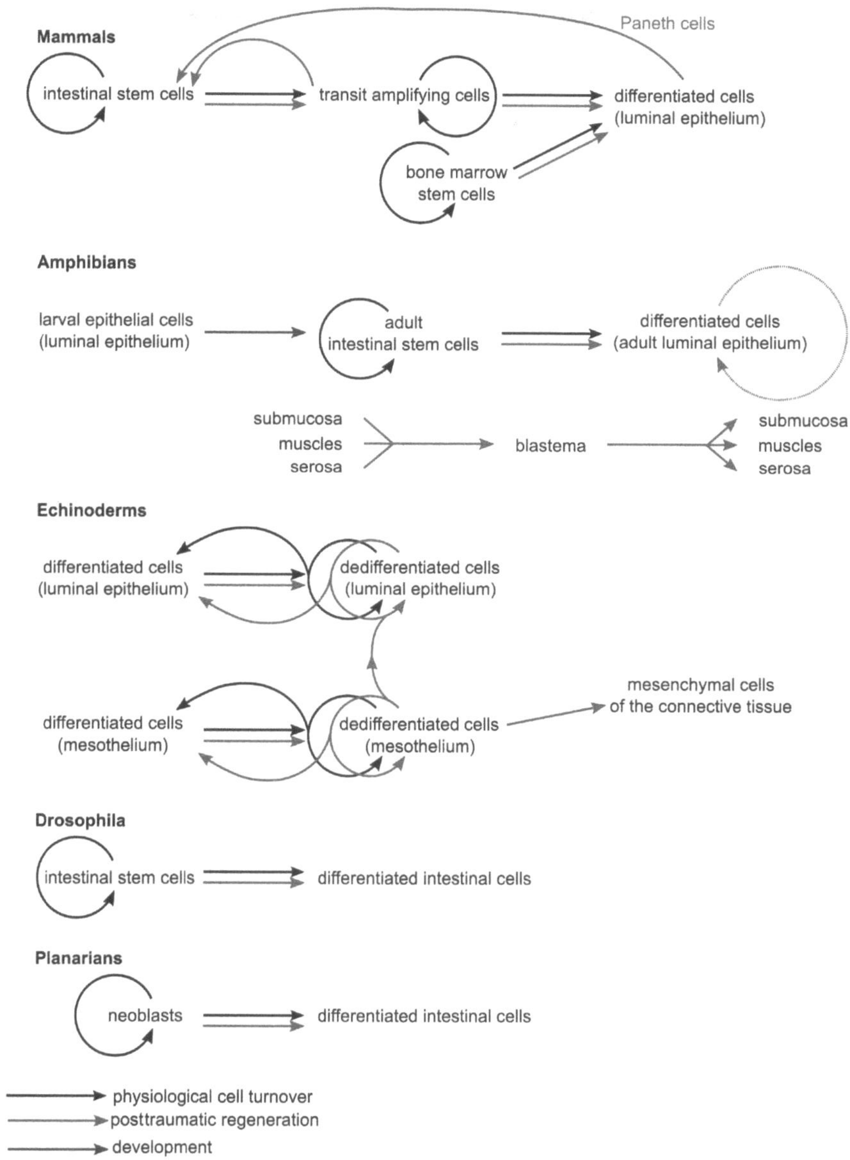

Figure 7.2 Diagram summarizing cell lineage relationships in normal (physiological) cell turnover and posttraumatic regeneration in the digestive tube of different bilaterian animal models. Straight arrows show cell lineage relationships in physiological cell turnover (black), posttraumatic regeneration (red), and development (blue). Circular arrows indicate self-renewal. (See color plate.)

and assume stem cell identity after $Lgr5^+$ stem cells had been ablated by sublethal doses of ionizing irradiation (Roth et al., 2012).

2.1.4 Contribution of nonintestinal cells to intestinal luminal epithelium repair

Some reports even suggest that differentiated cells of the digestive organs can originate from outside sources. Cell fate tracing in mice, as well as studies on human patients with transplanted sex-mismatched bone marrow or peripheral blood hematopoietic stem cells, have shown that the progeny of the adult stem cells from these grafted tissues can invade various organs and tissues of the recipient and differentiate into the respective specialized cell types, including mature epithelial cells of the gastrointestinal tract and liver hepatocytes (Fig. 7.2). Importantly, the number of bone marrow-derived epithelial cells increased after intestinal epithelium damage, suggesting that the bone marrow is recruited as an additional source of cells to spur tissue regeneration (Gupta, Dixit, Sales, Winslet, & Seifalian, 2006; Körbling et al., 2002; Krause et al., 2001; Okamoto, Matsumoto, & Watanabe, 2006; Okamoto et al., 2002). It is interesting to note here, however, that although these extraintestinal stem cells give rise to differentiated cells of the intestinal epithelium, they have never been reported to produce intestinal stem cells (Okamoto et al., 2006). This system of histogenetic relationships with bone marrow-derived stem cells localized outside of the digestive organs, whose cells they replenish, is reminiscent of the situation in planarians, whose neoblasts reside in the parenchyma and constitute the only proliferating universal pluripotent stem cell pool in the body (Reddien, 2013) (see below).

2.1.5 Molecular signaling in intestinal regeneration

As in normal cell turnover, Wnt signaling is one of the most essential signaling mechanisms required for regeneration of the intestinal epithelium. One of the key downstream genes activated by Wnt is c-Myc, whose elevated expression is required for regeneration to occur (Ashton et al., 2010; Bernal et al., 2005). It has been noted that not only the mere presence but also a significant increase in expression of the protein product of this gene is required for the biological effect. Forcing c-Myc expression to remain within the normal physiological range results in poor generation (Ashton et al., 2010). At the mechanistic level, it has been noted that Wnt signaling can affect cell fate plasticity, as exposure to Wnt3a can revert committed trans-amplifying secretory cell progenitors to the multipotent $Lgr5^+$ stem cell phenotype (van Es et al., 2012).

2.1.6 Regeneration of nonmucosal intestinal tissues

The mucosal epithelium is not the only intestinal component that is able to regenerate. To different degrees, all tissue layers of the digestive tube show regenerative capacities. This is the reason why functional reconnection takes place after the intestine is transected and then sutured back in the course of surgery. However, regeneration of nonmucosal tissue layers of the mammalian gut has received much less attention than that of the luminal epithelium. Occasional studies have shown that various components of the submucosa are able to regenerate following injury. For example, smooth musculature grows and reestablishes the anatomical integrity across the lesion site after surgical transection followed by end-to-end reanastomosis (Galligan, Furness, & Costa, 1989). However, the origin of the new intestinal muscle tissue has not been determined in mammals. It has been shown that the mesothelium, the outermost tissue layer of the bowel wall, is also capable of developing a regenerative response. The mesothelial cells surrounding the injury change their morphology (from flattened to cuboidal), proliferate, and migrate into the wounded area (Nagai et al., 2013). Even the enteric nervous system has been shown to regenerate. Both newly produced and previously existing surviving neurons can grow axons that expand within the injured area and reinnervate their targets (Galligan et al., 1989; Hanani et al., 2003; Laranjeira et al., 2011).

2.1.7 Limits to adult intestinal organogenesis

In spite of their extensive capacity to renew and regenerate the intestinal epithelium, mammals do not regrow their digestive tube at the whole-organ level, as they are not able to spontaneously regenerate full-thickness wounds and entire ablated regions. In human patients, the intestinal tissue can be lost to various disorders, such as intestinal ischemia, cancer, traumatic injury, and inflammatory bowel disease. All these pathologies often require extensive surgery and removal of considerable lengths of the digestive tube, resulting in the short bowel syndrome, a life-threatening condition, in which the remnant of the digestive tube fails to support the needs of the organism (Chen & Beierle, 2004; Levin & Grikscheit, 2012). Among the most common therapeutic solutions to treat patients with the short bowel syndrome is organ transplantation. However, application of this procedure is limited by donor availability, high incidence of graft rejection, toxicity of immune suppression, and relatively low success rates (Howell & Wells, 2011). Although alternative approaches involving tissue engineering are being actively sought, their development is at its infancy as their progress heavily depends on still incomplete fundamental biological knowledge of how to stimulate visceral regeneration in mammals at the whole-organ level (Chen & Beierle, 2004; Gupta et al., 2006; Howell & Wells, 2011; Jwo et al., 2013; Levin & Grikscheit, 2012; Ross et al., 2013).

2.2. Amphibians

2.2.1 *Intestinal regeneration following transection*

Intestinal regeneration in amphibians has been studied in two models: (a) posttraumatic regeneration after complete surgical transection or removal of the intestine in larval and adult animals (O'Steen, 1959; O'Steen & Walker, 1962) and (b) intestinal remodeling during metamorphosis (Fujimoto, Matsuura, Hu-Wang, Lu, & Shi, 2012; Ishizuya-Oka, 2007). Unfortunately, the studies of posttraumatic regeneration date back to about half a century ago (O'Steen, 1959; O'Steen & Walker, 1962), that is, well before modern imaging and molecular biology techniques had come into existence. Nevertheless, those studies did document some interesting phenomena. It was established that, unlike in mammals, the rostral and caudal cut ends of the injured intestine were able to grow across the injury gap to reconnect spontaneously and restore the anatomical integrity and physiological function of the digestive tube. The luminal epithelium (mucosa) was the first tissue layer to reform and it regenerated from its own cells. The serosa, muscular layer, and submucosa contributed cells to the blastema, which then underwent growth, remodeling, and differentiation to regenerate these tissue layers across the site of the original transection (Fig. 7.2).

2.2.2 *Intestinal remodeling during metamorphosis*

The second model focuses on the extensive reorganization of the digestive tube, which accompanies the switch from larval to adult diet during amphibian metamorphosis. At the onset of this transformation, most of the larval luminal epithelium is eliminated by extensive programmed cell death. Only some of the specialized epithelial cells are spared. They dedifferentiate, transform into adult stem cells, and then give rise to the new adult mucosa via extensive proliferation (Ishizuya-Oka, 2007; Ishizuya-Oka et al., 2009). There is a clear difference in histogenetic relationships in the intestine of the larval and adult amphibians. Before metamorphosis, the larval luminal epithelium does not contain undifferentiated stem cells and is composed of specialized epithelial cells capable of reentering the cell cycle, whereas the adult intestine shows a clear separation between the proliferative and differentiated domains, very much like in mammals (Ishizuya-Oka et al., 2009).

As to the molecular mechanisms underlying such a drastic tissue transformation, it has been shown that c-Myc plays a central role in gut remodeling during thyroid hormone-controlled metamorphosis in *Xenopus*. Its transcription is activated directly by the thyroid hormone receptor. In its turn, c-Myc regulates differential expression of a number of down-stream genes, including arginine methyltransferase 1, which is thought to play an important role in specification of adult intestinal stem cells (Fujimoto et al., 2012).

2.3. Tunicates

The highest regenerative capacities among the Tunicata, the subphylum of invertebrate chordates, are seen in those species that can reproduce asexually (Fujiwara, Isozaki, Mori, & Kawamura, 2011; Kaneko, Katsuyama, Kawamura, & Fujiwara, 2010; Kawamura, Sugino, Sunanaga, & Fujiwara, 2008). Depending on the species, one of several types of differentiated tissues, in budding or regenerating animals, is capable of undergoing dedifferentiation and giving rise to various organs, including the digestive tube. One of these tissues showing high developmental plasticity is the epicardium, an endodermally derived tubular or sac-like structure that surrounds the viscera (Kawamura et al., 2008). Another multipotent, but differentiated, tissue is the atrial epithelium, which constitutes the inner layer of the body wall in tunicates and has no homolog in vertebrates. The embryological origin of this structure is unclear, as conflicting results point to either the ectoderm or the endoderm as a possible source. In regenerating and asexually propagating animals, the atrial epithelium dedifferentiates to form a gut primordium. Interestingly, this process requires retinoic acid, since in the presence of inhibitors of retinoic acid synthesis, gut formation is suppressed (Kaneko et al., 2010). Gut formation from the atrial epithelium also depends on Myc expression, as injections of specific dsRNA caused defects in morphogenesis of the viscera (Fujiwara et al., 2011; Kawamura et al., 2008). The link between retinoic acid and Myc is not clearly established. Although investigators have proposed that while retinoic acid might contribute to the activation and maintenance of Myc transcription, it is not the factor that initiates the activation itself (Kaneko et al., 2010).

2.4. Echinoderms

The capacity of echinoderms to regenerate their visceral organs is among the highest among metazoan taxa. Most of what is known about organization and regeneration of the digestive tube in echinoderms has been learned from

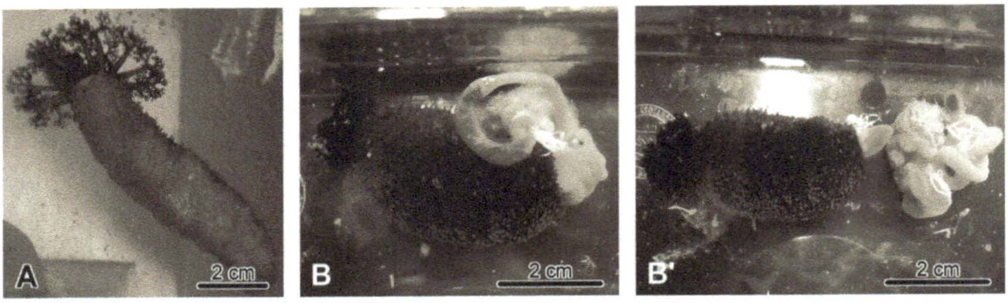

Figure 7.3 One of the best studied models of visceral regeneration: the sea cucumber *Holothuria glaberrima* Selenka. (A) Normal adult individual. (B) and (B′) An individual photographed early and late in the process of evisceration, respectively. (See color plate.)

experiments on sea cucumbers (holothurians) (Figs. 7.3 and 7.4) (reviewed in García-Arrarás & Greenberg, 2001; Mashanov & García-Arrarás, 2011), although occasional studies were also done on members of other classes, such as brittle stars (ophiuroids) (Frolova & Dolmatov, 2010) and sea lilies (crinoids) (Mozzi, Dolmatov, Bonasoro, & Candia Carnevali, 2006).

2.4.1 Anatomy of the adult sea cucumber intestine

The digestive tube of sea cucumbers consists of a number of specialized histologically distinct regions, including the pharynx, esophagus, long looped intestine, and the cloaca. It is attached to the inner surface of the body wall by the mesentery, a thin film made up of a central layer of connective tissue sandwiched between two layers of the contractile coelomic epithelium. The wall of the digestive tube is composed of an outer mesothelium, which is continuous with the coelomic epithelium of the mesentery, a middle connective tissue layer, and an inner luminal (digestive) epithelium (Fig. 7.4A). The luminal epithelium is made up mostly of vesicular enterocytes. Other less abundant cell types include granular enterocytes, goblet cells (mucocytes), and neurosecretory cells (Feral & Massin, 1982; Mashanov, Frolova, & Dolmatov, 2004). Like in many other metazoan animals, the luminal epithelium of echinoderms undergoes constant cell turnover under normal physiological conditions (Fig. 7.4B); however, unlike in vertebrates and insects, the proliferating cells are not localized to restricted specific locations, but rather seem to be scattered randomly throughout the luminal epithelium and mesothelium (Mashanov & García-Arrarás, 2011).

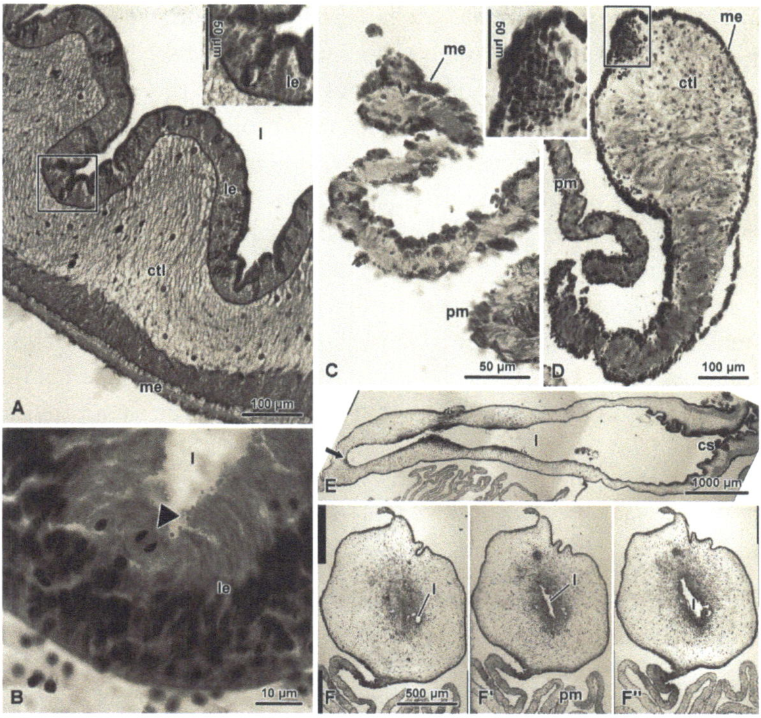

Figure 7.4 Histological organization of the digestive tube of the sea cucumber *H. glaberrima* in noneviscerated animals (A) and (B) and at different time points of regeneration (C–F″). (A) Cross-section of the intestine in a noneviscerated animal stained with alcian blue (pH 2.5) and nuclear fast red. The *inset* shows a higher magnification view of the luminal epithelium. (B) Dividing cell (anaphase, *arrowhead*) in the luminal epithelium of the normal intestine stained with hematoxylin and eosin. (C–F″) Regenerating digestive tube stained with Heindenhein's azan. (C and D) Cross-sections through the early gut rudiment on days 3 and 7 postevisceration, respectively. The *inset* in D shows a detailed view of the mesothelial cells ingressing into the underlying connective tissue. (E) Longitudinal section through the growing gut rudiment on day 14 postevisceration. The growing tip (*arrow*) is on the left, whereas the cloacal stump is on the right. (F–F″) Day 14 postevisceration; successive cross-sections of a posterior gut rudiment taken through the growing tip (F) and at distances of 40 μm (F′) and 110 μm (F″). *cs*, cloacal stump; *ctl*, connective tissue layer of the gut wall; *l*, gut lumen; *le*, luminal (mucosal) epithelium; *me*, mesothelium; *pm*, proximal mesentery. (See color plate.)

2.4.2 Stem cell status in the sea cucumber luminal epithelium

Morphological studies did not reveal any undifferentiated cells in the tissues of the gut wall, but instead identified mitotic cells of the luminal epithelium as vesicular enterocytes (Mashanov et al., 2004). Dividing enterocytes retained their secretory vacuoles and were thus capable of undergoing cell division while remaining in, at least, a partially differentiated state. It is currently unknown whether all enterocytes can divide or whether this capacity is restricted to only a subset of cells, thus separating the morphologically homogeneous population of enterocytes into proliferation-competent and permanently postmitotic cells, respectively. It is also unclear if the same cells can keep dividing in the long term and, thus, behave as bona fide stem cells or if the proliferative capacity is stochastically turned on and off in individual cells within the epithelium.

In order to clarify these issues, we have probed markers used to study intestinal stem cells in other animal models. One of the most basic techniques to mark putative stem cells in the tissue relies on their proposed ability to retain DNA labeling for extended periods of time after a pulse of a thymidine analog (e.g., BrdU) (Crosnier et al., 2006; Potten, 2004). After injecting holothurians with BrdU, randomly distributed, scattered, strongly immunoreactive label-retaining cells were observed in both the luminal epithelium and mesothelium of the intestine for at least 2–5 weeks after the last injection, suggesting the presence of cells with stem cell-like behavior (Mashanov & García-Arrarás, 2011). Further experiments are, however, needed to directly establish whether the label-retaining enterocytes actually act as stem cells (i.e., are capable of renewing themselves and producing specialized progeny in further cell cycles) or merely represent terminally postmitotic cells, which originate via only a limited number of cell divisions, and thus remain strongly labeled for an extended duration of time. Such long-lived label-retaining differentiated cells are not uncommon. Mammalian Paneth cells, for example, are able to retain labeling for up to 100 days (Roth et al., 2012).

We also studied expression of pluripotency factors, such as homologs of mammalian *Sox2*, *c-Myc*, *Lgr5*, and *BmiI* (Mashanov et al., in preparation), which were identified as intestinal stem cell markers in vertebrates. These genes are widely expressed in the intestinal luminal epithelium of *Holothuria glaberrima* without preferential localization to any particular region. This expression is, however, not ubiquitous, as there are always luminal cells, which show no *in situ* hybridization signal (Fig. 7.1B). Thus, the morphologically uniform enterocyte population is obviously heterogeneous in terms of expression of stem cell/multipotency markers. However, it remains to be established how these differences affect proliferative potential in individual enterocytes.

2.4.3 Cellular events in intestinal regeneration

Many sea cucumber species are well known for their impressive ability to spontaneously autotomize (eviscerate) their alimentary canal, which is then completely regenerated over the time course of a few weeks (Byrne, 1986; García-Arrarás & Greenberg, 2001; Hyman, 1955; Mashanov & García-Arrarás, 2011; Wilkie, 2001) (Figs. 7.3 and 7.4C–F″). In regenerating sea cucumbers, the first morphogenetic process observed after autotomy is remodeling of the torn free edge of the mesentery. The connective tissue partition in this region swells to form the primordium of the new intestine (compare Fig. 7.4C and D). The mesothelium, which covers this connective tissue swelling, undergoes deep dedifferentiation, that is, the cells lose specialized phenotypic features, enhancing their ability to proliferate and migrate. Nonetheless, it retains its epithelial integrity except at the distal (antimesenterial) side where the cells undergo an epithelial-mesenchymal transition and ingress into the underlying extracellular matrix (Fig. 7.4D inset) (Dolmatov & Mashanov, 2007; García-Arrarás et al., 2011).

Development of the solid connective tissue rudiment is followed by lumen formation, which can be accomplished through one of two mechanisms. The first one involves invasion of the luminal epithelium from the cloacal and/or esophageal stumps (Fig. 7.4E–F″) (García-Arrarás et al., 1998). In this case, only partial dedifferentiation is seen in the luminal epithelium. In spite of extensive cell division, mitotic enterocytes remain connected to their neighbors by intercellular junctions and even retain microvilli and secretory vacuoles (Mashanov, Dolmatov, & Heinzeller, 2005; Odintsova, Dolmatov, & Mashanov, 2005). A second path to lumen formation involves infolding of the mesothelium with subsequent detachment of the invaginated epithelial folds from the coelomic surface of the gut rudiment and transdifferentiation into the luminal epithelium (Mashanov et al., 2005; Mosher, 1956). The latter mechanism exemplifies a relatively rare phenomenon of direct transformation of one type of differentiated tissue into another one across the germ layer boundaries, since in embryogenesis the gut mesothelium and luminal epithelium originate from the mesoderm and endoderm, respectively (Mashanov & Dolmatov, 2004).

2.4.4 Molecular hallmarks of intestinal regeneration

The molecular basis of tissue plasticity in the echinoderm digestive tube has recently started to be elucidated. One of the first hypotheses to be tested was whether extensive dedifferentiation of specialized cells is related to increased expression of transcription factors, such as homologs of the vertebrate *Sox2*, *c-Myc*, *Klf4*, and *Oct3/4* genes, known to confer multipotency (Sridharan et al., 2009; Takahashi & Yamanaka, 2006). Of these, only *Myc* was found to be significantly upregulated in the regenerating tissues of the sea cucumber, and, importantly, this increase in expression correlated in time with commencing dedifferentiation. Surprisingly, *SoxB1*, the holothurian homolog of the mammalian *Sox2*, was almost completely absent from the dedifferentiating tissues, whereas the remaining two factors did not show any significant change in their expression values (Mashanov et al., in preparation). These observations suggest that the molecular signaling pathways that induce *in vivo* dedifferentiation of specialized cells in regenerating echinoderms might not be exactly the same as those employed to induce pluripotency in cultured mammalian cells. Nevertheless, the role of *Myc* in both processes seems to be conserved. Dedifferentiation in the gut mesothelium has also been shown to correlate with extensive expression of a number of genes known to be involved in both embryogenesis and cancer, including *survivin*, *mortalin*, *TCTP*, and *Bmp1/Tll* (Mashanov, Zueva, & García-Arrarás, 2012; Mashanov, Zueva, Rojas-Catagena, & García-Arrarás, 2010).

In echinoderms, Wnt signaling also potentially plays a role in intestinal regeneration, although its exact functions are not known yet. Expression of *Wnt9* was studied in the noninjured and regenerating sea cucumber *H. glaberrima* (Mashanov et al., 2012). No expression of this gene was detected in the uninjured digestive tube, suggesting that this ligand is not involved in the normal physiological cell turnover. However, *Wnt9* was extensively expressed in the mesothelium of the regenerating intestine. This contrasts with the expression pattern of *Wnt9* in the mammalian digestive system, where *Wnt9b*, one of the two paralogs that constitute the vertebrate Wnt9 subfamily, is specifically restricted to the epithelium of the crypt (Crosnier et al., 2006; Gregorieff et al., 2005). Of interest, induction of *Wnt9* expression in dedifferentiating tissues of the sea cucumber coincides with upregulation of *Myc*. As *Myc* homologs are known to be activated by Wnt signaling and are necessary to trigger intestinal regeneration in a number of phylogenetically distant species, including mammals and insects (Ashton et al., 2010; Bernal et al., 2005; Cordero, Stefanatos, Scopelliti, Vidal, & Sansom, 2012), overexpression of these genes might represent an evolutionary conserved signaling mechanism in visceral repair.

2.5. Insects

Most of what is known about insect intestinal regeneration has been learned from studies of *Drosophila* midgut. The midgut luminal epithelium consists of three cell types—columnar enterocytes, endocrine cells, and undifferentiated intestinal stem (regenerative) cells (ISCs). The *Drosophila*

ISCs were discovered less than a decade ago by two groups (Micchelli & Perrimon, 2006; Ohlstein & Spradling, 2006), and were demonstrated to produce the other two (differentiated) intestinal cell types (Fig. 7.2). Similar results have shown an undifferentiated intestinal cell giving rise to different cell phenotypes in the midgut of another insect, the grasshopper *Locusta migratoria* (Illa-Bochaca & Montuenga, 2006). In *Drosophila*, mitotic division of the stem cell gives rise to another daughter stem cell and a diploid progenitor enteroblast. Unlike in mammals, the enteroblasts do not divide under normal conditions, but, instead, differentiate directly into one of the two cell types. In 90% of cases, they become absorptive enterocytes. The remaining 10% give rise to small enteroendocrine cells (Cordero et al., 2012). The *Drosophila* adult intestine shows a robust regenerative response to epithelial injuries. In particular, the ISCs respond by switching to a rapid proliferative state to compensate for the cell loss (Cordero et al., 2012; Shaw et al., 2010). However, an alternative hypothesis that stem cells responding to injury originate from a different pool from those associated with normal homeostasis has also been proposed (Biteau, Hochmuth, & Jasper, 2011; Hochmuth, Biteau, Bohmann, & Jasper, 2011).

The *Drosophila* model system has provided important information on the genetic basis of intestinal stem cell regulation both in normal homeostasis and after injury (Biteau et al., 2011). It has been shown that at least part of the gene regulatory machinery that controls intestinal development, growth, and regeneration is highly conserved between mammals and *Drosophila* (Apidianakis & Rahme, 2011; Shaw et al., 2010). For example, in both systems, pathways activated by the insulin receptor and the epidermal growth factor receptor are essential for the proliferation of intestinal stem cells during normal maintenance, while Jun-N-terminal Kinase (JNK), JAK/Stat, and Hippo/Yorkie pathways are required for stress-related responses (Biteau et al., 2011), Moreover, like in mammals, wingless (Wg) signaling acting through Myc is required for damage-induced proliferation in the intestinal epithelium (Cordero et al., 2012). The induction of Wg occurs via the activation of the JNK pathway and takes place parallel to the activation of JAK/Stat signaling, which suggests that activation of multiple pathways is necessary for midgut regeneration in *Drosophila*. Regulation of cell differentiation has also been well studied in this system. For example, stem cells with Delta-rich vesicles activate Notch and produce daughter cells with downregulated Delta that become enterocytes, while stem cells with weak Delta activity produce progeny that become enteroendocrine cells (Ohlstein & Spradling, 2007). Won-Jae Lee and Masayuki Miura expand on the cellular and molecular basis of *Drosophila* regeneration in an accompanying review in this volume.

2.6. Planarians

The planarian digestive system is formed by a single-opening tube (pharynx) with extensive blind ramifications (intestine). Unlike in many other organisms, the intestinal luminal epithelium of flatworms consists entirely of developmentally static, postmitotic differentiated cells. Both physiological cell renewal and posttraumatic regeneration of the digestive system is accomplished via the involvement of neoblasts, the only dividing cell population in the animal (Fig. 7.2), which reside in the surrounding mesenchymal parenchyma and give rise to all differentiated cell types in the body (Forsthoefel, Park, & Newmark, 2011; Wagner, Wang, & Reddien, 2011). Although the lineage relationship between neoblasts and terminally differentiated cells is clearly unidirectional (differentiation cannot be reversed), the details of histogenetic hierarchy within this progenitor cell population have not been worked out yet. The emerging data suggest that individual neoblasts may differ in their developmental potential (Reddien, 2013). On the one hand, at least some of the neoblasts are true pluripotent cells (Wagner et al., 2011); on the other hand, the existence of more committed proliferating progenitors has also been demonstrated. The bona fide pluripotent neoblasts can, thus, give rise to lineage-restricted neoblasts, which, in turn, generate differentiated progeny (Reddien, 2013).

Recent findings have shown some of the genes that might be involved in intestinal homeostasis and regeneration (Forsthoefel et al., 2012). Among them are genes that regulate intestinal branching and differentiation of phagocytes (the main cell type in the planarian gut and the one responsible for digestion). Moreover, results from the same group provide evidence for the regulation of neoblast proliferation dynamics suggesting the possibility that intestinal phagocytes play a niche-like role for somatic stem cells.

2.7. Coelenterates

Coelenterates possess a simple gastric cavity, where they digest their food. The gastric cavity communicates with the external environment through a single opening and is lined by an endodermal epithelial layer, or gastrodermis, composed mainly of epitheliomuscular cells, and also gland cells that secrete digestive enzymes and mucous-secreting cells. Most of what is known about regeneration in this animal group has been studied in *Hydra* (Galliot, 2012) and, to a smaller level, in *Hydractinia* (Plickert, Frnak, & Müller, 2012).

The central body region of coelenterates, or the gastric column, contains a larger proportion of interstitial undifferentiated (stem) cells in contrast to the head or foot regions made up of terminally differentiated cells (Galliot, Welschof, Schuckert, Hoffmeister, & Schaller, 1995). Stem cells in the gastric column divide continuously and their progeny are displaced toward the apical and foot ends of the body, where they are eventually discarded (Bode, 1996). In *Hydra*, three independent cell lineages (ectoderm, endoderm, and interstitial cells) arise during development. Each lineage has its own set of "stem cells" that constantly replenishes the cells of the corresponding lineage during the animal's life and during regeneration (Hobmayer et al., 2012). The endodermal cell lineage consists of the epithelio-muscle cells that line the gastric cavity and, although differentiated, maintain the capability to divide (Plickert et al., 2012). Interstitial cells give rise to other cell types that are eventually incorporated into the endodermal epithelial layer. In addition to mucous and gland cells, they also produce other cell types such as nematocysts and neurons (David, 2012). The interstitial precursors undergo a limited number of divisions and differentiate into postmitotic cells (Bode, 1996). The differentiation of stem cells into nematocytes and neurons has been well studied, in comparison to the differentiation into mucous or gland cells (David, 2012). Nonetheless, it has been proposed that gland cells differentiate into mucous cells as they move from the body column and enter the head region (Bode, 1996; Siebert, Anton-Erxleben, & Bosch, 2008). The same authors also showed that mucous cells in the head region originate through direct differentiation of interstitial stem cells. Thus, in this simple digestive system, we find that many of the cellular processes are reminiscent of what takes place in intestinal regeneration of other metazoans. As in mammals, the interstitial stem cells within the body wall of *Hydra* show a compartmental separation between dividing and terminally differentiated cells. As in sea cucumbers, the differentiated cells of the endoderm (both epitheliomuscular and glands) retain their proliferative properties. Finally, similar to planarians, some of the cells within the gastrodermal layer (glands and mucous) originate from external stem cells.

Not only are the cellular processes reminiscent of what occurs in animals with a much more complex digestive tube, but similar molecules also appear to control these events. For example, Wnt signaling is associated with cell proliferation and maintenance of stemness during regeneration and budding (Hobmayer et al., 2012). Most interesting is the possible role of Myc family members, which, as in other metazoans, act downstream of Wnt. The expression of one of them (myc1) is specific to the proliferating interstitial stem cell derivatives and is absent in terminally differentiated cells (Hartl et al., 2010). This myc gene has been proposed to control proliferation of the rapidly dividing interstitial cellular lineage (Hobmayer et al., 2012).

3. EMERGING PRINCIPLES OF INTESTINAL REGENERATION ACROSS EUMETAZOANS

3.1. Regeneration strategies of the luminal epithelium

An overview of the literature shows that the luminal epithelium is the digestive system component whose regeneration has been best studied across animal phylogeny, while the connective tissue compartment, intestinal musculature and the outer mesothelium, has attracted much less attention. A comparative analysis suggests that two different strategies enable regeneration of the gut lining in metazoans (Fig. 7.5). The first strategy relies on resident epithelial cells, whose progeny give rise to differentiated cells that remain within the same tissue. The second strategy recruits cells from external sources located outside of the mucosal epithelium. Each strategy involves either stem cells or differentiated cells as the starting material. Unfortunately, the current state of knowledge precludes the drawing of any conclusions as to the evolution and maintenance of regeneration mechanisms.

These strategies are widely distributed across animal phylogeny and do not show any obvious correlation with the taxonomic position or with the complexity of the histological organization of the gut wall. For example,

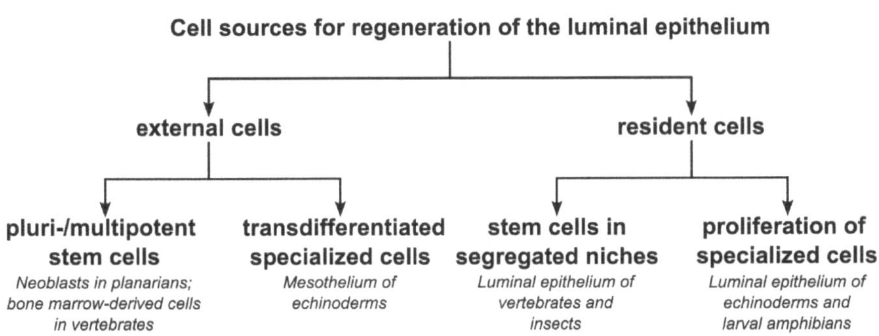

Figure 7.5 Diagram summarizing the different strategies employed by metazoan animals to regenerate their intestinal epithelium.

generation of the luminal epithelium from tissue-specific intestinal stem cells is seen in both vertebrates and insects. On the other hand, the reverse lineage relationship, when the cell pool within the tissue is replenished through induction of proliferation in differentiated cells, has also been reported for representatives of different taxonomic groups, such as amphibian tadpoles and sea cucumbers. Likewise, regeneration of the luminal epithelium from external sources of stem cells has been documented in phylogenetically distant taxa, such as planarians (neoblasts) and mammals (bone marrow-derived stem cells). *Hydra* makes use of both strategies: the epitheliomuscular cells of the endoderm propagate through self-renewal, while the gland cells that are incorporated into the gastrodermis originate from interstitial stem cells (David, 2012; Hobmayer et al., 2012). The only mechanism of gut regeneration that seems to be unique to one animal group in view of the current state of knowledge is regeneration of the luminal epithelium through direct transdifferentiation of mesothelial cells in echinoderms (Mashanov et al., 2005).

It is important to note that different mechanisms of intestinal regeneration are not necessarily mutually exclusive and can be employed sequentially or simultaneously by the same organ in the same organism. Thus, in humans, the damaged mucosal epithelium regenerates predominantly through migration of surviving differentiated cells and proliferation of the resident stem cells, but it can also incorporate bone marrow-derived cells (Gupta et al., 2006; Körbling et al., 2002; Krause et al., 2001; Okamoto et al., 2006, 2002). Sea cucumbers provide even more drastic examples of how different regenerative mechanisms, in spite of being based on dissimilar developmental events and cell sources, lead to the same final result—rebuilding of the lost regions of the digestive tube. Upon induction of visceral autotomy, the holothurian *Eupentacta fraudatrix* eviscerates in such a way that the only remnant of the original digestive tube is the cloacal stump at the posterior end of the body, while no endodermal tissues are present at the anterior end (Mashanov et al., 2005). The new intestinal rudiments that develop at the posterior and anterior ends of the body employ two different mechanisms of lumen formation: outgrowth of the luminal epithelium of the cloacal stump and invagination of the mesothelium, respectively. These observations lead to interesting questions as to which signaling pathways underlie induction of different coexisting regenerative mechanisms. For example, what makes the dedifferentiated mesothelial cells produce only mesothelium in the posterior gut rudiment and both the mesothelium and luminal epithelium in the anterior gut rudiment?

3.2. Posttraumatic regeneration and physiological cell turnover

There are extensive parallels between posttraumatic regeneration and other modes of development, including embryonic/postembryonic development and cell turnover under normal conditions (physiological regeneration) (Carlson, 2007). It has been suggested that regeneration might have originally evolved as an extension of normal developmental programs (Bely & Nyberg, 2010). Here, we discuss relationships between visceral regeneration and other forms of development that take place in the digestive system.

Under normal physiological conditions, as new cells are being continuously born and induced to differentiate and gradually replace the worn out counterparts, there must be a tightly regulated equilibrium between the number of new cells being produced and the number being lost. In contrast, posttraumatic regeneration and wound healing require a burst of proliferative activity and production of a large cohort of new cells over short time spans. This increase in cell proliferation, however, has to be reversible and also cease once the required number of cells has been generated. Furthermore, the new cells must differentiate and organize into a patch of functional tissue continuous with the surrounding (noninjured) regions of the body. Moreover, posttraumatic regeneration, instead of acting locally within a single particular tissue, often requires some reorganization of multiple tissues at the whole-organ level. This is directly related to the question of why it is that regeneration in mammals, including humans, is so limited in spite of the high rate of homeostatic cell turnover in the mucosal epithelium (Radtke & Clevers, 2005).

When one considers the differences in requirements that must be met by physiological cell turnover and posttraumatic regeneration in the interest of the multicellular body as a whole, it is not surprising that the two processes may not be necessarily identical in terms of their cell sources and regulatory mechanisms. For example, studies on mammals and *Drosophila* showed that regeneration of the luminal epithelium depends on the expression of sets of genes that are not strictly required for physiological cell turnover (Ashton et al., 2010; Cordero et al., 2012). Thus, high level of physiological regeneration does not necessarily imply extensive capacities for posttraumatic regeneration. The reverse is also true: Relatively quiescent tissues, such as liver, can readily regenerate after injury (Gilgenkrantz & Collin de l'Hortet, 2011). Posttraumatic visceral regeneration, therefore, cannot be interpreted simply as an extension of tissue self-renewal under normal physiological conditions (Cordero et al., 2012).

3.3. Regeneration and embryonic development

In general terms, posttraumatic regeneration is known to incorporate certain elements of embryonic and/or postembryonic development (Liozner, 1982). This is also true in the context of visceral regeneration. For example, the molecular mechanisms underpinning intestinal adaptation in mice (compensatory growth following massive small bowel resection) show significant similarity in the differential gene expression signature to the developing and immature intestine (Erwin et al., 2006). In some cases, the relationships between regeneration and development are extremely intimate, such as when regeneration becomes an integral component of the developmental program. This kind of "development-through-regeneration" is seen, for instance, in the intestinal epithelium remodeling during amphibian metamorphosis (Ishizuya-Oka, 2007; Ishizuya-Oka et al., 2009) and the formation of viscera in asexually propagating tunicates (Kawamura et al., 2008).

On the other hand, regeneration differs from embryonic development in that it often involves changes in cell fate specification and lineage relationships, that is, cells in regenerating tissues do not necessarily arise from other cells of the same kind or from their immediate developmental progenitors or resident tissue stem cells, but can originate from sources that are fundamentally different from those in normal tissue turnover and embryogenesis (Brockes & Kumar, 2008). Sometimes, exact recapitulation of embryonic mechanisms is rendered impossible by the very nature of the injury, such as when all endodermal tissues are lost in the anterior body region of some sea cucumbers (Mashanov & García-Arrarás, 2011). In these cases, the ability to induce developmental plasticity in adult tissues and to divert from embryonic developmental trajectories is a necessary prerequisite for regeneration.

Certain levels of cell fate plasticity are seen in many visceral regeneration models and retained even in the adult mammalian intestine. For example, intestinal fibroblasts and myofibroblasts may derive from nonmesenchymal cells, such as epithelial and endothelial cells, through transdifferentiation (Speca, Giusti, Rieder, & Latella, 2012). Furthermore, as mentioned earlier, recent lineage tracing experiments have shown that cell fate changes also occur within the luminal epithelium itself. These examples include injury-induced reversal of normally unidirectional lineage relationships from the stem cell through committed transit-amplifying progenitors to differentiated cells, when the ablated stem cell pool can be restored from the trans-amplifying or even differentiated (Paneth) cells (Doupé & Jones, 2013; Roth et al., 2012; Tian et al., 2011; van Es et al., 2012). Finally, the endodermally derived luminal epithelium of the human gastrointestinal tract can regenerate from mesodermal sources (bone marrow-derived stem cells) (Gupta et al., 2006; Körbling et al., 2002; Krause et al., 2001; Okamoto et al., 2006, 2002). Thus, posttraumatic visceral regeneration can incorporate, but is not based entirely on programs of normal development.

3.4. What can we learn from model animals to improve bowel regeneration in mammals?

What underlies the ability to fully regenerate the digestive tube in such animals as sea cucumbers, and what prevents mammals from doing so? At the level of individual tissues, at least some of the mechanisms and adaptations required for successful regeneration are already in place in mammals. These have been best demonstrated in the mucosal epithelium and include proliferation of resident intestinal stem cells and induction of stem cell properties in transit-amplifying progenitors and differentiated cells, as well as recruitment of extramucosal cells, that is, the same set of strategies that are used by animals capable of spontaneous regeneration of the digestive tube at the whole-organ level. The paradox then is the following. If all individual tissues can regenerate, why cannot the organ itself regenerate following major injury or grow back after resection? There are several possible answers to this question. First, tissues themselves might have restricted regenerative capabilities and overall organ regeneration might be limited by the capacities of the least regenerating component. Thus, although the mucosal epithelium is a highly dynamic self-sustaining and self-organizing tissue, which can recreate its typical crypt-villus structure from as little as a single $Lgr5^+$ stem cell (Merlos-Suárez et al., 2011; Sato et al., 2009), other tissue layers might have less prominent properties.

A second explanation for the low regenerative capacity of the mammalian bowel at the whole- organ level is the possible need for an additional factor that mediates intestinal lengthening. In this case, such a factor that would regulate the acquisition of the proper organ length might be lacking during the regenerative response. A number of signaling molecules affecting the lengthening of the intestine were identified in embryonic development, but have never been studied in the context of gut regeneration. They include epimorphin, FGF9, IGF-1, and Wnt 5a, which are factors that regulate cell proliferation and differentiation, particularly in the intestinal mesenchyme, as well as epithelial–mesenchymal interactions (Cervantes, Yamaguchi, & Hebrok, 2009; Geske, Zhang, Patel, Ornitz, & Stappenbeck, 2008; Wang et al., 2006; Williams, Fuller, Fagin, & Lund, 2002; Zhang et al., 2006).

Finally, the success of whole organ regeneration is determined not only by developmental processes at the level of discrete tissues, but also involves coordination and interaction between cellular and molecular events in individual components of the regenerating organ. For example, research on sea cucumbers has stressed the importance of timing and coordination of different events between tissue layers of the regenerating digestive tube (reviewed in García-Arrarás & Greenberg, 2001; Mashanov & García-Arrarás, 2011). The first structures to develop are the mesothelium and connective tissue of the new gut. These tissue layers initially form a solid rod-like structure, which only then is invaded by the luminal epithelium. Therefore, it is the outer, nonmucosal, layers of the gut wall that play the leading role during the early phases of regeneration in echinoderms. Depending on the organization and composition, the extracellular matrix (ECM) can either promote regeneration or suppress it. The importance of connective tissue remodeling for initializing the regrowth of the viscera was demonstrated by experimental inhibition of matrix metalloproteases, which either slowed down or completely abolished gut regeneration in holothurians, depending on the treatment protocol (Lamash & Dolmatov, 2013; Quiñones, Rosa, Ruiz, & García-Arrarás, 2002). On the other side of the spectrum is the observation that if acute posttraumatic inflammation of the mammalian bowel leads to formation of the fibrotic scar, regeneration is inhibited and the normal intestinal tissue architecture is permanently altered, resulting in inability of the affected region to resume normal organ function (Speca et al., 2012). Thus, since proper ECM remodeling is known to be required for the formation of various "new" functional organs (Daley, Peters, & Larsen, 2008), it is possible that regeneration-promoting ECM assemblage or degradation does not occur following resection or severe injury of the mammalian intestine.

The two contrasting properties of the mammalian digestive tube, namely, the lack of regeneration at the whole-organ level and high regenerative potential at the level of the mucosal epithelium, have long been taken into account by investigators in regenerative medicine and organ bioengineering. This has spurred the development of a variety of scaffolding materials to provide the neointestine with a three-dimensional supporting and organizing system to direct cell adhesion, proliferation, and proper differentiation. In addition, techniques have been developed that rely on seeding cells onto supporting scaffolds. Although it was possible to keep the bioengineered tissue viable, it has proved difficult to completely reproduce the organization of the entire organ. The engineered constructs often lacked critical components of the normal gut wall, such as nervous elements, blood and lymphatic vessels, and the muscular layer (Ross et al., 2013).

The gaps in the current knowledge about interactions between the epithelial cells and the underlying connective tissue layer are possibly one of the major obstacles for designing new bioengineering technologies (Ross et al., 2013). Further studies of the properties of the supporting scaffold that is formed during spontaneous gut regeneration in different animal groups can provide valuable insight into how posttraumatic remodeling of the bowel wall can be better managed in mammals and which properties should be enhanced or developed in artificial grafting materials.

4. CONCLUSIONS

Regeneration of the digestive epithelium in eumetazoans is accomplished through recruitment of cells from two types of sources. The first primary source is multi- or pluripotent stem cells, while the second one is reactivation of specialized cells. In both cases, the source cells can either reside within the luminal epithelium itself or be recruited from other tissues through migration and transdifferentiation. There is no clear correlation between the employment of either of these strategies and the phylogenetic position. Moreover, different strategies can be utilized within the same animal. Likewise, the capacity to regenerate the digestive tube does not necessarily correlate with the level of anatomical and histological complexity.

Although mammals are unable to regenerate lost regions of their digestive tube at the whole-organ level, there are reasons to think that their regenerative capacities can be significantly improved if the right conditions are found. First, even though mammals do not regrow the ablated segments of their gastrointestinal system after surgical resection, they still show a rudimentary regenerative response in the form of intestinal adaptation, which involves widening of the absorptive surface in the remnant bowel through extensive proliferation of intestinal stem and transit-amplifying cells resulting in increase in intestinal diameter and villus height (Bernal et al., 2005; Erwin et al., 2006). Second, regeneration of full thickness injuries or entire tubular segments of the digestive tube has been documented in other closely related groups, such as lower vertebrates, tunicates, and echinoderms, and, thus, occurs both in the chordate and nonchordate branches of the deuterostomian lineage. There is, therefore, a possibility that at least some of the components of the regenerative machinery might be evolutionarily conserved and have been preserved in mammals, although in a dormant state, as has been suggested to be the case in the central nervous system (Mashanov, Zueva, & García-Arrarás, 2013). Third, although the mammalian digestive tube does not regenerate at the whole-organ level, other endodermally derived organs (embryological derivatives of the digestive tube) retain a high regenerative capacity. These include the thymus, thyroid, liver, and the exocrine pancreas (Baddour, Sousounis, & Tsonis, 2012; Desai et al., 2007; Gilgenkrantz & Collin de l'Hortet, 2011). Therefore, whole-organ regeneration is not ubiquitously suppressed in the entire endodermal lineage of the adult mammals and possibly can be improved in the digestive tube *per se*.

Although a wide range of cellular mechanisms of visceral regeneration have evolved in animal phylogeny, at least some of the key components of the underlying molecular signaling machinery appear to be highly conserved. More comparative studies on spontaneously regenerating species are needed in order to distinguish between phylogenetically conserved and locally evolved components of the signaling cascades involved in regeneration of the digestive tube. This knowledge will help us understand if and how visceral injuries in poorly regenerating organisms, including humans, can be better managed and healed with minimal intervention.

ACKNOWLEDGMENTS

We thank Brigitte Galliot for her critical reading of the review, and members of the García-Arrarás lab for their contribution to understanding the phenomenon of intestinal regeneration in sea cucumbers. The work was supported by the NIH (Grants 1SC1GM084770-01, 1R03NS065275-01, 1R15NS081686-01), the NSF (Grant IOS-0842870), and the University of Puerto Rico.

REFERENCES

Apidianakis, Y., & Rahme, L. G. (2011). *Drosophila melanogaster* as a model for human intestinal infection and pathology. *Disease Models and Mechanisms, 4*, 21–30.

Ashton, G. H., Morton, J. P., Myant, K., Phesse, T. J., Ridgway, R. A., Marsh, V., et al. (2010). Focal adhesion kinase is required for intestinal regeneration and tumorigenesis downstream of Wnt/c-Myc signaling. *Developmental Cell, 19*, 259–269.

Baddour, J. A., Sousounis, K., & Tsonis, P. A. (2012). Organ repair and regeneration: An overview. *Birth Defects Research Part C, 96*, 1–29.

Barker, N., van Es, J. H., Kuipers, J., Kujala, P., van den Born, M., Cozijnsen, M., et al. (2007). Identification of stem cells in small intestine and colon by marker gene Lgr5. *Nature, 449*, 1003–1007.

Bely, A., & Nyberg, K. (2010). Evolution of animal regeneration: Re-emergence of a field. *Trends in Ecology & Evolution, 25*, 161–170.

Bernal, N. P., Stehr, W., Zhang, Y., Profitt, S., Erwin, C. R., & Warner, B. W. (2005). Evidence for active Wnt signaling during postresection intestinal adaptation. *Journal of Pediatric Surgeons, 40*, 1025–1029.

Biteau, B., Hochmuth, C. E., & Jasper, H. (2011). Maintaining tissue homeostasis: Dynamic control of somatic stem cell activity. *Cell Stem Cell, 9*, 402–411.
Bode, H. R. (1996). The interstitial cell lineage of hydra: A stem cell system that arose early in evolution. *Journal of Cell Science, 109*, 1155–1164.
Brockes, J. P., & Kumar, A. (2008). Comparative aspects of animal regeneration. *Annual Review of Cell and Developmental Biology, 24*, 525–549.
Buczacki, S. J. A., Zecchini, H. I., Nicholson, A. M., Russell, R., Vermeulen, L., Kemp, R., et al. (2013). Intestinal label-retaining cells are secretory precursors expressing Lgr5. *Nature, 495*, 65–69.
Byrne, M. (1986). Induction of evisceration in the holothurian *Eupentacta quinquesemita* and evidence for the existense of an endogenous evisceration factor. *The Journal of Experimental Biology, 120*, 25–39.
Carlson, B. M. (2007). *Principles of Regenerative Biology*. Amsterdam: Elsevier.
Cervantes, S., Yamaguchi, T. P., & Hebrok, M. (2009). Wnt5a is essential for intestinal elongation in mice. *Developmental Biology, 326*, 285–294.
Chen, M. K., & Beierle, E. A. (2004). Animal models for intestinal tissue engineering. *Biomaterials, 25*, 1675–1681.
Cordero, J. B., Stefanatos, R. K., Scopelliti, A., Vidal, M., & Sansom, O. J. (2012). Inducible progenitor-derived Wingless regulates adult midgut regeneration in *Drosophila*. *The EMBO Journal, 31*, 3901–3917.
Crosnier, C., Stamataki, D., & Lewis, J. (2006). Organizing cell renewal in the intestine: Stem cells, signals and combinatorial control. *Nature Reviews. Genetics*, 7, 349–359.
Daley, W. P., Peters, S. B., & Larsen, M. (2008). Extracellular matrix dynamics in development and regenerative medicine. *Journal of Cell Science, 121*, 255–264.
David, C. N. (2012). Interstitial stem cells in *Hydra*: Multipotency and decision-making. *The International Journal of Developmental Biology, 56*, 489–497.
Desai, B. M., Oliver-Krasinski, J., De Leon, D. D., Farzad, C., Hong, N., Leach, S. D., et al. (2007). Preexisting pancreatic acinar cells contribute to acinar cell, but not islet beta cell, regeneration. *The Journal of Clinical Investigation, 117*, 971–977.
Dolmatov, I. Y., & Mashanov, V. S. (2007). *Regeneration in holothurians*. Vladivostok: Dalnauka, (in Russian).
Doupé, D. P., & Jones, P. H. (2013). Cycling progenitors maintain epithelia while diverse cell types contribute to repair. *BioEssays, 35*, 443–451.
Erwin, C. R., Jarboe, M. D., Sartor, M. A., Medvedovic, M., Stringer, K. F., Warner, B. W., et al. (2006). Developmental characteristics of adapting mouse small intestine crypt cells. *Gastroenterology, 130*, 1324–1332.
Everhart, J. E., & Ruhl, C. E. (2009). Burden of digestive diseases in the United States part I: Overall and upper gastrointestinal diseases. *Gastroenterology, 136*, 376–386.
Feil, W., Lacy, E. R., Wong, Y. M., Burger, D., Wenzl, E., Starlinger, M., et al. (1989). Rapid epithelial restitution of human and rabbit colonic mucosa. *Gastroenterology, 97*, 685–701.
Feral, J., & Massin, C. (1982). Digestive system: Holothuroidea. In M. Jangoux & J. Lawrence (Eds.), *Echinoderm nutrition* (pp. 192–212). Rotterdam: Balkema.
Forsthoefel, D. J., James, N. P., Escobar, D. J., Stary, J. M., Vieira, A. P., Waters, F. A., et al. (2012). An RNAi screen reveals intestinal regulators of branching morphogenesis, differentiation, and stem cell proliferation in planarians. *Developmental Cell, 23*, 691–704.
Forsthoefel, D. J., Park, A. E., & Newmark, P. A. (2011). Stem cell-based growth, regeneration, and remodeling of the planarian intestine. *Developmental Biology, 356*, 445–459.
Frolova, L. T., & Dolmatov, I. Y. (2010). Microscopic anatomy of the digestive system in normal and regenerating specimens of the brittlestar *Amphipholis kochii*. *The Biological Bulletin, 218*, 303–316.

Fujimoto, K., Matsuura, K., Hu-Wang, E., Lu, R., & Shi, Y. (2012). Thyroid hormone activates protein arginine methyltransferase 1 expression by directly inducing c-Myc transcription during *Xenopus* intestinal stem cell development. *The Journal of Biological Chemistry, 287*, 10039–10050.

Fujiwara, S., Isozaki, T., Mori, K., & Kawamura, K. (2011). Expression and function of myc during asexual reproduction of the budding ascidian *Polyandrocarpa misakiensis*. *Development, Growth and Differentiation, 53*, 1004–1014.

Galligan, J. J., Furness, J. B., & Costa, M. (1989). Migration of the myoelectric complex after interruption of the myenteric plexus: Intestinal transection and regeneration of enteric nerves in the guinea pig. *Gastroenterology, 97*, 1135–1146.

Galliot, B. (2012). *Hydra*, a fruitful model system for 270 years. *The International Journal of Developmental Biology, 56*, 411–423.

Galliot, B., Welschof, M., Schuckert, O., Hoffmeister, S., & Schaller, H. C. (1995). The cAMP response element binding protein is involved in hydra regeneration. *Development, 121*, 1205–1216.

García-Arrarás, J. E., Estrada-Rodgers, L., Santiago, R., Torres, I. I., Díaz-Miranda, L., & Torres-Avillán, I. (1998). Cellular mechanisms of intestine regeneration in the sea cucumber, *Holothuria glaberrima* Selenka (Holothuroidea:Echinodermata). *The Journal of Experimental Zoology, 281*, 288–304.

García-Arrarás, J. E., & Greenberg, M. J. (2001). Visceral regeneration in holothurians. *Microscopy Research and Technique, 55*, 438–451.

García-Arrarás, J. E., Valentín-Tirado, G., Flores, J. E., Rosa, R. J., Rivera-Cruz, A., San Miguel-Ruiz, J. E., et al. (2011). Cell dedifferentiation and epithelial to mesenchymal transitions during intestinal regeneration in *H. glaberrima*. *BMC Developmental Biology, 11*, 61.

Garza-Garcia, A. A., Driscoll, P. C., & Brockes, J. P. (2010). Evidence for the local evolution of mechanisms underlying limb regeneration in salamanders. *Integrative and Comparative Biology, 50*, 528–535.

Geske, M. J., Zhang, X., Patel, K. K., Ornitz, D. M., & Stappenbeck, T. S. (2008). Fgf9 signaling regulates small intestinal elongation and mesenchymal development. *Development, 135*, 2959–2968.

Gilgenkrantz, H., & Collin de l'Hortet, A. (2011). New insights into liver regeneration. *Clinics and Research in Hepatology Gastroenterology, 35*, 623–629.

Goss, R. J. (1992). The evolution of regeneration: Adaptive or inherent? *Journal of Theoretical Biology, 159*, 241–260.

Gregorieff, A., Pinto, D., Begthel, H., Destrée, O., Kielman, M., & Clevers, H. (2005). Expression pattern of Wnt signaling components in the adult intestine. *Gastroenterology, 129*, 626–638.

Gupta, A., Dixit, A., Sales, K. M., Winslet, M. C., & Seifalian, A. M. (2006). Tissue engineering of small intestine—Current status. *Biomacromolecules, 7*, 2701–2709.

Hanani, M., Ledder, O., Yutkin, V., Abu-Dalu, R., Huang, T., Härtig, W., et al. (2003). Regeneration of myenteric plexus in the mouse colon after experimental denervation with benzalkonium chloride. *The Journal of Comparative Neurology, 462*, 315–327.

Hartl, M., Mitterstiller, A. M., Valoka, T., Breuker, K., Hobmayer, B., & Bister, K. (2010). Stem cell-specific activation of an ancestral myc proto- oncogene with conserved basic functions in the early metazoan *Hydra*. *Proceedings of the National Academy of Sciences of the United States of America, 107*, 4051–4056.

Hobmayer, B., Jenewein, M., Eder, D., Eder, M. K., Glasauer, S., Gufler, S., et al. (2012). Stemness in *Hydra*—A current perspective. *The International Journal of Developmental Biology, 56*, 509–517.

Hochmuth, C. E., Biteau, B., Bohmann, D., & Jasper, H. (2011). Redox regulation by Keap1 and Nrf2 controls intestinal stem cell proliferation in Drosophila. *Cell Stem Cell, 8*, 188–199.

Howell, J. C., & Wells, J. M. (2011). Generating intestinal tissue from stem cells: Potential for research and therapy. *Regenerative Medicine, 6*, 743–755.

Hyman, L. (1955). *The invertebrates. IV. Echinodermata. The celomate bilateria.* New York: McGraw-Hill Book Co. Inc.

Illa-Bochaca, I., & Montuenga, L. M. (2006). The regenerative nidi of the locust midgut as a model to study epithelial cell differentiation from stem cells. *The Journal of Experimental Biology, 209*, 2215–2223.

Ishizuya-Oka, A. (2007). Regeneration of the amphibian intestinal epithelium under the control of stem cell niche. *Development, Growth and Differentiation, 49*, 99–107.

Ishizuya-Oka, A., Hasebe, T., Buchholz, D. R., Kajita, M., Fu, L., & Shi, Y. (2009). Origin of the adult intestinal stem cells induced by thyroid hormone in *Xenopus laevis*. *The FASEB Journal, 23*, 2568–2575.

Jwo, S. C., Tang, S. J., Chen, J. R., Chiang, K. C., Huang, T. S., & Chen, H. Y. (2013). A novel model for simultaneous study of neointestinal regeneration and intestinal adaptation. *Wound Repair and Regeneration, 21*, 309–319.

Kaneko, N., Katsuyama, Y., Kawamura, K., & Fujiwara, S. (2010). Regeneration of the gut requires retinoic acid in the budding ascidian *Polyandrocarpa misakiensis*. *Development, Growth and Differentiation, 52*, 457–468.

Kawamura, K., Sugino, Y., Sunanaga, T., & Fujiwara, S. (2008). Multipotent epithelial cells in the process of regeneration and asexual reproduction in colonial tunicates. *Development, Growth and Differentiation, 50*, 1–11.

Körbling, M., Katz, R. L., Khanna, A., Ruifrok, A. C., Rondon, G., Albitar, M., et al. (2002). Hepatocytes and epithelial cells of donor origin in recipients of peripheral-blood stem cells. *The New England Journal of Medicine, 346*, 738–746.

Krause, D. S., Theise, N. D., Collector, M. I., Henegariu, O., Hwang, S., Gardner, R., et al. (2001). Multi-organ, multi-lineage engraftment by a single bone marrow-derived stem cell. *Cell, 105*, 369–377.

Lamash, N. E., & Dolmatov, I. Y. (2013). Proteases from the regenerating gut of the holothurian *Eupentacta fraudatrix*. *PLoS One, 8*, e58433.

Laranjeira, C., Sandgren, K., Kessaris, N., Richardson, W., Potocnik, A., Vanden Berghe, P., et al. (2011). Glial cells in the mouse enteric nervous system can undergo neurogenesis in response to injury. *The Journal of Clinical Investigation, 121*, 3412–3424.

Levin, D. E., & Grikscheit, T. C. (2012). Tissue-engineering of the gastrointestinal tract. *Current Opinion in Pediatrics, 24*, 365–370.

Liozner, L. D. (1982). *Regeneration and development.* Moscow: Nauka, (in Russian).

Mashanov, V. S., & Dolmatov, I. Y. (2004). Functional morphology of the developing alimentary canal in the holothurian *Eupentacta fraudatrix* (Holothuroidea, Dendrochirota). *Acta Zoologica, 85*, 29–39.

Mashanov, V. S., Dolmatov, I. Y., & Heinzeller, T. (2005). Transdifferentiation in holothurian gut regeneration. *The Biological Bulletin, 209*, 184–193.

Mashanov, V., Frolova, L., & Dolmatov, I. (2004). Structure of the digestive tube in the holothurian *Eupentacta fraudatrix* (Holothuroidea:Dendrochirota). *Russian Journal of Marine Biology, 30*, 314–322.

Mashanov, V. S., & García-Arrarás, J. E. (2011). Gut regeneration in holothurians: A snapshot of recent developments. *The Biological Bulletin, 221*, 93–109.

Mashanov, V. S., Zueva, O. R., & García-Arrarás, J. E. (2012). Expression of Wnt9, TCTP, and Bmp1/Tll in sea cucumber visceral regeneration. *Gene Expression Patterns, 12*, 24–35.

Mashanov, V. S., Zueva, O. R., & García-Arrarás, J. E. (2013). Radial glial cells play a key role in echinoderm neural regeneration. *BMC Biology*, *11*, 49.
Mashanov, V. S., Zueva, O. R., Rojas-Catagena, C., & García-Arrarás, J. E. (2010). Visceral regeneration in a sea cucumber involves extensive expression of survivin and mortalin homologs in the mesothelium. *BMC Developmental Biology*, *10*, 117.
Merlos-Suárez, A., Barriga, F. M., Jung, P., Iglesias, M., Céspedes, M. V., Rossell, D., et al. (2011). The intestinal stem cell signature identifies colorectal cancer stem cells and predicts disease relapse. *Cell Stem Cell*, *8*, 511–524.
Micchelli, C. A., & Perrimon, N. (2006). Evidence that stem cells reside in the adult Drosophila midgut epithelium. *Nature*, *439*, 475–479.
Mosher, C. (1956). Observation on evisceration and visceral regeneration in the sea cucumber Actinopyga agassizi Selenka. *Zoologica NY*, *41*, 17–26.
Mozzi, D., Dolmatov, I., Bonasoro, F., & Candia Carnevali, M. D. (2006). Visceral regeneration in the crinoid *Antedon mediterranea*: Basic mechanisms, tissues and cells involved in gut regrowth. *Central European Journal of Biology*, *1*, 609–635.
Nagai, H., Chew, S. H., Okazaki, Y., Funahashi, S., Namba, T., Kato, T., et al. (2013). Metamorphosis of mesothelial cells with active horizontal motility in tissue culture. *Scientific Reports*, *3*, 1144.
Odintsova, N., Dolmatov, I., & Mashanov, V. (2005). Regenerating holothurian tissues as a source of cells for long-time cell culture. *Marine Biology*, *146*, 915–921.
Ohlstein, B., & Spradling, A. (2006). The aduot Drosophila posterior midgut is maintained by pluripotent cells. *Nature*, *439*, 470–474.
Ohlstein, B., & Spradling, A. (2007). Multipotent Drosophila intestinal stem cells specify daughter cell fates by differential notch signaling. *Science*, *315*, 988–992.
Okamoto, R., Matsumoto, T., & Watanabe, M. (2006). Regeneration of the intestinal epithelia: Regulation of bone marrow-derived epithelial cell differentiation towards secretory lineage cells. *Human Cell*, *19*, 71–75.
Okamoto, R., Yajima, T., Yamazaki, M., Kanai, T., Mukai, M., Okamoto, S., et al. (2002). Damaged epithelia regenerated by bone marrow-derived cells in the human gastrointestinal tract. *Nature Medicine*, *8*, 1011–1017.
O'Steen, W. K. (1959). Regeneration and repair of the intestine in *Rana clamitans* larvae. *The Journal of Experimental Zoology*, *141*, 449–475.
O'Steen, W. K., & Walker, B. E. (1962). Radioautographic studies or regeneration in the common newt. III. Regeneration and repair of the intestine. *The Anatomical Record*, *142*, 179–187.
Plickert, G., Frnak, U., & Müller, W. A. (2012). *Hydractinia*, a pioneering model for stem cell biology and reprogramming somatic cells to pluripotency. *The International Journal of Developmental Biology*, *56*, 519–534.
Potten, C. S. (2004). Keratinocyte stem cells, label-retaining cells and possible genome protection mechanisms. *The Journal of Investigative Dermatology*, *9*, 183–195.
Quiñones, J. L., Rosa, R., Ruiz, D. L., & García-Arrarás, J. E. (2002). Extracellular matrix remodeling and metalloproteinase involvement during intestine regeneration in the sea cucumber *Holothuria glaberrima*. *Developmental Biology*, *250*, 181–197.
Radtke, F., & Clevers, H. (2005). Self-renewal and cancer of the gut: Two sides of a coin. *Science*, *307*, 1904–1909.
Reddien, P. W. (2013). Specialized progenitors and regeneration. *Development*, *140*, 951–957.
Ross, C. L., Booth, C., Sanders, B., Babbar, P., Bergman, C., Soker, T., et al. (2013). Regeneration and bioengineering of transplantable abdominal organs: Current status and future challenges. *Expert Opinion on Biological Therapy*, *13*, 103–113.
Roth, S., Franken, P., Sacchetti, A., Kremer, A., Anderson, K., Sansom, O., et al. (2012). Paneth cells in intestinal homeostasis and tissue injury. *PLoS One*, 7, e38965.

Sato, T., Vries, R. G., Snippert, H. J., van de Wetering, M., Barker, N., Stange, D. E., et al. (2009). Single Lgr5 stem cells build crypt-villus structures in vitro without a mesenchymal niche. *Nature*, *459*, 262–265.
Shaw, R. L., Kohlmaier, A., Polesello, C., Veelken, C., Edgar, B. A., & Tapon, N. (2010). The Hippo pathway regulates intestinal stem cell proliferation during *Drosophila* adult midgut regeneration. *Development*, *137*, 4147–4158.
Siebert, S., Anton-Erxleben, F., & Bosch, T. C. (2008). Cell type complexity in the basal metazoan *Hydra* is maintained by both stem cell based mechanisms and transdifferentiation. *Developmental Biology*, *313*, 13–24.
Speca, S., Giusti, I., Rieder, F., & Latella, G. (2012). Cellular and molecular mechanisms of intestinal fibrosis. *World Journal of Gastroenterology*, *18*, 3635–3661.
Sridharan, R., Tchieu, J., Mason, M. J., Yachechko, R., Kuoy, E., Horvath, S., et al. (2009). Role of the murine reprogramming factors in the induction of pluripotency. *Cell*, *136*, 364–377.
Sturm, A., & Dignass, A. U. (2008). Epithelial restitution and wound healing in inflammatory bowel disease. *World Journal of Gastroenterology*, *14*, 348–353.
Takahashi, K., & Yamanaka, S. (2006). Induction of pluripotent stem cells from mouse embryonic and adult fibroblast cultures by defined factors. *Cell*, *126*, 663–676.
Tian, H., Biehs, B., Warming, S., Leong, K. G., Rangell, L., Klein, O. D., et al. (2011). A reserve stem cell population in small intestine renders Lgr5-positive cells dispensable. *Nature*, *478*, 255–259.
van Es, J. H., Sato, T., van de Wetering, M., Lyubimova, A., Nee, A. N. Y., Gregorieff, A., et al. (2012). Dll1+ secretory progenitor cells revert to stem cells upon crypt damage. *Nature Cell Biology*, *14*, 1099–1104.
Wagner, D. E., Wang, I. E., & Reddien, P. W. (2011). Clonogenic neoblasts are pluripotent adult stem cells that underlie planarian regeneration. *Science*, *332*, 811–816.
Wang, Y., Wang, L., Iordanov, H., Swietlicki, E. A., Zheng, Q., Jiang, S., et al. (2006). Epimorphin(−/−) mice have increased intestinal growth, decreased susceptibility to dextran sodium sulfate colitis, and impaired spermatogenesis. *The Journal of Clinical Investigation*, *116*, 1535–1546.
Wilkie, I. C. (2001). Autotomy as a prelude to regeneration in echinoderms. *Microscopy Research and Technique*, *55*, 369–396.
Williams, K. L., Fuller, C. R., Fagin, J., & Lund, P. K. (2002). Mesenchymal IGF-I overexpression: Paracrine effects in the intestine, distinct from endocrine actions. *American Journal of Physiology. Gastrointestinal and Liver Physiology*, *283*, G875–G885.
Zhang, X., Stappenbeck, T. S., White, A. C., Lavine, K. J., Gordon, J. I., & Ornitz, D. M. (2006). Reciprocal epithelial-mesenchymal FGF signaling is required for cecal development. *Development*, *133*, 173–180.

CHAPTER EIGHT

Aging and Regeneration in Vertebrates

Konstantinos Sousounis*, Joelle A. Baddour†, Panagiotis A. Tsonis*,[1]
*Department of Biology and Center for Tissue Regeneration and Engineering, University of Dayton, Dayton, Ohio, USA
†Department of Chemical and Materials Engineering and Center for Tissue Regeneration and Engineering, University of Dayton, Dayton, Ohio, USA
[1]Corresponding author: e-mail address: ptsonis1@udayton.edu

Contents

Abstract

Aging is marked by changes that affect organs and resident stem cell function. Shorting of telomeres, DNA damage, oxidative stress, deregulation of genes and proteins, impaired cell–cell communication, and an altered systemic environment cause the eventual demise of cells. At the same time, reparative activities also decline. It is intriguing to correlate aging with the decline of regenerative abilities. Animal models with strong regenerative capabilities imply that aging processes might not be affecting regeneration. In this review, we selectively present age-dependent changes in stem/progenitor cells that are vital for tissue homeostasis and repair. In addition, the aging effect on regeneration following injury in organs such as lung, skeletal muscle, heart, nervous system, cochlear hair, lens, and liver are discussed. These tissues are also known

Current Topics in Developmental Biology, Volume 108
ISSN 0070-2153
http://dx.doi.org/10.1016/B978-0-12-391498-9.00008-5

for diseases such as heart attack, stroke, cognitive impairment, cataract, and hearing loss that occur mostly during aging in humans. Conclusively, vertebrate regeneration declines with age with the loss of stem/progenitor cell function. Future studies on improving the function of stem cells, along with studies in fish and amphibians where regeneration does not decline with age, will undoubtedly provide insights into both processes.

ABBREVIATIONS

BM bone marrow-derived
C/EBP CCAAT/enhancer-binding protein
EPCs endothelial progenitor cells
FGF fibroblast growth factor
HSCs hematopoietic stem cells
IGF insulin-like growth factor
MAPK mitogen-activated protein kinase
MSCs mesenchymal stem cells
NSCs neural stem cells
ROS reactive oxygen species
TGF transforming growth factor
VEGF vascular endothelial growth factor

1. AGING AND REGENERATION

It is generally accepted that aging is the gradual result of increased DNA damage, deregulation of gene expression and protein function, disruption of cell matrix and organelles, oxidative stress, and disrupted metabolism (Mattson & Magnus, 2006). Consequently, organ homeostasis and function are impaired. The aging process seems to be well programmed and all these changes take place according to the life span of organisms (Herbig, Jobling, Chen, Chen, & Sedivy, 2004; Vaziri & Benchimol, 1996). Among all these changes, it must be noted that reparative processes are also affected by aging. The players in this process are stem cells. These cells are supposed to repair damaged tissues, but as they are also subject to aging, their contribution to tissue damage repair diminishes as we age. On the other hand, regeneration in some animals is quite remarkable and seems to defy aging. Some invertebrates can regenerate entire body segments, while some vertebrates, such as amphibians, do so as well. Regeneration in these animals is made possible by reserved progenitor cells or even by transdifferentiation of somatic cells, which assumably revert to a stem cell-like status. In addition, recent research has indicated that regenerative abilities in newt and zebrafish do not decline with age (Eguchi et al., 2011; Itou, Kawakami, Burgoyne, & Kawakami, 2012). Thus, regeneration research can shed some light on the mechanisms of aging and vice versa.

In this review we summarize the effects of aging on stem/progenitor cells in the context of organ homeostasis and regeneration following injury in vertebrates. We focus on research that most profoundly demonstrates these effects in aged humans or animal models. We also present findings on reversal of the effects of aging resulting in better stem/progenitor cell function and regeneration.

2. EFFECTS OF AGING ON STEM CELLS AND CELL TURNOVER

2.1. Endothelial progenitor cells

Endothelial progenitor cells (EPCs), derived mainly from bone marrow, circulate in the blood and contribute to the repair of vascular tissue. Telomerase activity and telomere length are decreased, while apoptosis-related genes are increased in aged human EPCs (Kushner et al., 2009, 2011). In addition, tumor suppressor genes including $p16^{ink4a}/p19^{arf}$ (coding for cyclin-dependent kinase inhibitor 2A isoforms; CDKN2A) are upregulated because of an age-dependent downregulation of high mobility group protein HMGI-C (*Hmga2)* modulated by microRNA-10A and microRNA-21 (Zhu et al., 2013). Downregulation of these microRNAs could rescue the aging effects (Zhu et al., 2013). Furthermore, while human EPCs have a reduced expression level and activity of glutathione peroxidase 1 (GPX1) that renders them more sensitive to oxidative stress (He, Joyner, & Katusic, 2009), mice show a cytochrome b-245(Nox2)-dependent increase in reactive oxygen species (ROS) that can account for the loss of function of EPCs (Turgeon et al., 2012). In humans, the number of circulating EPCs is not reduced with age; however, their ability to incorporate during repair becomes limited (Heiss et al., 2005) because of changes in heparan sulfate in their membrane (Williamson et al., 2013). In mice, studies have shown that the ability of aged EPCs to protect against atherosclerosis, a disease that directly involves dysfunctional vascular tissue, is reduced (Rauscher et al., 2003). Impairment of the ability to recruit EPCs in the aged wound site is attributed to a reduced expression of stromal cell-derived factor 1 (SDF1) and vascular endothelial growth factor (VEGF) (Chang et al., 2007). In addition, insulin-like growth factor 1 (IGF1) expression is reduced during aging while its reintroduction can improve the function of the EPCs

(Thum et al., 2007). Other ways of improving the function of EPCs include transplantation of young bone marrow (Edelberg, Tang, Hattori, Lyden, & Rafii, 2002), physical exercise (Hoetzer et al., 2007; Xia et al., 2011; Yang et al., 2011), shear stress (Xia et al., 2012), and diet (Marin, Yubero-Serrano, Lopez-Miranda, & Perez-Jimenez, 2013), all of which contribute to enhanced migration, engraftment, and differentiation of the cells (for a review on EPCs, see also Felice, Barsotti, Poredos, Balbarini, & Di Stefano, 2013; Williamson, Stringer, & Alexander, 2012).

2.2. Hematopoietic stem cells

Hematopoietic stem cells (HSCs) are found mainly in the bone marrow and have the ability to differentiate to all blood cell lineages, including the myeloid and the lymphoid. During aging, HSCs experience a decline in function, epigenetic deregulation (Chambers et al., 2007), proliferation-dependent DNA methylation changes (Beerman et al., 2013) such as hypomethylation of differentiation genes (Bocker et al., 2011), reduced proliferation capability (de Haan, Nijhof, & Van Zant, 1997), telomere shortening (Vaziri et al., 1994), impaired DNA repair mechanisms (Nijnik et al., 2007; Rossi et al., 2007), and oxidative damage (Yahata et al., 2011), all of which affect their ability to self-renew and function. Mutations in the mitochondrial DNA have also been implicated in HSC aging (Norddahl et al., 2011; Yao et al., 2013). Moreover, upregulation of tumor suppressors has been shown to be responsible for HSC exhaustion. The activity of the tumor suppressor protein p53 (TP53) has been implicated in the rate of HSC aging (Dumble et al., 2007) and the regulation of cell quiescence (Liu et al., 2009), whereas *p53* null mice have been shown to have more functional HSC (TeKippe, Harrison, & Chen, 2003). Furthermore, seine/threonine-protein kinase mTOR (mTOR) signaling was found to be upregulated during aging and is associated with the activation of cyclin-dependent kinase inhibitors including CDKN1A and CDKN2A isoforms coded by $p21^{cip1}$, $p16^{ink4a}$, and $p19^{arf}$. In contrast, the decline in HSC function can be restored in the presence of two mTOR inhibitors rapamycin (Chen, Liu, & Zheng, 2009) and hamartin (TSC1) (Gan et al., 2008), or in the absence of Cdkn2a in $p16^{ink4a}$ null mice (Janzen et al., 2006). PTEN/PI3K/Akt/mTOR interplay has also been found to be responsible for regulating many aspects of HSC function (Magee et al., 2012). Phosphatidylinositol 3,4,5-triphosphate 3-phosphatase and dual-specificity protein phosphatase PTEN (PTEN) negatively regulate the PI3K/Akt pathway, which activates mTOR. Deletion of PTEN results in activation of mTOR (Lee et al., 2010) and HSC exhaustion (Zhang et al., 2006). Another downstream target of the PI3K/Akt pathway is Foxo3a, a forkhead transcriptional factor that has been shown to maintain HSC function after ROS-induced stress (Miyamoto et al., 2007). *Foxo3a* null mice show increased phosphorylation of p38 mitogen-activated protein kinase (MAPK) (Miyamoto et al., 2007), which results in elevated ROS, leading to HSC pool exhaustion (Ito et al., 2006) and increased expression of tumor suppressor genes including *p53* and *p21* (Yalcin et al., 2008). FoxO genes, in general, have been shown to play a role in oxidative stress in HSCs (Tothova et al., 2007). Foxo3 also upregulates the expression of serine-protein kinase ATM (ATM) (Yalcin et al., 2008), which regulates ROS (Ito et al., 2004) to maintain the normal self-renewal ability of HSCs. In addition, ATM phorsphorylates BH3-interacting domain death agonist (BID), which protects against irradiation and oxidative stress (Maryanovich et al., 2012). Normal HSC function under oxidative stress is maintained by E3 ubiquitin-protein ligase MDM2 (MDM2), which represses the activity of TP53 under these conditions (Abbas et al., 2010).

Also found to conserve normal HSC function under stress is NAD-dependent protein deacetylase sirtuin-1 (Sirt1), which protects the genomic content of the cell and regulates transcription (Singh et al., 2013). In addition, Sirt3 improves the function of aged HSCs (Brown et al., 2013), and both Sirt1 and Sirt3 are downregulated during aging. Ehz1, a histone-lysine N-methyltransferase, also maintains the ability of HSCs for proliferation (Hidalgo et al., 2012). Consistent with these results are the epigenetic changes that are found to be responsible for HSC aging. Wahlestedt and colleagues reprogrammed aged HSCs to induced pluripotent stem cells and redifferentiated them back to HSCs with functions that were found to be comparable to that of young cells (Wahlestedt et al., 2013). Furthermore, DNA methylation has been found to contribute to HSC aging by reducing cell proliferation (Beerman et al., 2013).

Other proteins that play a role in HSC function include Bmi-1 (Park et al., 2003) and Hif1a (Takubo et al., 2010), both of which regulate the tumor suppressor Cdkn2a. Atg7 (Mortensen et al., 2011) and Fip200 (Liu et al., 2010) play a role in the Foxo3a-dependent autophagy mechanism of the cells (Warr et al., 2013). Furthermore, a number of cytokines such as granulocyte colony-stimulating factor (G-CSF) have been shown to regulate the function of HSCs, and are deregulated by aging (Ju et al., 2007; Zhang & Lodish, 2008). Finally, dietary restriction (Chen, Astle, & Harrison, 2003) and telomerase induction (Allsopp, Morin, DePinho, Harley, & Weissman, 2003) can improve HSC function.

2.3. Muscle stem cells

Muscle stem cells, also known as satellite cells, reside in muscle tissue and contribute to muscle tissue renewal. The telomerase activity of human muscle stem cells is seized after birth, and the shortening of the telomere with age results in impaired proliferation (Decary et al., 1997), a process supported by CDKN2A (Zhu et al., 2007). Muscle stem cells also experience reduced proliferation (Decary et al., 1997; Zwetsloot, Childs, Gilpin, & Booth, 2013), reduced oxidative metabolism (Fulle et al., 2005), increased apoptosis (Jejurikar et al., 2006; Marzetti et al., 2008), reduced self-renewal (Day, Shefer, Shearer, & Yablonka-Reuveni, 2010), and reduced cell number (Gibson & Schultz, 1983; Renault, Thornell, Eriksson, Butler-Browne, & Mouly, 2002) during aging. Notch, transforming growth factor beta (TGF-beta), Wnt, and fibroblast growth factor (FGF) pathways have been shown to interact in order to maintain muscle stem cells' self-renewal ability and function. Notch signaling is important for the maintenance of muscle stem cells' quiescent state (Bjornson et al., 2012), whereas a balanced interplay between TGF-beta and Notch signaling pathways is essential for the cells' function (Carlson, Hsu, & Conboy, 2008). Cell quiescence is also regulated by FGF2, which is suppressed by Sprouty1 during aging (Chakkalakal, Jones, Basson, & Brack, 2012). Wnt signaling is increased during aging, leading to changes in muscle stem cell function and increased fibrosis (Brack et al., 2007). Increased fibrosis has been attributed to the age-dependent increase in TGF-beta signaling (Beggs et al., 2004). In addition, aged muscle stem cells show a reduced expression of pluripotency factors that contribute to their ability to proliferate and differentiate (Beccafico et al., 2011; Carlson & Conboy, 2007). Muscle stem cell function can be improved with caloric restriction (Cerletti, Jang, Finley, Haigis, & Wagers, 2012), IGF1 treatment (Barton-Davis, Shoturma, Musaro, Rosenthal, & Sweeney, 1998; Chakravarthy, Davis, & Booth, 2000), testosterone treatment (Kovacheva, Hikim, Shen, Sinha, & Sinha-Hikim, 2010), and exercise (Shefer, Rauner, Yablonka-Reuveni, & Benayahu, 2010). Improvement of cell function is due to regulation of age-related effects such as regulation of Notch and Akt pathway, increase in muscle stem cell number, and reduction of apoptosis and inflammation (Phillips & Leeuwenburgh, 2005). Most of these concepts have implications in regeneration of aged muscle tissue after injury (see next section).

2.4. Neural stem cells

Neural stem cells (NSCs) reside in the adult brain from where they can contribute to neurogenesis. During aging, NSC are reduced in number (Maslov,

Barone, Plunkett, & Pruitt, 2004), and they have a reduced ability for oxidative metabolism and an increased genome instability, with shorter telomeres that impair, in a p53-dependent fashion, proliferation, neurogenesis, and function (Ferron et al., 2009, 2004; Kuhn, Dickinson-Anson, & Gage, 1996; Stoll et al., 2011). Proliferation is also affected by an age-dependent decline in FGF2, IGF1, and VEGF (Shetty, Hattiangady, & Shetty, 2005), which, along with epidermal growth factor (EGF), are essential for NSCs (Tropepe et al., 1999). In addition, TGF-beta1 is increased during aging and contributes to NSC apoptosis (Pineda et al., 2013). Reversely, the introduction of FGF2, EGF, and brain-derived neurotrophic factor (BDNF), as well as the repression of TGF-beta1, can improve the function of aged NSCs (Jin et al., 2003; Pineda et al., 2013; Yang, Qiao, Sun, & Li, 2012). A study by Villeda and colleagues showed through mice parabiosis that an aged systemic environment can decrease NSC function (Villeda et al., 2011). Furthermore, let-7 microRNA, Hmga2, and Cdkn2a have been shown to regulate NSC function during aging. Age-dependent upregulation of let-7 microRNA represses Hmga2, a protein that promotes NSC function (Nishino, Kim, Chada, & Morrison, 2008) by repressing $p16^{ink4a}$ (codes for Cdkn2a), which, in turn, negatively affects NSC number and neurogenesis potential (Molofsky et al., 2006). Bmi-1 acts upstream of Cdkn2a by repressing it and also improves NSC function (Molofsky, He, Bydon, Morrison, & Pardal, 2005).

NSC survival is regulated by the transcriptional factor Sox-2, which has been shown to play an antiapoptotic role by regulating survivin expression (Feng et al., 2013). Other factors that contribute to NSC function are FoxOs (Paik et al., 2009; Renault et al., 2009), mTOR (Paliouras et al., 2012), p57 (Furutachi, Matsumoto, Nakayama, & Gotoh, 2013), ATP (Suyama et al., 2012), PR domain zinc finger protein 16 (Prdm16) (Chuikov, Levi, Smith, & Morrison, 2010), choline acetyltransferase (Park et al., 2013), and Wnt signaling (Lie et al., 2005; Miranda et al., 2012). Aged NSC function can be restored by inactivation of Dickkopf-related protein 1 (DKK1), a negative regulator of Wnt signaling (Seib et al., 2013). In addition, dietary restriction has also been found to improve NSC function (Lee, Duan, Long, Ingram, & Mattson, 2000).

2.5. Mesenchymal stem cells

Mesenchymal stem cells (MSCs) include many cell types that reside in the mesenchyme. The most studied among them are bone marrow-, adipose-, and umbilical cord-derived MSCs. These cells differentiate to many different cell types and contribute to the repair of damaged tissues (Baddour, Sousounis, & Tsonis, 2012). During aging, bone marrow-derived MSCs (BM-MSCs) are reduced in number, less proliferative, impaired in their differentiation potential, and show an increased expression of tumor suppressors and apoptosis genes such as *p53*, *p21,* and *BAX* (Zheng, Martin, Duwayri, Falcon, & Buckwalter, 2007; Zhou et al., 2008). The niche where the cells reside is important for their function (Chen, 2010). It has been shown that FGF2 treatment maintains the chondrogenic potential of BM-MSCs in prolonged cultures (Solchaga, Penick, Goldberg, Caplan, & Welter, 2010), in addition to promoting cell growth by the downregulation of tumor suppressor genes including *p53*, $p21^{Cip1}$ (codes for CDKN1A), and $p16^{ink4a}$ (codes for CDKN2A) and the time-dependent expression of TGF-beta2 (Ito, Sawada, Fujiwara, Seyama, & Tsuchiya, 2007). Similarly, FGF2 and FGF receptors 1/2 have been shown to regulate proliferation and function of bone MSCs (Coutu, Francois, & Galipeau, 2011; Ou et al., 2010). Interestingly, introduction of the homeobox protein and pluripotency factor Nanog to aged BM-MSCs can improve their proliferation and differentiation potential (Han et al., 2012). Similarly, SOX2 has been found to be important for osteoblasts' function (Basu-Roy et al., 2010). Osteoblast and osteoclast functions are associated with estrogen through their receptor, an effect that declines with age, especially in postmenopausal women, which can lead to osteoporosis (empty bones). Estrogen treatment can rescue this effect (Cortet, 2011; Jakob et al., 2013).

Adipose-derived MSCs are subjected to similar aging-induced changes. Their number and differentiation potential are reduced, while tumor suppressor genes including *CHEK1* and $p16^{ink4a}$ are upregulated (Alt et al., 2012). During aging, there is also a decrease in adipogenesis via interruption of the CCAAT/enhancer-binding protein (C/EBP) family gene expression (Karagiannides et al., 2001), as well as impaired osteoblastic (de Girolamo et al., 2009) and angiogenic differentiation, and proliferation (Efimenko, Starostina, Kalinina, & Stolzing, 2011; Madonna et al., 2011). Furthermore, DNA methylation and repressive histone modifications (Schellenberg et al., 2011), in addition to caloric restriction (Schmuck, Mulligan, & Saupe, 2011), are involved in the decreased proliferation of adipose-derived MSCs. On the other hand, the ROS-dependent impaired angiogenic capability of aged adipose-derived MSCs can be restored after hypoxic preconditioning (De Barros et al., 2013).

3. AGE-DEPENDENT CHANGES OF REGENERATION POTENTIAL AFTER INJURY

3.1. Lung

Lungs are the organs that aid gas exchange. Lung regeneration is usually studied by treatment with bleomycin, naphthalene, viruses or other infectious-related reagents, and pneumonectomy. Failure to regenerate following injury often causes pulmonary fibrosis. During aging, bleomycin-induced (Xu et al., 2009) or virus-induced (Torres-Gonzalez et al., 2012) lung injury causes increased fibrosis and apoptosis. Regeneration of the lung after penumonectomy is successful in young mice, incomplete in middle-aged ones, and absent in old mice. These differences were related to myofibroblast differentiation (Paxson et al., 2011). Impaired regeneration after naphthalene treatment has been attributed to senescence of Clara cells and increased p38 MAPK-dependent inflammation (Zhou, Onizawa, Nagai, & Aoshiba, 2011), which are common effects of lung aging (Aoshiba & Nagai, 2009). The roles of aging in lung regeneration have been discussed in studies on lung disease, smoking, and inflammation (Gan, Man, & Sin, 2005; Kapetanaki, Mora, & Rojas, 2013; Teramoto & Ishii, 2007), though the direct effect of these aspects on the regeneration potential of aged injured lungs requires further investigation.

3.2. Skeletal muscle and heart

Muscle has the ability to contract and, as such, mediates the movement of organs, including the blood pumping of the heart. Skeletal muscle and heart tissue regeneration are usually studied by subjecting the tissue to mechanical or chemical injury. Muscle stem cells contribute to the regeneration of mammalian skeletal muscle. Research on the effects of aging on muscle regeneration is primarily conducted because of the importance of the ability of aged muscle stem cells to retain their regenerative capability in the context of function improvement. Aging affects muscle stem cells (as discussed earlier) by reducing their function and, as such, by impairing their regenerative capacity. Muscle stem cells from aged mice show a delayed proliferation after mechanical injury (McGeachie & Grounds, 1995). Similarly, the thyroarytenoid muscle regeneration in rats has been found to decline with age (Lee et al., 2012). Regulation of TGF-beta1, Wnt, complement C1q, and Notch pathways are found to be important for aged muscle stem cells and their regeneration ability after injury. Elevated TGF-beta1 contributes to muscle fibrosis, whereas its inhibition enhances aged muscle regeneration after cardiotoxin damage (Carlson et al., 2009). In addition, induction of Notch pathway, a pathway that has TGF-beta antagonistic effects (Carlson et al., 2008), can improve regeneration of aged muscle (Conboy, Conboy, Smythe, & Rando, 2003). Complement C1q has been shown to activate Wnt signaling and reduces aged muscle regeneration, an effect that can be reversed by C1q blocking (Naito et al., 2012). Moreover, induced IGF1 expression (Musaro et al., 2001) or testosterone treatment (Serra et al., 2013) in the muscle enhances regeneration in aged animals, while inhibition of osteopontin enhances muscle regeneration after cardiotoxin-induced injury (Paliwal, Pishesha, Wijaya, & Conboy, 2012). Notably, transplantation experiments in old to young animals clearly indicate that a young or aged systemic environment contributes to the regeneration potential of the muscle (Carlson & Faulkner, 1989; Gutmann & Carlson, 1976; Smythe et al., 2008). In humans, it has recently been shown that reduced exercise contributes to a reduced regeneration potential of aged muscle (Buford et al., 2013). Muscle regeneration in relation to aged muscle stem cell function has been discussed in recent reviews (Barberi et al., 2013; Garcia-Prat, Sousa-Victor, & Munoz-Canoves, 2013; Jang, Sinha, Cerletti, Dall'Osso, & Wagers, 2011).

The mammalian heart has the ability to regenerate after injury, as shown in neonatal mice, but this ability is lost when the animal is 1 week old (Porrello et al., 2011). Regeneration in neonatal mice is carried out by the preexisting cardiomyocytes as during tissue homeostasis (Senyo et al., 2013). Since the regeneration ability of mice is lost early in their early, studies have been focusing on reducing the effects of aging on vascular tissue along with reducing fibrosis after injury, which can cause heart failure. Cardiomyocytes, cardiac stem cells, EPCs, and MSCs contribute to the homeostasis of the heart as well as its diseased state. Studies that enhance the function of these cells also increase the heart function following stress (for reviews, see Ballard, 2010; Mohsin, Siddiqi, Collins, & Sussman, 2011; Jakob et al., 2012). Notably, a membrane-bound stem cell factor expressed in the heart can enhance its function in aged mice after myocardial infarction (Sun et al., 2012). In addition, a recent report by Boon and colleagues showed an increased regeneration potential of the heart by inhibition of the age-dependent increase in microRNA-34 expression (Boon et al., 2013). In contrast to mammals, the zebrafish has been shown to retain the regeneration ability of the heart throughout its life (Itou et al., 2012).

3.3. Nervous system

Neurons transfer electrical signals mainly to and from the brain and the spinal cord. Aging affects this process, leading to several diseases including impairment of cognition and movement. Interestingly, recent studies have shown that aging in the nervous system is programmed, and this might also be the case for regeneration (Zou et al., 2013). There are several reports showing a reduced regeneration capability of the nervous system including the chorda tympani nerve (Saito, Yamada, Okamoto, & Narita, 2012), the spinal cord (Jaerve, Kruse, Malik, Hartung, & Muller, 2012), motorneurons (Yuan et al., 2012), and the gustatory nerve (He, Yadgarov, Sharif, & McCluskey, 2012) following injury. In addition, a stroke damages the brain, especially during aging, and several studies have linked the regeneration ability after a stroke to age and potential therapeutic approaches (see review Buga, Vintilescu, Pop, & Popa-Wagner, 2011, and references therein). The regeneration potential can be affected by the type of damage that is incurred by the nervous system (Rupp, Cunningham, Yao, Furukawa, & Willison, 2013) and the ability of stem/progenitor cells to maintain their proper functions (aspects discussed previously). A young systemic environment (Ruckh et al., 2012), antiscarring treatment (Jaerve et al., 2012), expression of EphrinB3 (Duffy et al., 2012), and FGF treatment (Hirano, Tateya, Kishimoto, Kanemaru, & Ito, 2012) have all been shown to enhance nervous system regeneration during aging.

3.4. Cochlear hair cell

The cochlea mediates hearing ability. Sensory hair as well as supporting cells in the organ of Corti located in the cochlea is responsible for the conversion of sound. In mammals, hair cells lose their regenerative capability early in development (Walters & Zuo, 2013), whereas newts, fishes, and birds retain this ability throughout adulthood (Brignull, Raible, & Stone, 2009; Taylor & Forge, 2005). The focus in this field has been to increase the window that hair cells can regenerate in mammals, and to study the mechanism of regeneration in fishes and birds. The regeneration capability of hair cells has been attributed to stem/progenitor cells in the cochlea and the transdifferentiation of supporting cells. Methods to injure the inner ear include ototoxins, sound, and mechanical manipulation. Factors that improve proliferation and regeneration of mammalian hair cells include inhibition of $p27^{Kip1}$ (Chen & Segil, 1999; Lowenheim et al., 1999), retinoblastoma (*Rb*) (Weber et al., 2008; Yu et al., 2010), retinoblastoma-like protein 2

(*Rbl2)* (Rocha-Sanchez et al., 2011), $p21^{Cip1}$ and $p19^{Ink4d}$ (Laine et al., 2007), and overexpression of beta-catenin (Chai et al., 2012) and protein atonal homolog 1 (Atoh1) (Kelly, Chang, Pan, Lin, & Chen, 2012). However, in the animal models, these effects do not extend to more than one to three months of age (Huang et al., 2011; Liu et al., 2012, 2012; Oesterle, Chien, Campbell, Nellimarla, & Fero, 2011). Studies on fishes and birds, as well as *in vitro* mammalian cultures from the inner ear implicate FGF (Pickles & van Heumen, 1997; Sweet, Vemaraju, & Riley, 2011), EGF receptor (White, Stone, Groves, & Segil, 2012), and retinoic acid signaling (Kelley, Xu, Wagner, Warchol, & Corwin, 1993; Lee & Cotanche, 1996; Lefebvre, Malgrange, Staecker, Moonen, & Van de Water, 1993) in hair cell regeneration. These pathways are seized early after birth in mammals (Hayashi, Ray, Younkins, & Bermingham-McDonogh, 2010; Kelley et al., 1993; Luo, Koutnouyan, Baird, & Ryan, 1993; Zine, Nyffeler, & de Ribaupierre, 2000). Other potential mechanisms and approaches for hair regeneration are discussed in this recent review (Walters & Zuo, 2013).

3.5. Lens

The eye lens is a clear organ that mediates vision by focusing the incoming light to areas of the retina. The continuous exposure of the lens to the UV light from the sun makes it very vulnerable to aging effects. The most common human lens disease is cataract or the opacification of the lens, whereby light can no longer pass through and the lens appears cloudy. Many signaling pathways and genes, especially crystallins, have been implicated in cataract formation as a result of age (Dawes, Duncan, & Wormstone, 2013; Hejtmancik & Kantorow, 2004; Sousounis & Tsonis, 2012). While the mammalian lens cannot regenerate if fully removed, amphibian lens regeneration remains the area of focus of most studies. Frogs can regenerate the lens in a small developmental window when they are tadpoles (Henry & Tsonis, 2010). Similarly, a recent report shows that axololts can regenerate the lens during a period of 2 weeks posthatching although regeneration is seized thereafter (Suetsugu-Maki et al., 2012). The most exciting results came from a long-lasting experiment on the red-spotted newt, *Notophthalmus viridescens*; Eguchi and colleagues performed a 16-year experiment in which the lens was removed eighteen times from the same animals and was regenerated every time without any sign of cataract. In addition, the lenses from these aged animals were compared to those of young ones and, strikingly and in contrast to other organs discussed earlier, they were similar

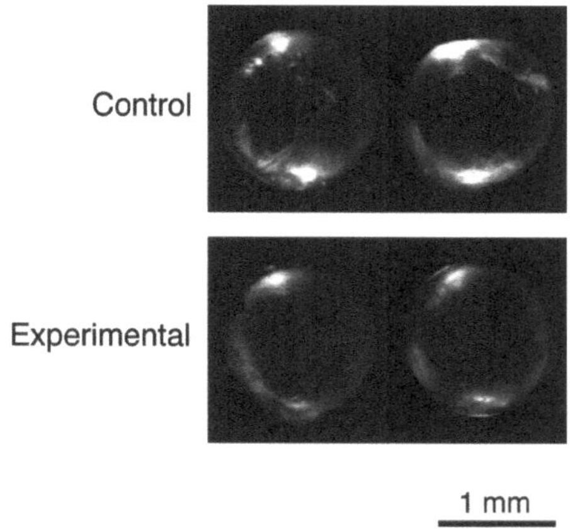

Figure 8.1 Newt lens regeneration during aging. Control newt lenses taken from 14-year-old newts are similar in size, shape, and morphology to experimental newt lenses obtained from approximately 30-year=old newts that had regenerated the lens 17 times over a period of 16 years. Bar indicates size. (See color plate.)

in structure and gene expression (Fig. 8.1) (Eguchi et al., 2011). The transcriptome of the aged lenses from these animals is an ongoing investigation; nevertheless, recent studies on gene expression during newt lens regeneration show upregulation of DNA repair and ROS-related genes, which might counteract the effects of aging at the site where regeneration occurs (Sousounis et al., 2013a, 2013b).

3.6. Liver

Liver is one of the best known organs having the ability to regenerate after injury, even in humans. Liver has a variety of functions including detoxification and lipid catabolism. Mammalian liver regeneration declines with age (Biondo-Simoes Mde et al., 2006; Timchenko, 2009). This age-dependent delay in regeneration has been attributed to the reduced proliferation and function of EGF, EGF receptors, and ERK (Hutter et al., 2000; Ishigami, Reed, & Roth, 1993; Kamat et al., 2008; Marti, 1993; Palmer, Tuzon, & Paulson, 1999; Sawada, 1989) (for review, see Schmucker & Sanchez, 2011). In addition, $p66^{Shc}$ has recently been implicated in the age-dependent decline in liver regeneration by inducing oxidative stress and apoptosis following injury (Haga et al., 2010). Increased hepatocyte proliferation has been achieved with expression of FoxM1B (Wang et al., 2001), a transcriptional factor that downregulates tumor suppressor $p27^{Kip1}$ and activates cell cycle regulator Cdc25B (Wang et al., 2002). Growth hormone (Krupczak-Hollis, Wang, Dennewitz, & Costa, 2003) and Farnesoid X receptor (Chen et al., 2010) have been shown to regulate FoxM1B and, subsequently, they can improve proliferation of cells in the aged liver. Moreover, serotonin acting through the VEGF pathway (Furrer et al., 2011) and ablation of interferon gamma signaling, and the subsequent reduction in inflammation (Singh, Goode, Dean, Awad, & Darlington, 2011) have showed enhanced liver regeneration during aging. In addition, the age-dependent repression of Sirt1 by C/EBPbeta-HDAC1 complex has been shown to play a role in the impaired ability of the liver to regenerate, while expression of Sirt1 can reverse the effect (Jin, Iakova, Jiang, Medrano, & Timchenko, 2011).

4. CONCLUDING REMARKS

Regeneration and aging are two vastly opposite processes. Adult stem cells are subjected to the effects of aging, which limit their ability to self-renew, differentiate, and maintain tissue homeostasis, leading to loss of function (Conboy & Rando, 2012; Signer & Morrison, 2013). Short telomeres, oxidative stress, DNA damage, impaired metabolism, and deregulation of gene expression are all "hallmarks" of aging that account for the impaired stem/progenitor cell function (Lopez-Otin, Blasco, Partridge, Serrano, & Kroemer, 2013). On the other hand, regeneration requires cells to proliferate and differentiate to the lost cell types. This process requires increased metabolism, regulation of tumor suppressors, and regulation of the cell cycle, all the while the cells having to tackle the DNA damage and ROS stress that is generated. We conclude by highlighting some major concepts (Fig. 8.2).

Cell division is necessary for cell replacement and regeneration. In every cycle, the telomeres are reduced in length, an effect that imposes a direct age-dependent limitation. In the absence of telomerase, DNA integrity is compromised and division is stopped. Most of the aged tissues reviewed here have short telomeres during aging with demonstrated negative effects on stem/progenitor cell function in several tissues (Hao et al., 2005). Strikingly, activation of telomerase in aged animal models reversed the aging effects (Jaskelioff et al., 2011), showing the importance of telomeres and DNA integrity during aging.

As time passes and the organism advances in age, there begins a battle between maintaining the proliferation and differentiation ability of stem/progenitor cells and pool exhaustion and impaired tissue turnover. These events are modulated by tumor suppressor genes, with the most important being the interplay between TP53, CDKN1A, CDKN2A, ATM, FOXOs,

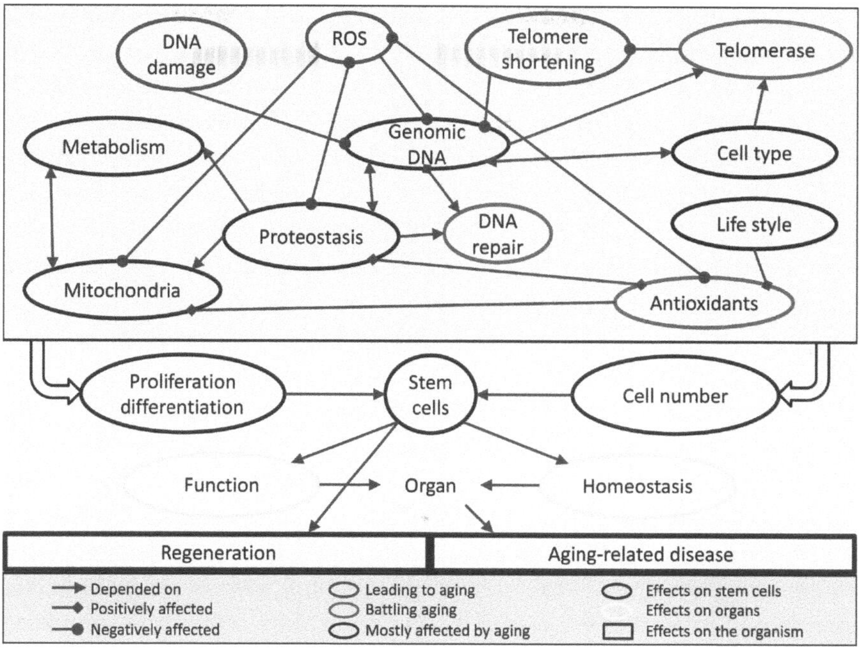

Figure 8.2 Molecular interactions during aging affecting stem cell function and the ability for organ regeneration, and susceptibility to disease. During aging, all the components of a cell are deregulated while natural defenses are weakened. DNA damage, reactive oxygen species (ROS)-induced stress, and telomere shortening are some of the most often encountered events during aging. The integrity of the DNA, proteostasis, metabolism, and mitochondria are all compromised. In addition, the negative effects are enhanced because of negative feedback interactions between these components. These negative effects on the cellular components also disrupt the mechanisms that could alleviate the aging damage such as DNA repair. Due to the nature of their role, certain cell types express proteins that can battle aging and are, thus, more resilient to these negative effects. The lifestyle of the organism is another parameter that can change the rate of aging. Healthy food and exercise are the most common factors that are known to slow aging. All these interacting elements (in light green background) affect the stem cells' ability for proliferation and differentiation as well as their number. The function of these cells is related to their regeneration ability and the degree of age-dependent decline. In addition, stem cells regulate the function and homeostasis of the organs that they reside in, and regulate the onset of aging-related diseases such as stroke and heart attack. Arrows show the type of interaction between the different elements. Boxes or circles are color-coded depending on the role of the element in aging. (For color version of this figure, the reader is referred to the online version of this chapter.)

and the PTEN/PI3K/AKT/mTOR pathway, which are also responsible for the after-effect responses of aging (e.g., DNA damage). Interestingly, inhibition of CDKN1A or CDKN2A leads to enhanced stem/progenitor cell function in many tissues (Baker et al., 2011; Choudhury et al., 2007), which implies that these genes play a major aging role in the entire organism.

In all the systems discussed earlier, there are important signaling pathways such as FGF, EGF, IGF, Wnt, and TGF that play various roles in tissue homeostasis and have been found to be deregulated during aging. It seems that restoring the correct signaling of these pathways can improve stem cell function in the particular tissue, for example, the Wnt pathway in NSCs (Seib et al., 2013) and Notch in muscle stem cells (Carlson et al., 2008). However, this balance between the signaling pathways is tissue-specific and depends primarily on the embryonic origin. In addition, the TGF-beta pathway is mostly shown to be a negative regulator of stem cell function, as in NSCs (Pineda et al., 2013), and interferes with other signaling pathways, as in muscle stem cells (Carlson et al., 2008). Furthermore, this pathway plays a role during the disease state and following injury, two instances linked to increased fibrosis, as in the case of cataract formation in the lens (Liu, Hales, Chamberlain, & McAvoy, 1994; Roberts et al., 1986).

The epigenetic state of the cells is another major aspect of age-dependent regeneration. Cell identity and cell response to injury via expression of certain genes is mediated through DNA methylation and histone marks, which can activate or inhibit gene expression with their potential roles in the regeneration response. Changes in DNA methylation and histone marks have been shown in most of the systems studied, influencing pluripotency-, proliferation-, and differentiation-related genes. We found Sirt1, a histone deacetylase, to play a role in the young epigenetic state in many of the systems reviewed while in some cases HDCA1 played the opposite role. Pluripotency factors such as Nanog and SOX2 also decline with the aging process since they are associated with pluripotency and proliferation mostly during development. The altered epigenetic state caused by changes in proteins and protein complexes that regulate it during aging in the stem/progenitor cells contributes to their impaired proliferation and differentiation ability.

Finally, a major study that was conducted by Conboy and colleagues in 2005 clearly showed that the environment surrounding the cells is critical for their function. These researchers used heterochronic parabiosis of a young and aged mouse and showed that it could rescue the aging effects of muscle and liver (Conboy et al., 2005). More recently, growth differentiation factor

11 was found to be responsible for rejuvenating the aged mouse heart with hypertrophy (Loffredo et al., 2013), a result that has potential applications in improving regeneration during aging. These studies, and others discussed previously concerning the systemic milieu that bathes the cells, contribute another parameter to the aging process. Is the systemic environment upstream to the other aging effects? If that holds true, is there a central tissue or organ that orchestrates aging in a damage-accumulating way or a strictly timed developmental way or a combination ofboth (de Magalhaes & Church, 2005)? Or, are cells affected by aging, change their behavior, and contribute differently to the systemic environment, which, in turn, qualitatively affects the other cells? While these aspects are true in mammals, fish heart and fin (Itou et al., 2012), and amphibian lens (Eguchi et al., 2011) show their regenerative abilities throughout life. While more studies are needed to further understand the differences in mammalian systems, more robust mechanisms tackling the age-related effects in combination with differences in tumor suppressor genes might be indicative of this discrepancy (Pajcini, Corbel, Sage, Pomerantz, & Blau, 2010; Sanchez Alvarado, 2012).

Taken together, this review provides a summary of age-dependent effects on cell turnover and regeneration after injury in tissues and organs. Aging affects all the cells, and the organism seems to age as a whole driven by similar mechanisms. As part of the natural cycle, aging is, as of today, inevitable and irreversible. However, studies that show well-controlled and developmentally triggered aging effects, and improved stem/progenitor function, as well as studies on lower vertebrates continue to light the way to reversing the core of the aging mechanisms that compromise organ function. This review also underlines the importance of studying animal models with good regenerative ability in which aging does not seem to be a factor. Such animal models can provide fundamental insights into the process of aging and regeneration.

ACKNOWLEDGMENTS

We apologize to the authors whose work has not been cited. Our summary is not exhaustive, but focuses on studies that connect aging and regeneration. Supported by NIH Grant EY10540 to P. A. T.

REFERENCES

Abbas, H. A., Maccio, D. R., Coskun, S., Jackson, J. G., Hazen, A. L., Sills, T. M., et al. (2010). Mdm2 is required for survival of hematopoietic stem cells/progenitors via dampening of ROS-induced p53 activity. *Cell Stem Cell*, 7, 606–617.

Allsopp, R. C., Morin, G. B., DePinho, R., Harley, C. B., & Weissman, I. L. (2003). Telomerase is required to slow telomere shortening and extend replicative lifespan of HSCs during serial transplantation. *Blood, 102*, 517–520.

Alt, E. U., Senst, C., Murthy, S. N., Slakey, D. P., Dupin, C. L., Chaffin, A. E., et al. (2012). Aging alters tissue resident mesenchymal stem cell properties. *Stem Cell Research, 8*, 215–225.

Aoshiba, K., & Nagai, A. (2009). Senescence hypothesis for the pathogenetic mechanism of chronic obstructive pulmonary disease. *Proceedings of the American Thoracic Society, 6*, 596–601.

Baddour, J. A., Sousounis, K., & Tsonis, P. A. (2012). Organ repair and regeneration: An overview. *Birth Defects Research. Part C, Embryo Today, 96*, 1–29.

Baker, D. J., Wijshake, T., Tchkonia, T., LeBrasseur, N. K., Childs, B. G., van de Sluis, B., et al. (2011). Clearance of p16Ink4a-positive senescent cells delays ageing-associated disorders. *Nature, 479*, 232–236.

Ballard, V. L. (2010). Stem cells for heart failure in the aging heart. *Heart Failure Reviews, 15*, 447–456.

Barberi, L., Scicchitano, B. M., De Rossi, M., Bigot, A., Duguez, S., Wielgosik, A., et al. (2013). Age-dependent alteration in muscle regeneration: The critical role of tissue niche. *Biogerontology, 14*, 273–292.

Barton-Davis, E. R., Shoturma, D. I., Musaro, A., Rosenthal, N., & Sweeney, H. L. (1998). Viral mediated expression of insulin-like growth factor I blocks the aging-related loss of skeletal muscle function. *Proceedings of the National Academy of Sciences of the United States of America, 95*, 15603–15607.

Basu-Roy, U., Ambrosetti, D., Favaro, R., Nicolis, S. K., Mansukhani, A., & Basilico, C. (2010). The transcription factor Sox2 is required for osteoblast self-renewal. *Cell Death and Differentiation, 17*, 1345–1353.

Beccafico, S., Riuzzi, F., Puglielli, C., Mancinelli, R., Fulle, S., Sorci, G., et al. (2011). Human muscle satellite cells show age-related differential expression of S100B protein and RAGE. *Age (Dordrecht, Netherlands), 33*, 523–541.

Beerman, I., Bock, C., Garrison, B. S., Smith, Z. D., Gu, H., Meissner, A., et al. (2013). Proliferation-dependent alterations of the DNA methylation landscape underlie hematopoietic stem cell aging. *Cell Stem Cell, 12*, 413–425.

Beggs, M. L., Nagarajan, R., Taylor-Jones, J. M., Nolen, G., Macnicol, M., & Peterson, C. A. (2004). Alterations in the TGFbeta signaling pathway in myogenic progenitors with age. *Aging Cell, 3*, 353–361.

Biondo-Simoes Mde, L., Matias, J. E., Montibeller, G. R., Siqueira, L. C., Nunes Eda, S., & Grassi, C. A. (2006). Effect of aging on liver regeneration in rats. *Acta Cirúrgica Brasileira, 21*, 197–202.

Bjornson, C. R., Cheung, T. H., Liu, L., Tripathi, P. V., Steeper, K. M., & Rando, T. A. (2012). Notch signaling is necessary to maintain quiescence in adult muscle stem cells. *Stem Cells, 30*, 232–242.

Bocker, M. T., Hellwig, I., Breiling, A., Eckstein, V., Ho, A. D., & Lyko, F. (2011). Genome-wide promoter DNA methylation dynamics of human hematopoietic progenitor cells during differentiation and aging. *Blood, 117*, e182–e189.

Boon, R. A., Iekushi, K., Lechner, S., Seeger, T., Fischer, A., Heydt, S., et al. (2013). MicroRNA-34a regulates cardiac ageing and function. *Nature, 495*, 107–110.

Brack, A. S., Conboy, M. J., Roy, S., Lee, M., Kuo, C. J., Keller, C., et al. (2007). Increased Wnt signaling during aging alters muscle stem cell fate and increases fibrosis. *Science, 317*, 807–810.

Brignull, H. R., Raible, D. W., & Stone, J. S. (2009). Feathers and fins: Non-mammalian models for hair cell regeneration. *Brain Research, 1277*, 12–23.

Brown, K., Xie, S., Qiu, X., Mohrin, M., Shin, J., Liu, Y., et al. (2013). SIRT3 reverses aging-associated degeneration. *Cell Reports, 3*, 319–327.
Buford, T. W., Macneil, R. G., Clough, L. G., Dirain, M., Sandesara, B., Pahor, M., et al. (2013). Active muscle regeneration following eccentric contraction-induced injury is similar between healthy young and older adults. *Journal of Applied Physiology*, ahead of print.
Buga, A. M., Vintilescu, R., Pop, O. T., & Popa-Wagner, A. (2011). Brain aging and regeneration after injuries: An organismal approach. *Aging and Disease, 2*, 64–79.
Carlson, M. E., & Conboy, I. M. (2007). Loss of stem cell regenerative capacity within aged niches. *Aging Cell, 6*, 371–382.
Carlson, M. E., Conboy, M. J., Hsu, M., Barchas, L., Jeong, J., Agrawal, A., et al. (2009). Relative roles of TGF-beta1 and Wnt in the systemic regulation and aging of satellite cell responses. *Aging Cell, 8*, 676–689.
Carlson, B. M., & Faulkner, J. A. (1989). Muscle transplantation between young and old rats: Age of host determines recovery. *The American Journal of Physiology, 256*, C1262–C1266.
Carlson, M. E., Hsu, M., & Conboy, I. M. (2008). Imbalance between pSmad3 and Notch induces CDK inhibitors in old muscle stem cells. *Nature, 454*, 528–532.
Cerletti, M., Jang, Y. C., Finley, L. W., Haigis, M. C., & Wagers, A. J. (2012). Short-term calorie restriction enhances skeletal muscle stem cell function. *Cell Stem Cell, 10*, 515–519.
Chai, R., Kuo, B., Wang, T., Liaw, E. J., Xia, A., Jan, T. A., et al. (2012). Wnt signaling induces proliferation of sensory precursors in the postnatal mouse cochlea. *Proceedings of the National Academy of Sciences of the United States of America, 109*, 8167–8172.
Chakkalakal, J. V., Jones, K. M., Basson, M. A., & Brack, A. S. (2012). The aged niche disrupts muscle stem cell quiescence. *Nature, 490*, 355–360.
Chakravarthy, M. V., Davis, B. S., & Booth, F. W. (2000). IGF-I restores satellite cell proliferative potential in immobilized old skeletal muscle. *Journal of Applied Physiology, 89*, 1365–1379.
Chambers, S. M., Shaw, C. A., Gatza, C., Fisk, C. J., Donehower, L. A., & Goodell, M. A. (2007). Aging hematopoietic stem cells decline in function and exhibit epigenetic dysregulation. *PLoS Biology, 5*, e201.
Chang, E. I., Loh, S. A., Ceradini, D. J., Lin, S. E., Bastidas, N., Aarabi, S., et al. (2007). Age decreases endothelial progenitor cell recruitment through decreases in hypoxia-inducible factor 1alpha stabilization during ischemia. *Circulation, 116*, 2818–2829.
Chen, X. D. (2010). Extracellular matrix provides an optimal niche for the maintenance and propagation of mesenchymal stem cells. *Birth Defects Research Part C, Embryo Today, 90*, 45–54.
Chen, J., Astle, C. M., & Harrison, D. E. (2003). Hematopoietic senescence is postponed and hematopoietic stem cell function is enhanced by dietary restriction. *Experimental Hematology, 31*, 1097–1103.
Chen, C., Liu, Y., & Zheng, P. (2009). mTOR regulation and therapeutic rejuvenation of aging hematopoietic stem cells. *Science Signaling, 2*, ra75.
Chen, P., & Segil, N. (1999). p27(Kip1) links cell proliferation to morphogenesis in the developing organ of Corti. *Development, 126*, 1581–1590.
Chen, W. D., Wang, Y. D., Zhang, L., Shiah, S., Wang, M., Yang, F., et al. (2010). Farnesoid X receptor alleviates age-related proliferation defects in regenerating mouse livers by activating forkhead box m1b transcription. *Hepatology, 51*, 953–962.
Choudhury, A. R., Ju, Z., Djojosubroto, M. W., Schienke, A., Lechel, A., Schaetzlein, S., et al. (2007). Cdkn1a deletion improves stem cell function and lifespan of mice with dysfunctional telomeres without accelerating cancer formation. *Nature Genetics, 39*, 99–105.
Chuikov, S., Levi, B. P., Smith, M. L., & Morrison, S. J. (2010). Prdm16 promotes stem cell maintenance in multiple tissues, partly by regulating oxidative stress. *Nature Cell Biology, 12*, 999–1006.

Conboy, I. M., Conboy, M. J., Smythe, G. M., & Rando, T. A. (2003). Notch-mediated restoration of regenerative potential to aged muscle. *Science, 302*, 1575–1577.

Conboy, I. M., Conboy, M. J., Wagers, A. J., Girma, E. R., Weissman, I. L., & Rando, T. A. (2005). Rejuvenation of aged progenitor cells by exposure to a young systemic environment. *Nature, 433*, 760–764.

Conboy, I. M., & Rando, T. A. (2012). Heterochronic parabiosis for the study of the effects of aging on stem cells and their niches. *Cell Cycle, 11*, 2260–2267.

Cortet, B. (2011). Bone repair in osteoporotic bone: Postmenopausal and cortisone-induced osteoporosis. *Osteoporosis International, 22*, 2007–2010.

Coutu, D. L., Francois, M., & Galipeau, J. (2011). Inhibition of cellular senescence by developmentally regulated FGF receptors in mesenchymal stem cells. *Blood, 117*, 6801–6812.

Dawes, L. J., Duncan, G., & Wormstone, I. M. (2013). Age-related differences in signaling efficiency of human lens cells underpin differential wound healing response rates following cataract surgery. *Investigative Ophthalmology & Visual Science, 54*, 333–342.

Day, K., Shefer, G., Shearer, A., & Yablonka-Reuveni, Z. (2010). The depletion of skeletal muscle satellite cells with age is concomitant with reduced capacity of single progenitors to produce reserve progeny. *Developmental Biology, 340*, 330–343.

De Barros, S., Dehez, S., Arnaud, E., Barreau, C., Cazavet, A., Perez, G., et al. (2013). Aging-related decrease of human ASC angiogenic potential is reversed by hypoxia preconditioning through ROS production. *Molecular Therapy, 21*, 399–408.

de Girolamo, L., Lopa, S., Arrigoni, E., Sartori, M. F., Baruffaldi Preis, F. W., & Brini, A. T. (2009). Human adipose-derived stem cells isolated from young and elderly women: Their differentiation potential and scaffold interaction during in vitro osteoblastic differentiation. *Cytotherapy, 11*, 793–803.

de Haan, G., Nijhof, W., & Van Zant, G. (1997). Mouse strain-dependent changes in frequency and proliferation of hematopoietic stem cells during aging: Correlation between lifespan and cycling activity. *Blood, 89*, 1543–1550.

de Magalhaes, J. P., & Church, G. M. (2005). Genomes optimize reproduction: Aging as a consequence of the developmental program. *Physiology (Bethesda), 20*, 252–259.

Decary, S., Mouly, V., Hamida, C. B., Sautet, A., Barbet, J. P., & Butler-Browne, G. S. (1997). Replicative potential and telomere length in human skeletal muscle: Implications for satellite cell-mediated gene therapy. *Human Gene Therapy, 8*, 1429–1438.

Duffy, P., Wang, X., Siegel, C. S., Tu, N., Henkemeyer, M., Cafferty, W. B., et al. (2012). Myelin-derived ephrinB3 restricts axonal regeneration and recovery after adult CNS injury. *Proceedings of the National Academy of Sciences of the United States of America, 109*, 5063–5068.

Dumble, M., Moore, L., Chambers, S. M., Geiger, H., Van Zant, G., Goodell, M. A., et al. (2007). The impact of altered p53 dosage on hematopoietic stem cell dynamics during aging. *Blood, 109*, 1736–1742.

Edelberg, J. M., Tang, L., Hattori, K., Lyden, D., & Rafii, S. (2002). Young adult bone marrow-derived endothelial precursor cells restore aging-impaired cardiac angiogenic function. *Circulation Research, 90*, E89–E93.

Efimenko, A., Starostina, E., Kalinina, N., & Stolzing, A. (2011). Angiogenic properties of aged adipose derived mesenchymal stem cells after hypoxic conditioning. *Journal of Translational Medicine, 9*, 10.

Eguchi, G., Eguchi, Y., Nakamura, K., Yadav, M. C., Millan, J. L., & Tsonis, P. A. (2011). Regenerative capacity in newts is not altered by repeated regeneration and ageing. *Nature Communications, 2*, 384.

Felice, F., Barsotti, M. C., Poredos, P., Balbarini, A., & Di Stefano, R. (2013). Effect of aging on metabolic pathways in endothelial progenitor cells. *Current Pharmaceutical Design, 19*, 2351–2365.

Feng, R., Zhou, S., Liu, Y., Song, D., Luan, Z., Dai, X., et al. (2013). Sox2 protects neural stem cells from apoptosis via up-regulating survivin expression. *The Biochemical Journal, 450*, 459–468.

Ferron, S. R., Marques-Torrejon, M. A., Mira, H., Flores, I., Taylor, K., Blasco, M. A., et al. (2009). Telomere shortening in neural stem cells disrupts neuronal differentiation and neuritogenesis. *The Journal of Neuroscience, 29*, 14394–14407.

Ferron, S., Mira, H., Franco, S., Cano-Jaimez, M., Bellmunt, E., Ramirez, C., et al. (2004). Telomere shortening and chromosomal instability abrogates proliferation of adult but not embryonic neural stem cells. *Development, 131*, 4059–4070.

Fulle, S., Di Donna, S., Puglielli, C., Pietrangelo, T., Beccafico, S., Bellomo, R., et al. (2005). Age-dependent imbalance of the antioxidative system in human satellite cells. *Experimental Gerontology, 40*, 189–197.

Furrer, K., Rickenbacher, A., Tian, Y., Jochum, W., Bittermann, A. G., Kach, A., et al. (2011). Serotonin reverts age-related capillarization and failure of regeneration in the liver through a VEGF-dependent pathway. *Proceedings of the National Academy of Sciences of the United States of America, 108*, 2945–2950.

Furutachi, S., Matsumoto, A., Nakayama, K. I., & Gotoh, Y. (2013). p57 controls adult neural stem cell quiescence and modulates the pace of lifelong neurogenesis. *The EMBO Journal, 32*, 970–981.

Gan, W. Q., Man, S. F., & Sin, D. D. (2005). The interactions between cigarette smoking and reduced lung function on systemic inflammation. *Chest, 127*, 558–564.

Gan, B., Sahin, E., Jiang, S., Sanchez-Aguilera, A., Scott, K. L., Chin, L., et al. (2008). mTORC1-dependent and -independent regulation of stem cell renewal, differentiation, and mobilization. *Proceedings of the National Academy of Sciences of the United States of America, 105*, 19384–19389.

Garcia-Prat, L., Sousa-Victor, P., & Munoz-Canoves, P. (2013). Functional dysregulation of stem cells during aging: A focus on skeletal muscle stem cells. *The FEBS Journal, 280*, 4051–4062.

Gibson, M. C., & Schultz, E. (1983). Age-related differences in absolute numbers of skeletal muscle satellite cells. *Muscle & Nerve, 6*, 574–580.

Gutmann, E., & Carlson, B. M. (1976). Regeneration and transplantation of muscles in old rats and between young and old rats. *Life Sciences, 18*, 109–114.

Haga, S., Morita, N., Irani, K., Fujiyoshi, M., Ogino, T., Ozawa, T., et al. (2010). p66(Shc) has a pivotal function in impaired liver regeneration in aged mice by a redox-dependent mechanism. *Laboratory Investigation, 90*, 1718–1726.

Han, J., Mistriotis, P., Lei, P., Wang, D., Liu, S., & Andreadis, S. T. (2012). Nanog reverses the effects of organismal aging on mesenchymal stem cell proliferation and myogenic differentiation potential. *Stem Cells, 30*, 2746–2759.

Hao, L. Y., Armanios, M., Strong, M. A., Karim, B., Feldser, D. M., Huso, D., et al. (2005). Short telomeres, even in the presence of telomerase, limit tissue renewal capacity. *Cell, 123*, 1121–1131.

Hayashi, T., Ray, C. A., Younkins, C., & Bermingham-McDonogh, O. (2010). Expression patterns of FGF receptors in the developing mammalian cochlea. *Developmental Dynamics, 239*, 1019–1026.

He, T., Joyner, M. J., & Katusic, Z. S. (2009). Aging decreases expression and activity of glutathione peroxidase-1 in human endothelial progenitor cells. *Microvascular Research, 78*, 447–452.

He, L., Yadgarov, A., Sharif, S., & McCluskey, L. P. (2012). Aging profoundly delays functional recovery from gustatory nerve injury. *Neuroscience, 209*, 208–218.

Heiss, C., Keymel, S., Niesler, U., Ziemann, J., Kelm, M., & Kalka, C. (2005). Impaired progenitor cell activity in age-related endothelial dysfunction. *Journal of the American College of Cardiology, 45*, 1441–1448.

Hejtmancik, J. F., & Kantorow, M. (2004). Molecular genetics of age-related cataract. *Experimental Eye Research, 79*, 3–9.
Henry, J. J., & Tsonis, P. A. (2010). Molecular and cellular aspects of amphibian lens regeneration. *Progress in Retinal and Eye Research, 29*, 543–555.
Herbig, U., Jobling, W. A., Chen, B. P., Chen, D. J., & Sedivy, J. M. (2004). Telomere shortening triggers senescence of human cells through a pathway involving ATM, p53, and p21(CIP1), but not p16(INK4a). *Molecular Cell, 14*, 501–513.
Hidalgo, I., Herrera-Merchan, A., Ligos, J. M., Carramolino, L., Nunez, J., Martinez, F., et al. (2012). Ezh1 Is required for hematopoietic stem cell maintenance and prevents senescence-like cell cycle arrest. *Cell Stem Cell, 11*, 649–662.
Hirano, S., Tateya, I., Kishimoto, Y., Kanemaru, S., & Ito, J. (2012). Clinical trial of regeneration of aged vocal folds with growth factor therapy. *Laryngoscope, 122*, 327–331.
Hoetzer, G. L., Van Guilder, G. P., Irmiger, H. M., Keith, R. S., Stauffer, B. L., & DeSouza, C. A. (2007). Aging, exercise, and endothelial progenitor cell clonogenic and migratory capacity in men. *Journal of Applied Physiology, 102*, 847–852.
Huang, M., Sage, C., Tang, Y., Lee, S. G., Petrillo, M., Hinds, P. W., et al. (2011). Overlapping and distinct pRb pathways in the mammalian auditory and vestibular organs. *Cell Cycle, 10*, 337–351.
Hutter, D., Yo, Y., Chen, W., Liu, P., Holbrook, N. J., Roth, G. S., et al. (2000). Age-related decline in Ras/ERK mitogen-activated protein kinase cascade is linked to a reduced association between Shc and EGF receptor. *The Journals of Gerontology. Series A, Biological Sciences and Medical Sciences, 55*, B125–B134.
Ishigami, A., Reed, T. D., & Roth, G. S. (1993). Effect of aging on EGF stimulated DNA synthesis and EGF receptor levels in primary cultured rat hepatocytes. *Biochemical and Biophysical Research Communications, 196*, 181–186.
Ito, K., Hirao, A., Arai, F., Matsuoka, S., Takubo, K., Hamaguchi, I., et al. (2004). Regulation of oxidative stress by ATM is required for self-renewal of haematopoietic stem cells. *Nature, 431*, 997–1002.
Ito, K., Hirao, A., Arai, F., Takubo, K., Matsuoka, S., Miyamoto, K., et al. (2006). Reactive oxygen species act through p38 MAPK to limit the lifespan of hematopoietic stem cells. *Nature Medicine, 12*, 446–451.
Ito, T., Sawada, R., Fujiwara, Y., Seyama, Y., & Tsuchiya, T. (2007). FGF-2 suppresses cellular senescence of human mesenchymal stem cells by down-regulation of TGF-beta2. *Biochemical and Biophysical Research Communications, 359*, 108–114.
Itou, J., Kawakami, H., Burgoyne, T., & Kawakami, Y. (2012). Life-long preservation of the regenerative capacity in the fin and heart in zebrafish. *Biol Open, 1*, 739–746.
Jaerve, A., Kruse, F., Malik, K., Hartung, H. P., & Muller, H. W. (2012). Age-dependent modulation of cortical transcriptomes in spinal cord injury and repair. *PLoS One, 7*, e49812.
Jakob, F., Ebert, R., Ignatius, A., Matsushita, T., Watanabe, Y., Groll, J., et al. (2013). Bone tissue engineering in osteoporosis. *Maturitas, 75*, 118–124.
Jakob, F., Ebert, R., Rudert, M., Noth, U., Walles, H., Docheva, D., et al. (2012). In situ guided tissue regeneration in musculoskeletal diseases and aging: Implementing pathology into tailored tissue engineering strategies. *Cell and Tissue Research, 347*, 725–735.
Jang, Y. C., Sinha, M., Cerletti, M., Dall'Osso, C., & Wagers, A. J. (2011). Skeletal muscle stem cells: Effects of aging and metabolism on muscle regenerative function. *Cold Spring Harbor Symposia on Quantitative Biology, 76*, 101–111.
Janzen, V., Forkert, R., Fleming, H. E., Saito, Y., Waring, M. T., Dombkowski, D. M., et al. (2006). Stem-cell ageing modified by the cyclin-dependent kinase inhibitor p16INK4a. *Nature, 443*, 421–426.
Jaskelioff, M., Muller, F. L., Paik, J. H., Thomas, E., Jiang, S., Adams, A. C., et al. (2011). Telomerase reactivation reverses tissue degeneration in aged telomerase-deficient mice. *Nature, 469*, 102–106.

Jejurikar, S. S., Henkelman, E. A., Cederna, P. S., Marcelo, C. L., Urbanchek, M. G., & Kuzon, W. M., Jr. (2006). Aging increases the susceptibility of skeletal muscle derived satellite cells to apoptosis. *Experimental Gerontology, 41*, 828–836.

Jin, J., Iakova, P., Jiang, Y., Medrano, E. E., & Timchenko, N. A. (2011). The reduction of SIRT1 in livers of old mice leads to impaired body homeostasis and to inhibition of liver proliferation. *Hepatology, 54*, 989–998.

Jin, K., Sun, Y., Xie, L., Batteur, S., Mao, X. O., Smelick, C., et al. (2003). Neurogenesis and aging: FGF-2 and HB-EGF restore neurogenesis in hippocampus and subventricular zone of aged mice. *Aging Cell, 2*, 175–183.

Ju, Z., Jiang, H., Jaworski, M., Rathinam, C., Gompf, A., Klein, C., et al. (2007). Telomere dysfunction induces environmental alterations limiting hematopoietic stem cell function and engraftment. *Nature Medicine, 13*, 742–747.

Kamat, A., Ghosh, P. M., Glover, R. L., Zhu, B., Yeh, C. K., Choudhury, G. G., et al. (2008). Reduced expression of epidermal growth factor receptors in rat liver during aging. *The Journals of Gerontology Series A, Biological Sciences and Medical Sciences, 63*, 683–692.

Kapetanaki, M. G., Mora, A. L., & Rojas, M. (2013). Influence of age on wound healing and fibrosis. *The Journal of Pathology, 229*, 310–322.

Karagiannides, I., Tchkonia, T., Dobson, D. E., Steppan, C. M., Cummins, P., Chan, G., et al. (2001). Altered expression of C/EBP family members results in decreased adipogenesis with aging. *American Journal of Physiology. Regulatory, Integrative and Comparative Physiology, 280*, R1772–R1780.

Kelley, M. W., Xu, X. M., Wagner, M. A., Warchol, M. E., & Corwin, J. T. (1993). The developing organ of Corti contains retinoic acid and forms supernumerary hair cells in response to exogenous retinoic acid in culture. *Development, 119*, 1041–1053.

Kelly, M. C., Chang, Q., Pan, A., Lin, X., & Chen, P. (2012). Atoh1 directs the formation of sensory mosaics and induces cell proliferation in the postnatal mammalian cochlea in vivo. *The Journal of Neuroscience, 32*, 6699–6710.

Kovacheva, E. L., Hikim, A. P., Shen, R., Sinha, I., & Sinha-Hikim, I. (2010). Testosterone supplementation reverses sarcopenia in aging through regulation of myostatin, c-Jun NH2-terminal kinase, Notch, and Akt signaling pathways. *Endocrinology, 151*, 628–638.

Krupczak-Hollis, K., Wang, X., Dennewitz, M. B., & Costa, R. H. (2003). Growth hormone stimulates proliferation of old-aged regenerating liver through forkhead box m1b. *Hepatology, 38*, 1552–1562.

Kuhn, H. G., Dickinson-Anson, H., & Gage, F. H. (1996). Neurogenesis in the dentate gyrus of the adult rat: Age-related decrease of neuronal progenitor proliferation. *The Journal of Neuroscience, 16*, 2027–2033.

Kushner, E. J., MacEneaney, O. J., Weil, B. R., Greiner, J. J., Stauffer, B. L., & DeSouza, C. A. (2011). Aging is associated with a proapoptotic endothelial progenitor cell phenotype. *Journal of Vascular Research, 48*, 408–414.

Kushner, E. J., Van Guilder, G. P., Maceneaney, O. J., Cech, J. N., Stauffer, B. L., & DeSouza, C. A. (2009). Aging and endothelial progenitor cell telomere length in healthy men. *Clinical Chemistry and Laboratory Medicine, 47*, 47–50.

Laine, H., Doetzlhofer, A., Mantela, J., Ylikoski, J., Laiho, M., Roussel, M. F., et al. (2007). p19(Ink4d) and p21(Cip1) collaborate to maintain the postmitotic state of auditory hair cells, their codeletion leading to DNA damage and p53-mediated apoptosis. *The Journal of Neuroscience, 27*, 1434–1444.

Lee, K. H., & Cotanche, D. A. (1996). Potential role of bFGF and retinoic acid in the regeneration of chicken cochlear hair cells. *Hearing Research, 94*, 1–13.

Lee, J., Duan, W., Long, J. M., Ingram, D. K., & Mattson, M. P. (2000). Dietary restriction increases the number of newly generated neural cells, and induces BDNF expression, in the dentate gyrus of rats. *Journal of Molecular Neuroscience, 15*, 99–108.

Lee, K., Kletzien, H., Connor, N. P., Schultz, E., Chamberlain, C. S., & Bless, D. M. (2012). Effects of aging on thyroarytenoid muscle regeneration. *Laryngoscope, 122*, 2800–2807.
Lee, J. Y., Nakada, D., Yilmaz, O. H., Tothova, Z., Joseph, N. M., Lim, M. S., et al. (2010). mTOR activation induces tumor suppressors that inhibit leukemogenesis and deplete hematopoietic stem cells after Pten deletion. *Cell Stem Cell, 7*, 593–605.
Lefebvre, P. P., Malgrange, B., Staecker, H., Moonen, G., & Van de Water, T. R. (1993). Retinoic acid stimulates regeneration of mammalian auditory hair cells. *Science, 260*, 692–695.
Lie, D. C., Colamarino, S. A., Song, H. J., Desire, L., Mira, H., Consiglio, A., et al. (2005). Wnt signalling regulates adult hippocampal neurogenesis. *Nature, 437*, 1370–1375.
Liu, Z., Dearman, J. A., Cox, B. C., Walters, B. J., Zhang, L., Ayrault, O., et al. (2012). Age-dependent in vivo conversion of mouse cochlear pillar and Deiters' cells to immature hair cells by Atoh1 ectopic expression. *The Journal of Neuroscience, 32*, 6600–6610.
Liu, Y., Elf, S. E., Miyata, Y., Sashida, G., Huang, G., Di Giandomenico, S., et al. (2009). p53 regulates hematopoietic stem cell quiescence. *Cell Stem Cell, 4*, 37–48.
Liu, J., Hales, A. M., Chamberlain, C. G., & McAvoy, J. W. (1994). Induction of cataract-like changes in rat lens epithelial explants by transforming growth factor beta. *Investigative Ophthalmology & Visual Science, 35*, 388–401.
Liu, F., Lee, J. Y., Wei, H., Tanabe, O., Engel, J. D., Morrison, S. J., et al. (2010). FIP200 is required for the cell-autonomous maintenance of fetal hematopoietic stem cells. *Blood, 116*, 4806–4814.
Liu, Z., Walters, B. J., Owen, T., Brimble, M. A., Steigelman, K. A., Zhang, L., et al. (2012). Regulation of p27Kip1 by Sox2 maintains quiescence of inner pillar cells in the murine auditory sensory epithelium. *The Journal of Neuroscience, 32*, 10530–10540.
Loffredo, F. S., Steinhauser, M. L., Jay, S. M., Gannon, J., Pancoast, J. R., Yalamanchi, P., et al. (2013). Growth differentiation factor 11 is a circulating factor that reverses age-related cardiac hypertrophy. *Cell, 153*, 828–839.
Lopez-Otin, C., Blasco, M. A., Partridge, L., Serrano, M., & Kroemer, G. (2013). The hallmarks of aging. *Cell, 153*, 1194–1217.
Lowenheim, H., Furness, D. N., Kil, J., Zinn, C., Gultig, K., Fero, M. L., et al. (1999). Gene disruption of p27(Kip1) allows cell proliferation in the postnatal and adult organ of corti. *Proceedings of the National Academy of Sciences of the United States of America, 96*, 4084–4088.
Luo, L., Koutnouyan, H., Baird, A., & Ryan, A. F. (1993). Acidic and basic FGF mRNA expression in the adult and developing rat cochlea. *Hearing Research, 69*, 182–193.
Madonna, R., Renna, F. V., Cellini, C., Cotellese, R., Picardi, N., Francomano, F., et al. (2011). Age-dependent impairment of number and angiogenic potential of adipose tissue-derived progenitor cells. *European Journal of Clinical Investigation, 41*, 126–133.
Magee, J. A., Ikenoue, T., Nakada, D., Lee, J. Y., Guan, K. L., & Morrison, S. J. (2012). Temporal changes in PTEN and mTORC2 regulation of hematopoietic stem cell self-renewal and leukemia suppression. *Cell Stem Cell, 11*, 415–428.
Marin, C., Yubero-Serrano, E. M., Lopez-Miranda, J., & Perez-Jimenez, F. (2013). Endothelial aging associated with oxidative stress can be modulated by a healthy mediterranean diet. *International Journal of Molecular Sciences, 14*, 8869–8889.
Marti, U. (1993). Handling of epidermal growth factor and number of epidermal growth factor receptors are changed in aged male rats. *Hepatology, 18*, 1432–1436.
Maryanovich, M., Oberkovitz, G., Niv, H., Vorobiyov, L., Zaltsman, Y., Brenner, O., et al. (2012). The ATM-BID pathway regulates quiescence and survival of haematopoietic stem cells. *Nature Cell Biology, 14*, 535–541.
Marzetti, E., Wohlgemuth, S. E., Lees, H. A., Chung, H. Y., Giovannini, S., & Leeuwenburgh, C. (2008). Age-related activation of mitochondrial caspase-independent apoptotic signaling in rat gastrocnemius muscle. *Mechanisms of Ageing and Development, 129*, 542–549.

Maslov, A. Y., Barone, T. A., Plunkett, R. J., & Pruitt, S. C. (2004). Neural stem cell detection, characterization, and age-related changes in the subventricular zone of mice. *The Journal of Neuroscience, 24,* 1726–1733.

Mattson, M. P., & Magnus, T. (2006). Ageing and neuronal vulnerability. *Nature Reviews. Neuroscience, 7,* 278–294.

McGeachie, J. K., & Grounds, M. D. (1995). Retarded myogenic cell replication in regenerating skeletal muscles of old mice: An autoradiographic study in young and old BALBc and SJL/J mice. *Cell and Tissue Research, 280,* 277–282.

Miranda, C. J., Braun, L., Jiang, Y., Hester, M. E., Zhang, L., Riolo, M., et al. (2012). Aging brain microenvironment decreases hippocampal neurogenesis through Wnt-mediated survivin signaling. *Aging Cell, 11,* 542–552.

Miyamoto, K., Araki, K. Y., Naka, K., Arai, F., Takubo, K., Yamazaki, S., et al. (2007). Foxo3a is essential for maintenance of the hematopoietic stem cell pool. *Cell Stem Cell, 1,* 101–112.

Mohsin, S., Siddiqi, S., Collins, B., & Sussman, M. A. (2011). Empowering adult stem cells for myocardial regeneration. *Circulation Research, 109,* 1415–1428.

Molofsky, A. V., He, S., Bydon, M., Morrison, S. J., & Pardal, R. (2005). Bmi-1 promotes neural stem cell self-renewal and neural development but not mouse growth and survival by repressing the p16Ink4a and p19Arf senescence pathways. *Genes & Development, 19,* 1432–1437.

Molofsky, A. V., Slutsky, S. G., Joseph, N. M., He, S., Pardal, R., Krishnamurthy, J., et al. (2006). Increasing p16INK4a expression decreases forebrain progenitors and neurogenesis during ageing. *Nature, 443,* 448–452.

Mortensen, M., Soilleux, E. J., Djordjevic, G., Tripp, R., Lutteropp, M., Sadighi-Akha, E., et al. (2011). The autophagy protein Atg7 is essential for hematopoietic stem cell maintenance. *The Journal of Experimental Medicine, 208,* 455–467.

Musaro, A., McCullagh, K., Paul, A., Houghton, L., Dobrowolny, G., Molinaro, M., et al. (2001). Localized Igf-1 transgene expression sustains hypertrophy and regeneration in senescent skeletal muscle. *Nature Genetics, 27,* 195–200.

Naito, A. T., Sumida, T., Nomura, S., Liu, M. L., Higo, T., Nakagawa, A., et al. (2012). Complement C1q activates canonical Wnt signaling and promotes aging-related phenotypes. *Cell, 149,* 1298–1313.

Nijnik, A., Woodbine, L., Marchetti, C., Dawson, S., Lambe, T., Liu, C., et al. (2007). DNA repair is limiting for haematopoietic stem cells during ageing. *Nature, 447,* 686–690.

Nishino, J., Kim, I., Chada, K., & Morrison, S. J. (2008). Hmga2 promotes neural stem cell self-renewal in young but not old mice by reducing p16Ink4a and p19Arf Expression. *Cell, 135,* 227–239.

Norddahl, G. L., Pronk, C. J., Wahlestedt, M., Sten, G., Nygren, J. M., Ugale, A., et al. (2011). Accumulating mitochondrial DNA mutations drive premature hematopoietic aging phenotypes distinct from physiological stem cell aging. *Cell Stem Cell, 8,* 499–510.

Oesterle, E. C., Chien, W. M., Campbell, S., Nellimarla, P., & Fero, M. L. (2011). p27 (Kip1) is required to maintain proliferative quiescence in the adult cochlea and pituitary. *Cell Cycle, 10,* 1237–1248.

Ou, G., Charles, L., Matton, S., Rodner, C., Hurley, M., Kuhn, L., et al. (2010). Fibroblast growth factor-2 stimulates the proliferation of mesenchyme-derived progenitor cells from aging mouse and human bone. *The Journals of Gerontology. Series A, Biological Sciences and Medical Sciences, 65,* 1051–1059.

Paik, J. H., Ding, Z., Narurkar, R., Ramkissoon, S., Muller, F., Kamoun, W. S., et al. (2009). FoxOs cooperatively regulate diverse pathways governing neural stem cell homeostasis. *Cell Stem Cell, 5,* 540–553.

Pajcini, K. V., Corbel, S. Y., Sage, J., Pomerantz, J. H., & Blau, H. M. (2010). Transient inactivation of Rb and ARF yields regenerative cells from postmitotic mammalian muscle. *Cell Stem Cell*, *7*, 198–213.

Paliouras, G. N., Hamilton, L. K., Aumont, A., Joppe, S. E., Barnabe-Heider, F., & Fernandes, K. J. (2012). Mammalian target of rapamycin signaling is a key regulator of the transit-amplifying progenitor pool in the adult and aging forebrain. *The Journal of Neuroscience*, *32*, 15012–15026.

Paliwal, P., Pishesha, N., Wijaya, D., & Conboy, I. M. (2012). Age dependent increase in the levels of osteopontin inhibits skeletal muscle regeneration. *Aging (Albany NY)*, *4*, 553–566.

Palmer, H. J., Tuzon, C. T., & Paulson, K. E. (1999). Age-dependent decline in mitogenic stimulation of hepatocytes. Reduced association between Shc and the epidermal growth factor receptor is coupled to decreased activation of Raf and extracellular signal-regulated kinases. *The Journal of Biological Chemistry*, *274*, 11424–11430.

Park, I. K., Qian, D., Kiel, M., Becker, M. W., Pihalja, M., Weissman, I. L., et al. (2003). Bmi-1 is required for maintenance of adult self-renewing haematopoietic stem cells. *Nature*, *423*, 302–305.

Park, D., Yang, Y. H., Bae, D. K., Lee, S. H., Yang, G., Kyung, J., et al. (2013). Improvement of cognitive function and physical activity of aging mice by human neural stem cells over-expressing choline acetyltransferase. *Neurobiology of Aging*, *34*, 2639–2646.

Paxson, J. A., Gruntman, A., Parkin, C. D., Mazan, M. R., Davis, A., Ingenito, E. P., et al. (2011). Age-dependent decline in mouse lung regeneration with loss of lung fibroblast clonogenicity and increased myofibroblastic differentiation. *PLoS One*, *6*, e23232.

Phillips, T., & Leeuwenburgh, C. (2005). Muscle fiber specific apoptosis and TNF-alpha signaling in sarcopenia are attenuated by life-long calorie restriction. *The FASEB Journal*, *19*, 668–670.

Pickles, J. O., & van Heumen, W. R. (1997). The expression of messenger RNAs coding for growth factors, their receptors, and eph-class receptor tyrosine kinases in normal and ototoxically damaged chick cochleae. *Developmental Neuroscience*, *19*, 476–487.

Pineda, J. R., Daynac, M., Chicheportiche, A., Cebrian-Silla, A., Sii Felice, K., Garcia-Verdugo, J. M., et al. (2013). Vascular-derived TGF-beta increases in the stem cell niche and perturbs neurogenesis during aging and following irradiation in the adult mouse brain. *EMBO Molecular Medicine*, *5*, 548–562.

Porrello, E. R., Mahmoud, A. I., Simpson, E., Hill, J. A., Richardson, J. A., Olson, E. N., et al. (2011). Transient regenerative potential of the neonatal mouse heart. *Science*, *331*, 1078–1080.

Rauscher, F. M., Goldschmidt-Clermont, P. J., Davis, B. H., Wang, T., Gregg, D., Ramaswami, P., et al. (2003). Aging, progenitor cell exhaustion, and atherosclerosis. *Circulation*, *108*, 457–463.

Renault, V. M., Rafalski, V. A., Morgan, A. A., Salih, D. A., Brett, J. O., Webb, A. E., et al. (2009). FoxO3 regulates neural stem cell homeostasis. *Cell Stem Cell*, *5*, 527–539.

Renault, V., Thornell, L. E., Eriksson, P. O., Butler-Browne, G., & Mouly, V. (2002). Regenerative potential of human skeletal muscle during aging. *Aging Cell*, *1*, 132–139.

Roberts, A. B., Sporn, M. B., Assoian, R. K., Smith, J. M., Roche, N. S., Wakefield, L. M., et al. (1986). Transforming growth factor type beta: Rapid induction of fibrosis and angiogenesis in vivo and stimulation of collagen formation in vitro. *Proceedings of the National Academy of Sciences of the United States of America*, *83*, 4167–4171.

Rocha-Sanchez, S. M., Scheetz, L. R., Contreras, M., Weston, M. D., Korte, M., McGee, J., et al. (2011). Mature mice lacking Rbl2/p130 gene have supernumerary inner ear hair cells and supporting cells. *The Journal of Neuroscience*, *31*, 8883–8893.

Rossi, D. J., Bryder, D., Seita, J., Nussenzweig, A., Hoeijmakers, J., & Weissman, I. L. (2007). Deficiencies in DNA damage repair limit the function of haematopoietic stem cells with age. *Nature*, *447*, 725–729.

Ruckh, J. M., Zhao, J. W., Shadrach, J. L., van Wijngaarden, P., Rao, T. N., Wagers, A. J., et al. (2012). Rejuvenation of regeneration in the aging central nervous system. *Cell Stem Cell*, *10*, 96–103.

Rupp, A., Cunningham, M. E., Yao, D., Furukawa, K., & Willison, H. J. (2013). The effects of age and ganglioside composition on the rate of motor nerve terminal regeneration following antibody-mediated injury in mice. *Synapse*, *67*, 382–389.

Saito, T., Yamada, T., Okamoto, M., & Narita, N. (2012). Comparison of regeneration of the chorda tympani nerve and gustatory function recovery after severing the nerve between pediatric and adult patients. *Otology & Neurotology*, *33*, 1567–1572.

Sanchez Alvarado, A. (2012). Cellular hyperproliferation and cancer as evolutionary variables. *Current Biology*, *22*, R772–R778.

Sawada, N. (1989). Hepatocytes from old rats retain responsiveness of c-myc expression to EGF in primary culture but do not enter S phase. *Experimental Cell Research*, *181*, 584–588.

Schellenberg, A., Lin, Q., Schuler, H., Koch, C. M., Joussen, S., Denecke, B., et al. (2011). Replicative senescence of mesenchymal stem cells causes DNA-methylation changes which correlate with repressive histone marks. *Aging (Albany NY)*, *3*, 873–888.

Schmuck, E. G., Mulligan, J. D., & Saupe, K. W. (2011). Caloric restriction attenuates the age-associated increase of adipose-derived stem cells but further reduces their proliferative capacity. *Age (Dordrecht, Netherlands)*, *33*, 107–118.

Schmucker, D. L., & Sanchez, H. (2011). Liver regeneration and aging: A current perspective. *Curr Gerontol Geriatr Res*, *2011*, 526379.

Seib, D. R., Corsini, N. S., Ellwanger, K., Plaas, C., Mateos, A., Pitzer, C., et al. (2013). Loss of Dickkopf-1 restores neurogenesis in old age and counteracts cognitive decline. *Cell Stem Cell*, *12*, 204–214.

Senyo, S. E., Steinhauser, M. L., Pizzimenti, C. L., Yang, V. K., Cai, L., Wang, M., et al. (2013). Mammalian heart renewal by pre-existing cardiomyocytes. *Nature*, *493*, 433–436.

Serra, C., Tangherlini, F., Rudy, S., Lee, D., Toraldo, G., Sandor, N. L., et al. (2013). Testosterone improves the regeneration of old and young mouse skeletal muscle. *The Journals of Gerontology. Series A, Biological Sciences and Medical Sciences*, *68*, 17–26.

Shefer, G., Rauner, G., Yablonka-Reuveni, Z., & Benayahu, D. (2010). Reduced satellite cell numbers and myogenic capacity in aging can be alleviated by endurance exercise. *PLoS One*, *5*, e13307.

Shetty, A. K., Hattiangady, B., & Shetty, G. A. (2005). Stem/progenitor cell proliferation factors FGF-2, IGF-1, and VEGF exhibit early decline during the course of aging in the hippocampus: Role of astrocytes. *Glia*, *51*, 173–186.

Signer, R. A., & Morrison, S. J. (2013). Mechanisms that regulate stem cell aging and life span. *Cell Stem Cell*, *12*, 152–165.

Singh, P., Goode, T., Dean, A., Awad, S. S., & Darlington, G. J. (2011). Elevated interferon gamma signaling contributes to impaired regeneration in the aged liver. *The Journals of Gerontology. Series A, Biological Sciences and Medical Sciences*, *66*, 944–956.

Singh, S. K., Williams, C. A., Klarmann, K., Burkett, S. S., Keller, J. R., & Oberdoerffer, P. (2013). Sirt1 ablation promotes stress-induced loss of epigenetic and genomic hematopoietic stem and progenitor cell maintenance. *The Journal of Experimental Medicine*, *210*, 987–1001.

Smythe, G. M., Shavlakadze, T., Roberts, P., Davies, M. J., McGeachie, J. K., & Grounds, M. D. (2008). Age influences the early events of skeletal muscle regeneration: Studies of whole muscle grafts transplanted between young (8 weeks) and old (13-21 months) mice. *Experimental Gerontology*, *43*, 550–562.

Solchaga, L. A., Penick, K., Goldberg, V. M., Caplan, A. I., & Welter, J. F. (2010). Fibroblast growth factor-2 enhances proliferation and delays loss of chondrogenic potential in human adult bone-marrow-derived mesenchymal stem cells. *Tissue Engineering. Part A*, *16*, 1009–1019.

Sousounis, K., Looso, M., Maki, N., Ivester, C. J., Braun, T., & Tsonis, P. A. (2013). Transcriptome analysis of newt lens regeneration reveals distinct gradients in gene expression patterns. *PLoS One*, *8*, e61445.

Sousounis, K., Michel, C. S., Bruckskotten, M., Maki, N., Borchardt, T., Braun, T., et al. (2013). A microarray analysis of gene expression patterns during early phases of newt lens regeneration. *Molecular Vision*, *19*, 135–145.

Sousounis, K., & Tsonis, P. A. (2012). Patterns of gene expression in microarrays and expressed sequence tags from normal and cataractous lenses. *Human Genomics*, *6*, 14.

Stoll, E. A., Cheung, W., Mikheev, A. M., Sweet, I. R., Bielas, J. H., Zhang, J., et al. (2011). Aging neural progenitor cells have decreased mitochondrial content and lower oxidative metabolism. *The Journal of Biological Chemistry*, *286*, 38592–38601.

Suetsugu-Maki, R., Maki, N., Nakamura, K., Sumanas, S., Zhu, J., Del Rio-Tsonis, K., et al. (2012). Lens regeneration in axolotl: New evidence of developmental plasticity. *BMC Biology*, *10*, 103.

Sun, Z., Lee, C. J., Mejia-Guerrero, S., Zhang, Y., Higuchi, K., Li, R. K., et al. (2012). Neonatal transfer of membrane-bound stem cell factor improves survival and heart function in aged mice after myocardial ischemia. *Human Gene Therapy*, *23*, 1280–1289.

Suyama, S., Sunabori, T., Kanki, H., Sawamoto, K., Gachet, C., Koizumi, S., et al. (2012). Purinergic signaling promotes proliferation of adult mouse subventricular zone cells. *The Journal of Neuroscience*, *32*, 9238–9247.

Sweet, E. M., Vemaraju, S., & Riley, B. B. (2011). Sox2 and Fgf interact with Atoh1 to promote sensory competence throughout the zebrafish inner ear. *Developmental Biology*, *358*, 113–121.

Takubo, K., Goda, N., Yamada, W., Iriuchishima, H., Ikeda, E., Kubota, Y., et al. (2010). Regulation of the HIF-1alpha level is essential for hematopoietic stem cells. *Cell Stem Cell*, *7*, 391–402.

Taylor, R. R., & Forge, A. (2005). Hair cell regeneration in sensory epithelia from the inner ear of a urodele amphibian. *The Journal of Comparative Neurology*, *484*, 105–120.

TeKippe, M., Harrison, D. E., & Chen, J. (2003). Expansion of hematopoietic stem cell phenotype and activity in Trp53-null mice. *Experimental Hematology*, *31*, 521–527.

Teramoto, S., & Ishii, M. (2007). Aging, the aging lung, and senile emphysema are different. *American Journal of Respiratory and Critical Care Medicine*, *175*, 197–198, author reply 198.

Thum, T., Hoeber, S., Froese, S., Klink, I., Stichtenoth, D. O., Galuppo, P., et al. (2007). Age-dependent impairment of endothelial progenitor cells is corrected by growth-hormone-mediated increase of insulin-like growth-factor-1. *Circulation Research*, *100*, 434–443.

Timchenko, N. A. (2009). Aging and liver regeneration. *Trends in Endocrinology and Metabolism*, *20*, 171–176.

Torres-Gonzalez, E., Bueno, M., Tanaka, A., Krug, L. T., Cheng, D. S., Polosukhin, V. V., et al. (2012). Role of endoplasmic reticulum stress in age-related susceptibility to lung fibrosis. *American Journal of Respiratory Cell and Molecular Biology*, *46*, 748–756.

Tothova, Z., Kollipara, R., Huntly, B. J., Lee, B. H., Castrillon, D. H., Cullen, D. E., et al. (2007). FoxOs are critical mediators of hematopoietic stem cell resistance to physiologic oxidative stress. *Cell*, *128*, 325–339.

Tropepe, V., Sibilia, M., Ciruna, B. G., Rossant, J., Wagner, E. F., & van der Kooy, D. (1999). Distinct neural stem cells proliferate in response to EGF and FGF in the developing mouse telencephalon. *Developmental Biology*, *208*, 166–188.

Turgeon, J., Haddad, P., Dussault, S., Groleau, J., Maingrette, F., Perez, G., et al. (2012). Protection against vascular aging in Nox2-deficient mice: Impact on endothelial progenitor cells and reparative neovascularization. *Atherosclerosis*, *223*, 122–129.

Vaziri, H., & Benchimol, S. (1996). From telomere loss to p53 induction and activation of a DNA-damage pathway at senescence: The telomere loss/DNA damage model of cell aging. *Experimental Gerontology*, *31*, 295–301.
Vaziri, H., Dragowska, W., Allsopp, R. C., Thomas, T. E., Harley, C. B., & Lansdorp, P. M. (1994). Evidence for a mitotic clock in human hematopoietic stem cells: Loss of telomeric DNA with age. *Proceedings of the National Academy of Sciences of the United States of America*, *91*, 9857–9860.
Villeda, S. A., Luo, J., Mosher, K. I., Zou, B., Britschgi, M., Bieri, G., et al. (2011). The ageing systemic milieu negatively regulates neurogenesis and cognitive function. *Nature*, *477*, 90–94.
Wahlestedt, M., Norddahl, G. L., Sten, G., Ugale, A., Frisk, M. A., Mattsson, R., et al. (2013). An epigenetic component of hematopoietic stem cell aging amenable to reprogramming into a young state. *Blood*, *121*, 4257–4264.
Walters, B. J., & Zuo, J. (2013). Postnatal development, maturation and aging in the mouse cochlea and their effects on hair cell regeneration. *Hearing Research*, *297*, 68–83.
Wang, X., Krupczak-Hollis, K., Tan, Y., Dennewitz, M. B., Adami, G. R., & Costa, R. H. (2002). Increased hepatic Forkhead Box M1B (FoxM1B) levels in old-aged mice stimulated liver regeneration through diminished p27Kip1 protein levels and increased Cdc25B expression. *The Journal of Biological Chemistry*, *277*, 44310–44316.
Wang, X., Quail, E., Hung, N. J., Tan, Y., Ye, H., & Costa, R. H. (2001). Increased levels of forkhead box M1B transcription factor in transgenic mouse hepatocytes prevent age-related proliferation defects in regenerating liver. *Proceedings of the National Academy of Sciences of the United States of America*, *98*, 11468–11473.
Warr, M. R., Binnewies, M., Flach, J., Reynaud, D., Garg, T., Malhotra, R., et al. (2013). FOXO3A directs a protective autophagy program in haematopoietic stem cells. *Nature*, *494*, 323–327.
Weber, T., Corbett, M. K., Chow, L. M., Valentine, M. B., Baker, S. J., & Zuo, J. (2008). Rapid cell-cycle reentry and cell death after acute inactivation of the retinoblastoma gene product in postnatal cochlear hair cells. *Proceedings of the National Academy of Sciences of the United States of America*, *105*, 781–785.
White, P. M., Stone, J. S., Groves, A. K., & Segil, N. (2012). EGFR signaling is required for regenerative proliferation in the cochlea: Conservation in birds and mammals. *Developmental Biology*, *363*, 191–200.
Williamson, K. A., Hamilton, A., Reynolds, J. A., Sipos, P., Crocker, I., Stringer, S. E., et al. (2013). Age-related impairment of endothelial progenitor cell migration correlates with structural alterations of heparan sulfate proteoglycans. *Aging Cell*, *12*, 139–147.
Williamson, K., Stringer, S. E., & Alexander, M. Y. (2012). Endothelial progenitor cells enter the aging arena. *Frontiers in Physiology*, *3*, 30.
Xia, W. H., Li, J., Su, C., Yang, Z., Chen, L., Wu, F., et al. (2011). Physical exercise attenuates age-associated reduction in endothelium-reparative capacity of endothelial progenitor cells by increasing CXCR4/JAK-2 signaling in healthy men. *Aging Cell*, *11*, 111–119.
Xia, W. H., Yang, Z., Xu, S. Y., Chen, L., Zhang, X. Y., Li, J., et al. (2012). Age-related decline in reendothelialization capacity of human endothelial progenitor cells is restored by shear stress. *Hypertension*, *59*, 1225–1231.
Xu, J., Gonzalez, E. T., Iyer, S. S., Mac, V., Mora, A. L., Sutliff, R. L., et al. (2009). Use of senescence-accelerated mouse model in bleomycin-induced lung injury suggests that bone marrow-derived cells can alter the outcome of lung injury in aged mice. *The Journals of Gerontology. Series A, Biological Sciences and Medical Sciences*, *64*, 731–739.
Yahata, T., Takanashi, T., Muguruma, Y., Ibrahim, A. A., Matsuzawa, H., Uno, T., et al. (2011). Accumulation of oxidative DNA damage restricts the self-renewal capacity of human hematopoietic stem cells. *Blood*, *118*, 2941–2950.

Yalcin, S., Zhang, X., Luciano, J. P., Mungamuri, S. K., Marinkovic, D., Vercherat, C., et al. (2008). Foxo3 is essential for the regulation of ataxia telangiectasia mutated and oxidative stress-mediated homeostasis of hematopoietic stem cells. *The Journal of Biological Chemistry, 283*, 25692–25705.

Yang, Z., Qiao, H., Sun, Z., & Li, X. (2012). Effect of BDNF-plasma-collagen matrix controlled delivery system on the behavior of adult rats neural stem cells. *Journal of Biomedical Materials Research. Part A, 101*, 599–606.

Yang, Z., Xia, W. H., Su, C., Wu, F., Zhang, Y. Y., Xu, S. Y., et al. (2011). Regular exercise-induced increased number and activity of circulating endothelial progenitor cells attenuates age-related decline in arterial elasticity in healthy men. *International Journal of Cardiology, 165*, 247–254.

Yao, Y. G., Kajigaya, S., Feng, X., Samsel, L., McCoy, J. P., Jr., Torelli, G., et al. (2013). Accumulation of mtDNA variations in human single CD34+ cells from maternally related individuals: Effects of aging and family genetic background. *Stem Cell Research, 10*, 361–370.

Yu, Y., Weber, T., Yamashita, T., Liu, Z., Valentine, M. B., Cox, B. C., et al. (2010). In vivo proliferation of postmitotic cochlear supporting cells by acute ablation of the retinoblastoma protein in neonatal mice. *The Journal of Neuroscience, 30*, 5927–5936.

Yuan, Q., Su, H., Guo, J., Tsang, K. Y., Cheah, K. S., Chiu, K., et al. (2012). Decreased c-Jun expression correlates with impaired spinal motoneuron regeneration in aged mice following sciatic nerve crush. *Experimental Gerontology, 47*, 329–336.

Zhang, J., Grindley, J. C., Yin, T., Jayasinghe, S., He, X. C., Ross, J. T., et al. (2006). PTEN maintains haematopoietic stem cells and acts in lineage choice and leukaemia prevention. *Nature, 441*, 518–522.

Zhang, C. C., & Lodish, H. F. (2008). Cytokines regulating hematopoietic stem cell function. *Current Opinion in Hematology, 15*, 307–311.

Zheng, H., Martin, J. A., Duwayri, Y., Falcon, G., & Buckwalter, J. A. (2007). Impact of aging on rat bone marrow-derived stem cell chondrogenesis. *The Journals of Gerontology. Series A, Biological Sciences and Medical Sciences, 62*, 136–148.

Zhou, S., Greenberger, J. S., Epperly, M. W., Goff, J. P., Adler, C., Leboff, M. S., et al. (2008). Age-related intrinsic changes in human bone-marrow-derived mesenchymal stem cells and their differentiation to osteoblasts. *Aging Cell, 7*, 335–343.

Zhou, F., Onizawa, S., Nagai, A., & Aoshiba, K. (2011). Epithelial cell senescence impairs repair process and exacerbates inflammation after airway injury. *Respiratory Research, 12*, 78.

Zhu, S., Deng, S., Ma, Q., Zhang, T., Jia, C., Zhuo, D., et al. (2013). MicroRNA-10A* and MicroRNA-21 modulate endothelial progenitor cell senescence via suppressing high-mobility group A2. *Circulation Research, 112*, 152–164.

Zhu, C. H., Mouly, V., Cooper, R. N., Mamchaoui, K., Bigot, A., Shay, J. W., et al. (2007). Cellular senescence in human myoblasts is overcome by human telomerase reverse transcriptase and cyclin-dependent kinase 4: Consequences in aging muscle and therapeutic strategies for muscular dystrophies. *Aging Cell, 6*, 515–523.

Zine, A., Nyffeler, M., & de Ribaupierre, F. (2000). Spatial expression patterns of epidermal growth factor receptor gene transcripts in the postnatal mammalian cochlea. *Hearing Research, 141*, 19–27.

Zou, Y., Chiu, H., Zinovyeva, A., Ambros, V., Chuang, C. F., & Chang, C. (2013). Developmental decline in neuronal regeneration by the progressive change of two intrinsic timers. *Science, 340*, 372–376.

Zwetsloot, K. A., Childs, T. E., Gilpin, L. T., & Booth, F. W. (2013). Non-passaged muscle precursor cells from 32-month old rat skeletal muscle have delayed proliferation and differentiation. *Cell Proliferation, 46*, 45–57.

CHAPTER NINE

Skeletal Muscle Degeneration and Regeneration in Mice and Flies

Mamta Rai*, Upendra Nongthomba*, Miranda D. Grounds†,[1]
*Department of Molecular Reproduction, Development and Genetics, Indian Institute of Science, Bangalore, India
†School of Anatomy, Physiology and Human Biology, The University of Western Australia, Perth, Australia
[1]Corresponding author: e-mail address: miranda.grounds@uwa.edu.au

Contents

Abstract

Many aspects of skeletal muscle biology are remarkably similar between mammals and tiny insects, and experimental models of mice and flies (*Drosophila*) provide powerful tools to understand factors controlling the growth, maintenance, degeneration (atrophy and necrosis), and regeneration of normal and diseased muscles, with potential

Current Topics in Developmental Biology, Volume 108
ISSN 0070-2153
http://dx.doi.org/10.1016/B978-0-12-391498-9.00007-3

applications to the human condition. This review compares the limb muscles of mice and the indirect flight muscles of flies, with respect to the mechanisms of adult myofiber formation, homeostasis, atrophy, hypertrophy, and the response to muscle degeneration, with some comment on myogenic precursor cells and common gene regulatory pathways. There is a striking similarity between the species for events related to muscle atrophy and hypertrophy, without contribution of any myoblast fusion. Since the flight muscles of adult flies lack a population of reserve myogenic cells (equivalent to satellite cells), this indicates that such cells are not required for maintenance of normal muscle function. However, since satellite cells are essential in postnatal mammals for myogenesis and regeneration in response to myofiber necrosis, the extent to which such regeneration might be possible in flight muscles of adult flies remains unclear. Common cellular and molecular pathways for both species are outlined related to neuromuscular disorders and to age-related loss of skeletal muscle mass and function (sarcopenia). The commonality of events related to skeletal muscles in these disparate species (with vast differences in size, growth duration, longevity, and muscle activities) emphasizes the combined value and power of these experimental animal models.

ABBREVIATIONS

AMP(s) adult muscle precursor(s)
APF after puparium formation
DLM(s) dorsal longitudinal muscle(s)
DVM(s) dorsal ventral muscle(s)
ECM extracellular matrix
IFM(s) indirect flight muscle(s)
NMJ neuromuscular junction

1. INTRODUCTION

This review compares skeletal muscle maintenance, degeneration, and regeneration in the widely used laboratory models of the mouse (*Mus musculus* representing mammals) and the fruit fly (*Drosophila melanogaster* representing insects). These diverse models are vastly different in their life spans, body design, and the way muscles are formed, yet there are remarkable similarities between these species with respect to skeletal muscle formation (myogenesis), function, and also mechanisms of muscle degeneration and regeneration. Accordingly, the mass of data generated from genetic studies in flies has greatly informed research into mammalian models and vice versa (Daczewska, Picchio, Jagla, Figeac, & Jagla, 2010; Taylor, 2006). This review will discuss mechanisms for modulation of muscle size and degeneration and restoration/regeneration of muscles in various situations, with respect to the possible involvement of myogenic precursor cells, with a focus on limb muscles of the mouse and the indirect flight muscles (IFMs) of *Drosophila*. First, we present a broad overview of muscle formation (myogenesis) and maturation in the mouse and fly before discussing the issues of degeneration and regeneration.

2. SKELETAL MUSCLE FORMATION (MYOGENESIS) AND MATURATION IN MICE AND FLIES

Skeletal muscle is formed in a wide range of species from a pool of muscle precursor cells (myoblasts) that proliferate and then differentiate and fuse together to form multinucleated muscle cells (known as myotubes). These multinucleated cells increase in size, become innervated, and mature into myofibers.

2.1. The mouse

In mammals such as the mouse, myofibers are formed by myoblast fusion during *in utero* embryogenesis. There are considerable differences between skeletal muscles of the limbs, head, and body, with respect to the source of precursor cells and expression patterns of regulatory genes (Bismuth & Relaix, 2010). This discussion will focus mainly on the limb muscles that arise from the somites; however, there is much in common for a wide range of muscles for mechanisms of myogenesis and the regulatory genes involved between mice and flies (Relaix & Buckingham, 1999). In mouse limb muscle development, the myoblasts first fuse to generate primary myotubes. The primary myotube then becomes innervated and provides a scaffold for assembly of secondary myotubes that form in close association with the site of innervation (Duxson & Sheard, 1995; Paul, Sheard, & Duxson, 2004). In larger species with much bigger muscles (e.g., sheep), tertiary myotubes are also required (Paul et al., 2004). The myofiber number is generally fixed by birth and this is followed by postnatal growth and muscle maturation.

In mature mice and other mammals, a small population of quiescent myoblasts persists on the surface of the mature myofiber (outside the sarcolemma and beneath the basement membrane); these are called satellite cells and often referred to as myogenic stem cells. These quiescent satellite cells are activated in response to disturbances of myofiber homeostasis, and especially in response to necrosis of myofibers in order to form new muscle to replace the damaged segment (reviewed in Relaix & Zammit, 2012; Yablonka-Reuveni, 2011).

The diameter (and cross-sectional area, CSA) of mature myofibers is reasonably constant across species as this is limited physically by diffusion of nutrients and oxygen, and penetration of action potentials (generated from nerves) throughout the myofiber: the largest muscles tend to be those that generate more power (up to around 100 μm diameter or 6000 μm^2 CSA). In marked contrast, the length of mature myofibers varies dramatically within one animal and across species, dependent on the anatomical location of the specific muscle and size of the animal (Grounds & Shavlakadze, 2011). For example, in flies the length of largest adult mature myofibers ranges from 0.5 to 0.8 mm (U.N., unpublished data); whereas in humans the length of an individual myofiber can range from about 2 mm (small stapedius muscle in the ear) up to 500 mm for the sartorius muscle in the thigh. In large animals, increasing the length of myofibers can necessitate multiple sites of innervation to effectively propagate the electrical stimulation throughout the entire myofiber (Paul et al., 2004).

2.2. The fly

The basic cellular and developmental processes involved in muscle biology are largely conserved from insects to mammals (Sink, 2006). In insects such as *Drosophila*, the eggs are laid outside the body and muscle development is biphasic, from the embryo/larva (through the transition of pupation) to the adult form. Embryonic muscle formation takes place in less than a day and the body muscle pattern is retained in the larva. However, most of these muscles degenerate at pupal metamorphosis by a process similar to necrosis. Adult muscle formation occurs after pupal metamorphosis and is characterized by marked remodeling of larval muscles and their formation takes place in less than 4 days (Bate, Rushton, & Currie, 1991; Fernandes, Bate, & Vijayraghavan, 1991). *Drosophila* has a set of 30 larval muscles in each abdominal hemisegment (Bate, 1990) and at least 78 muscles in the adult thorax (Miller, 1950). Of all the adult thoracic muscles, IFMs are the largest. Embryonic muscles are single myofibers containing 3–25 myonuclei (Bate, 1990). In contrast, IFMs that are similar to vertebrate skeletal muscles (Taylor, 2006) contain hundreds of nuclei and have intricate organization very similar to that of vertebrate skeletal muscles (Rai & Nongthomba, 2013). IFMs are the only fibrillar muscles in *Drosophila* (Crossley, 1978) and are comparable to vertebrate skeletal muscles in structure (Fig. 9.1).

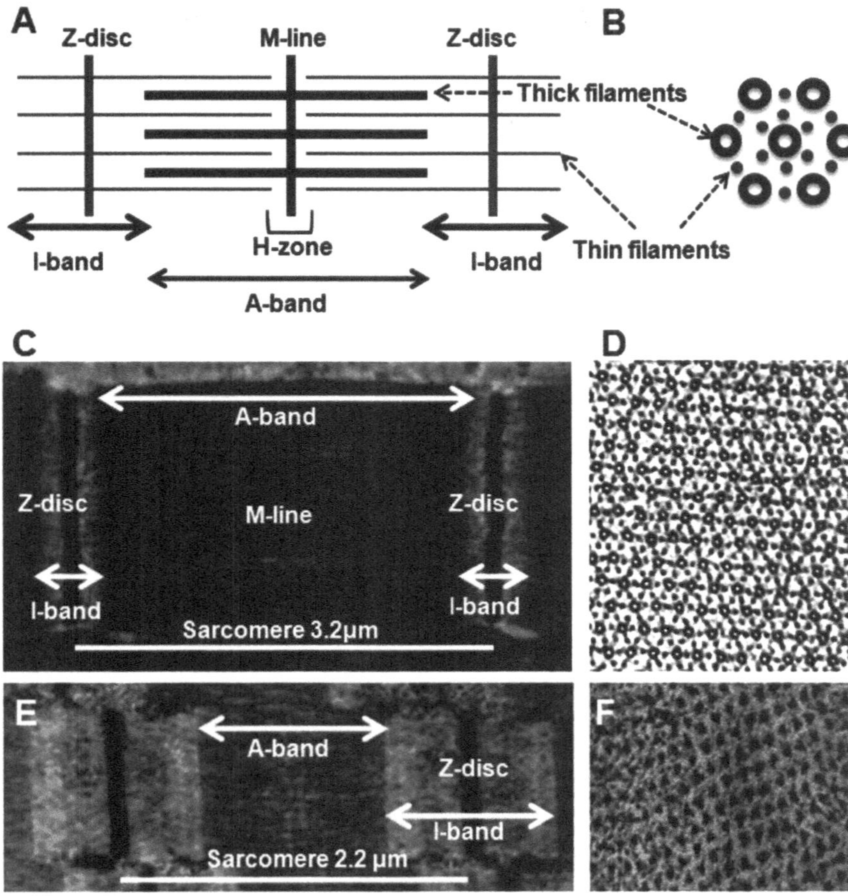

Figure 9.1 Comparison of *Drosophila* IFM and human skeletal muscle sarcomere organization. (A) Diagram of a single sarcomere viewed in longitudinal and (B) in transverse section showing hexagonal arrangement. (C) Electron micrograph of *Drosophila* IFM sarcomere shown in longitudinal section and (D) in transverse section. (E) Electron micrograph of human skeletal muscles sarcomere shown in longitudinal and (F) in transverse section. *(Courtesy Prof. Gayathri N., NIMHANS, Bangalore, for E, and F)*. (See color plate.)

Much of the review will thus compare the *Drosophila* IFMs to vertebrate skeletal muscles.

The bulk of the thorax of the adult fly is occupied by IFMs, which are divided into two groups. One contains the dorsal longitudinal muscles (DLMs) which require larval muscles as templates for their formation and pattern (Crossley, 1978; Fernandes et al., 1991). The other group contains dorsal ventral muscles (DVMs), which form by *de novo* fusion of myoblasts

(Fernandes et al., 1991). During pupal metamorphosis embryonic/larval muscles and motoneurons are reconstructed to provide the functional adult motor system (Consoulas, Restifo, & Levine, 2002; Fernandes et al., 1991; Fernandes & VijayRaghavan, 1993; Tissot & Stocker, 2000). At the onset of metamorphosis, in the prepupal and pupal stages, most of the larval musculature undergoes histolysis (Crossley, 1978), apart from the three mesothoracic muscles called larval oblique muscles (LOMs) (Fernandes et al., 1991). These spared larval muscles form the template for future DLM formation (Fig. 9.2). This mechanism of formation of flight muscles from larval muscles during development in *Drosophila* is also seen in other insects such as the Colorado beetle *Leprinorarsa decemliwata* (Smit & Velzing, 1986). Both embryonic and adult muscles are derived from mesodermal cells (Bate, 1993). Progenitors for the adult muscle formation called adult muscle precursors (AMPs) are set aside (in quiescence) during embryonic muscle

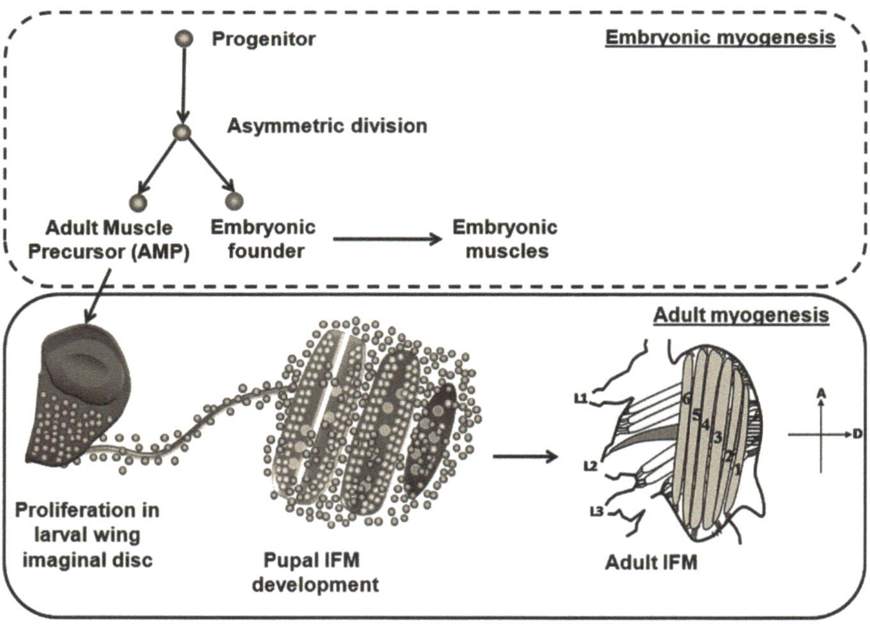

Figure 9.2 Biphasic muscle development in *Drosophila*. In the embryo, muscle progenitors are specified to form both embryonic and adult muscles. The AMPs that will form the IFMs get segregated in the wing imaginal disc where they proliferate during larval stages and at pupal metamorphosis, they then migrate toward, and fuse with, the three larval template muscles to form the six adult DLMs, as part of the IFMs. Green balls: myoblasts; Yellow balls: fused myoblasts or nuclei; large yellow balls: larval nuclei; adult DLMs are labeled 1–6; DVMs are present beneath the DLM. (See color plate.)

development (Fernandes et al., 1991). These AMPs subsequently undergo proliferation in wing and leg imaginal discs during the larval life and persistently express Twist (Bate et al., 1991; Currie & Bate, 1991; Fernandes et al., 1991). Their undifferentiated nature makes them similar to vertebrate satellite cells and hence offers a model to understand the biology of satellite cells in regeneration of muscles (Daczewska et al., 2010; Figeac, Daczewska, Marcelle, & Jagla, 2007; Taylor, 2006). Involvement of *Drosophila* AMPs is particularly well understood for adult DLMs, which form by fusion of AMPs to larval muscle templates each of which split to eventually form six adult DLMs in each hemithorax (Fig. 9.2). The motoneurons supplying the larval muscles also retract (Currie & Bate, 1991; Fernandes & VijayRaghavan, 1993), leaving a partial framework of peripheral nerves which provide a track for migration of AMPs. *De novo* fusion of AMPs beneath the DLMs gives rise to DVMs (Fernandes et al., 1991). In this way, adult IFMs develop from remnant larval muscle using neuronal cues from remodeled larval innervations.

2.3. Genetic regulation in mice and flies

Mice are a very valuable mammalian research model due to many sophisticated tools available for modulation of gene expression in this species (Schiaffino & Mammucari, 2011). A similar and highly complementary situation exists for genetic manipulation of *Drosophila*, with the advantage that this technology is quicker and cheaper in flies (Sparrow, Hughes, & Segalat, 2008). Research into the molecular and cellular regulation of IFM development has identified roles for many genes in *Drosophila*, which have equivalents in vertebrate muscle development and function, with comparable phenotype alterations across species (Schnorrer et al., 2010). There are many examples of this and a few are indicated below. For instance, when the human cell cycle inhibitor p21 is overexpressed in *Drosophila* myoblasts that will contribute to IFMs, the myoblasts fail to proliferate and fail to fuse with the DLM templates. As a result, the DLMs degenerate (Roy & VijayRaghavan, 1998) suggesting a common role for cell cycle genes during the muscle development. Persistent expression of dedifferentiating factors like Twist (Anant, Roy, & VijayRaghavan, 1998; Bernard et al., 2003) and Notch (Anant et al., 1998; Soler & Taylor, 2009) leads to apoptotic degeneration of IFMs. The counterparts to integrins which play important roles during IFM development and attachment (Fernandes, Celniker, & VijayRaghavan, 1996) are similarly affected (alpha7beta1) in congenital muscular dystrophy due to laminin deficiency in humans and mice (Vachon et al., 1997). The IFMs provide a unique system to genetically dissect the pathways of new genes and of genes whose phenotypes are clearly represented but whose pathways are not fully understood.

2.4. Comparison of life span, muscle growth, and body size

The duration of the life span and the overall size of the animal will affect the kinetics of growth and tissue maintenance and also the overall loading of different muscles. A comparison of the time line and scale of muscle growth for flies, mice, and humans is indicated in Fig. 9.3. Another factor to consider is the marked impact of different forms of locomotion on specific muscles: for example, the bipedal posture of humans means far greater loading on muscles of the legs compared with arms, in contrast with the relatively shared load on all limb muscles in quadrupedal animals such as mice, and the IFMs that aid rapidly moving wings in *Drosophila* are distinctly different again.

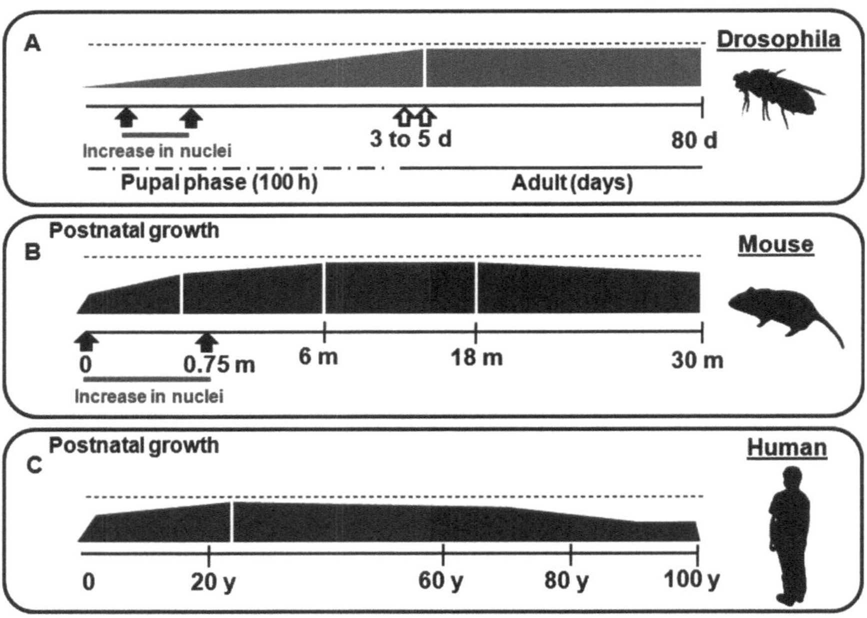

Figure 9.3 Indication of growth pattern over time for increasing myofiber size, maintenance of adult muscles, and longevity in the three species: not to scale. (A) In *Drosophila*, IFM development starts at pupal metamorphosis, with myoblast fusion starting as early as 8 h (h) "After Puparium Formation" (APF) with increasing numbers of myonuclei continuing up to approximately 36 h APF; the duration of this hyperplasia phase is indicated by black arrows (Fernandes et al., 1991). The muscles then undergo hypertrophic growth until 3–5 days posteclosion; shown by white arrows, with 1d indicating the start

In addition to understanding the formation and architecture of myofibers and the contractile properties of myofilaments, it is pertinent to make some brief comments about other factors that impact the normal maintenance of mass and function of skeletal muscles.

3. HOMEOSTASIS AND A RANGE OF FACTORS ESSENTIAL FOR OPTIMAL MUSCLE FUNCTION

Once skeletal muscles are formed, have finished growth, and are mature, their homeostasis and function depends not only on the size of the myofiber and the contractile properties of the myofilaments, but on many other factors both within and outside the myofibers. These include modulation of metabolism, vascular supply, the extracellular matrix composition, and efficacy of neuronal stimulation.

Within the myofiber, the sensitivity of response to a wide range of growth factors (including hormone levels that can vary throughout life) can alter the capacity for protein synthesis and degradation to result in net loss of protein, and there may also be changes to many other aspects of metabolism that include energy generation. These can be influenced by aging and disease. For example, young growing muscles of mammals are more sensitive to growth stimuli such as IGF-1 signaling that results in ready hypertrophy of neonatal/growing compared with adult myofibers (Davis & Fiorotto, 2009; Shavlakadze et al., 2010) and young mice more rapidly switch on protein degradation (catabolic) pathways in response to food deprivation (fasting), compared with adult mice (Shavlakadze, Soffe, Anwari, Cozens, & Grounds, 2013). Such turnover of proteins by signaling molecules

of the adult muscle phase (Reedy & Beall, 1993), and myofibers retain their size throughout the life of the fly (Miller et al., 2008). (B) In mice, rapid postnatal growth of hindlimb muscles occurs for about 6 weeks, then growth (including elongation) slows and has almost ceased by 3 months. Initially, there is extensive fusion of myoblasts (satellite cells) with the growing myofibers (hyperplasia) but this stops around 3 weeks (0.75 month), with growth after this time being due almost entirely to hypertrophy (White et al., 2010). Age-related loss of muscle mass (sarcopenia) is evident by 24 months of age and is more pronounced by 29 months. (C) In humans, the muscles increase in size until after adolescence (around 18 years of age); after this time when elongation has ceased in adults, muscles may increase in cross-sectional area (especially in males) due to hypertrophy and retain their mass until about 50–60 years of age. After this age, sarcopenia can become apparent with muscle mass decreased by about 30% by 70 years (d, m, and y represents day(s), month(s), and year(s), respectively). (See color plate.)

involved in both the proteasomal and autophagosomal proteolysis pathways is an essential part of normal homeostasis: however, impaired protein removal can result in accumulation of damaged proteins that progressively compromise normal cell function. The maximum metabolic rate of *Drosophila* IFM is approximately 10 times greater than the metabolic rate of human skeletal muscles (Crossley, 1978), suggesting extensive muscle usage. Similar to mouse skeletal muscles, both autophagosomal and proteasomal pathways are important for homeostasis and proper functioning of *Drosophila* IFMs (Demontis & Perrimon, 2010).

The mechanism used to generate energy for myofibrillar contraction is influenced by the myofiber type, with fast (glycolytic) myofibers depending mainly on glycogen metabolism. In contrast, slow (oxidative) myofibers (including *Drosophila* IFMs) have a greater dependency on mitochondria and lipid oxidation and thus the numbers of healthy mitochondria can influence the capacity to maintain muscle mass and function. Many aspects of metabolism may be altered in old mammalian muscles, as discussed further under Sarcopenia (Section 8): a detailed discussion of the metabolic complexities is beyond the scope of this review.

Outside the myofiber, a range of factors must be considered related to homeostasis; only three are mentioned here and these are associated with variables within the muscle tissue. A healthy blood supply and capillary network is essential to deliver oxygen and nutrients required for energy and to remove wastes. Disruption to the blood supply will compromise muscle contraction and in severe cases of ischemia will result in myofiber necrosis (Blaisdell, 2002; Labbe, Lindsay, & Walker, 1987). The extracellular matrix (ECM) plays a vital role in transmission of contractile force to move parts of the body. The force generated by the contractile myofilaments within the myofibers is transferred at the myotendinous junctions and also laterally along the length of the myofiber and across the sarcolemma to the ECM: initially to the basement membrane (a special form of ECM) and then to the interstitial connective tissue where collagens play a major role in strength and contraction (Grounds, Sorokin, & White, 2005). The ECM is a highly complex and dynamic environment. Many roles for the multitude of interacting components (proteins, sugars, enzymes, receptors) have been described during mammalian myogenesis/embryogenesis and in normal and regenerating muscles, with some important changes that affect 'stiffness' and strength also well documented during aging (reviewed in Grounds, 2008). Equivalent roles have already been identified for some ECM components in *Drosophila*, for example, for integrins (Fernandes et al., 1996; Pines et al., 2012) and for thrombospondin in myotendinous junction formation (Subramanian, Wayburn, Bunch, & Volk, 2007). Innervation is clearly fundamental for initiating myofiber contraction and thus any factors that influence the quality of the nerve supply (some of these may be in the central or peripheral nervous system) will modify the contractile properties and function. It is well documented that denervation and unloading/ disuse all result in loss of muscle mass—known as atrophy or wasting (as does starvation). Some of these situations of muscle atrophy are reversible whereas others are not. Denervation can be transient if the damaged nerve (or another motor-neuron axon) can sprout and reconnect with the denervated myofiber; however, severing of major nerves may be irreversible with permanent loss of the neuromuscular junction (NMJ) connections. In aging, there is a well-documented heterogeneous loss of nerves and alterations to NMJs in some muscles and such denervation will ultimately lead to impaired muscle function. This denervation may become far more severe over time in animals with a long life span, such as humans that age for many years, compared with only about months of aging in mice and rats, and days in *Drosophila* (Fig. 9.3).

Having established the mode of muscle formation and multitude of factors that influence the normal mass and function of adult skeletal muscles across a range of species, we will now consider the mechanisms involved in increasing (hypertrophy) and decreasing (atrophy) muscle mass and regeneration of skeletal muscles. Again, remarkable similarities are apparent between the experimental models of mice and *Drosophila*.

4. MECHANISMS OF MUSCLE HYPERTROPHY

During growth, there is a tremendous increase in myofiber size with respect to both length and cross-sectional area and this has many implications for properties of the expanding sarcolemma (Grounds & Shavlakadze, 2011). Once growth ceases, that is, in mammals the bones and hence muscles are of a fixed length, any increase in muscle mass (hypertrophy) is due to the expanded cross-sectional area of individual myofibers. Increased mass of mature muscles is not due to addition of new myofibers, since the number of myofibers is generally considered to be fixed at birth. The increase in myofiber size can involve a combination of two mechanisms (Fig. 9.4). In some situations, myoblasts fuse with the growing/hypertrophic myofibers and this process of addition of myonuclei, which is called "hyperplasia," certainly happens during normal muscle growth (Fig. 9.4B.1). In the context of multinucleated skeletal muscle cells, the term hyperplasia can be confusing as it can refer to an increase in numbers of myonuclei within an individual

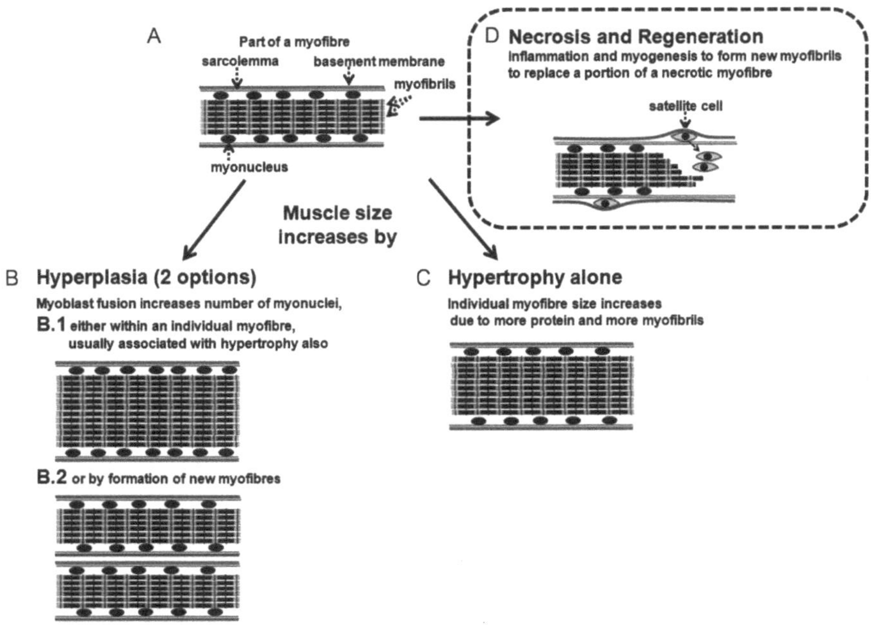

Figure 9.4 Mechanisms of skeletal muscle growth and regeneration. (A) represents a myofiber. Two mechanisms are shown for increasing muscle size: (B) hyperplasia involves fusion of myoblasts and can increase (B.1) numbers of myonuclei within individual myofibers (seen in neonatal) or (B.2) can generate new myofibers (occurs during embryogenesis and *Drosophila* mid-pupal DLM development). (C) Hypertrophy increases myofiber size without a change in number of myonuclei and is largely due to increased net protein content (this occurs during restoration after atrophy, postnatal muscle, and late pupal DLM development). A combination of B.1 and C can occur within an individual myofiber. (D) In response to myofiber necrosis, classic regeneration with myogenesis (involving satellite cells) occurs to replace the damaged myofiber.

myofiber, rather than an increase in the number of cells (mononucleated or myofibers) *per se*: this issue and many other aspects of factors controlling muscle mass are discussed elsewhere (Shavlakadze & Grounds, 2006). In many situations, an increase in myofiber size can be due purely to increased net protein content without any addition of myonuclei (Fig. 9.4C): this results in an altered myonuclear/cytoplasmic domain (Liu et al., 2009). It is of interest to consider the contribution of these two mechanisms to increased myofiber size in different situations.

In neonatal mice, there is steep increase in muscle size for about 6 weeks and then the growth rate slows and almost ceases by 12 weeks of age (White, Bierinx, Gnocchi, & Zammit, 2010). The number of myoblasts is high at birth and progressively decreases until 3 weeks of age (Beilharz, Lareu,

Garrett, Grounds, & Fletcher, 1992). During this stage there is extensive fusion of myoblasts into the elongating growing myofibers (hyperplasia), but after this time the increasing myofiber size is due mainly to hypertrophy (with minimal change in myonuclear number) (Fig. 9.3B) (White et al., 2010). The reason for this striking transition from hyperplasia to hypertrophy during growth is not known and has not been widely investigated for other species. However, an interesting recent paper (Delhaas, Van der Meer, Schaart, Degens, & Drost, 2013) investigated this aspect of myonuclear domain size in growing thoracoabdominal muscles of humans (aged 0.2 months to 15 years) and concluded that satellite cell density steeply declines in early infancy. The skeletal muscle growth between birth and the age of 15 years is due to an increase in the size of the fibers without a proportional increase in myonuclear number per fiber. As a result, the myonuclear domain area [μm^2] increases rapidly in the first months after birth and reaches adult values in the young adolescent phase (Delhaas et al., 2013). This rapid postnatal decrease in satellite cell numbers and steep increase in myonuclear domain size over the first year indicates a very early switch between hyperplasia and hypertrophic during growth of these human muscles (although this might differ for limb muscles with greater elongation) and supports great plasticity in the size of the myonuclear domain (Van der Meer, Jaspers, & Degens, 2011).

In *Drosophila* IFM development, hyperplasic growth takes place before the initiation of sarcomere assembly, when hundreds of myoblasts fuse to existing larval templates and grow in size (Fernandes et al., 1991). After fusion, the myonuclei realign into columns to form myofibers (M.R. and U.N. unpublished). Moreover, in a recent study, it has been found that reducing the pool of myoblasts available for fusion affects the number of myofibers constituting a DLM, suggesting an increase in myofiber number with development (Rai & Nongthomba, 2013). This implies a hyperplastic muscle growth where number of myofibers increase during the early DLM development (Fig. 9.4B.2). Myoblast fusion takes place until 36 h after puparium formation (h APF) after which the fusion rate slows and ceases by 40 h APF (DeSimone, Coelho, Roy, VijayRaghavan, & White, 1996). Once the fusion and differentiation process is complete, myofibers grow by increased protein synthesis/hypertrophy (Fig. 9.3A) (Rai & Nongthomba, 2013; Reedy & Beall, 1993), as documented for vertebrate skeletal muscles (Fig. 9.4C).

While nutrition and hormones are essential during the growth phase for muscles to grow and reach maturity, in healthy adult skeletal muscle an increase in mass is primarily driven by mechanical factors due to exercise and physical loading. Plus there is the restoration of muscle by hypertrophy in response to a prior loss of muscle mass (atrophy). A wide range of conditions including starvation, disuse, denervation, disease, and aging can result in muscle atrophy (wasting) and associated loss of function, across many species. The molecular signaling associated with these changes in muscle size is complex and has been widely studied, especially in mouse models, and is generally conserved across species (reviewed in Shavlakadze & Grounds, 2006; Schiaffino & Mammucari, 2011).

4.1. Are satellite cells required for mouse muscle homeostasis and hypertrophy?

Satellite cells (that give rise to myoblasts) are considered essential in mammals for postnatal muscle growth and also for myogenesis and regeneration in response to muscle necrosis (reviewed in Relaix & Zammit, 2012; Yablonka-Reuveni, 2011). However, it seems that satellite cells may play little or no role in the homeostasis of normal uninjured mature muscle. This is controversial. It seems to be widely assumed that there is some constant turnover of myofibers resulting from regeneration and this in turn assumes some form of ongoing damage in normal muscles. Yet there is little direct evidence for these in sedentary individuals and even in humans or mice subjected to moderate physiological exercise (Grounds & Shavlakadze, 2011). Instead, it may be that there is no routine need for satellite cells in normal muscle function: they might just remain quiescent for years (only to be used in case of an emergency). This proposal is supported by the fact that *Drosophila* appears to have no satellite cells in adult IFMs, yet clearly these muscles function perfectly well in their absence (in the homeostatic uninjured state). Skeletal muscle regeneration is clearly essential for long-lived large animals including humans and mice where there is much opportunity for severe injury to occur during their life span. It is emphasized that satellite cells play a crucial role for muscle regeneration in mammals (Relaix & Zammit, 2012), although the need for this may be infrequent in many situations of a moderate lifestyle.

The question then arises of how much hypertrophy can occur in the absence of a contribution by satellite cells? Clearly there is extensive hypertrophy without any fusion of myoblasts during later stages of postnatal growth (as discussed earlier) with merely a redistribution of the existing myonuclei and an adjustment of myonuclear/cytoplasmic ratio (White et al., 2010). Recent studies in *Drosophila* also indicate that the myoblast number dictates the initial hypertrophy of IFMs (Rai & Nongthomba, 2013). A similar situation of marked changes in myonuclear domain size has been reported in adult rats, where hindlimb suspension did not result in apoptosis of myonuclei (although nuclei of many interstitial cells succumbed to apoptosis) and there was no change in numbers of myonuclei during either atrophy or subsequent hypertrophy (Bruusgaard et al., 2012). The same conclusion was reached from studies with mice that were genetically engineered to deplete satellite cells, where muscle regrowth with hypertrophy occurred over 2 weeks in response to muscle atrophy (Jackson et al., 2012). Other studies of mature muscle in mice lacking satellite cells observed hypertrophy in response to mechanical overload (McCarthy et al., 2011) or to activation of hypertrophic signaling pathways (Lee et al., 2012). These studies confirm that hypertrophy of adult muscle can certainly occur initially in the absence of satellite cells. The extent to which the myonuclear domain might then become a limiting factor and restrict more extensive further hypertrophy is not clear. Significant increases in normal muscle mass (e.g., by resistance exercise training) may indeed require activation of satellite cells and their fusion with the expanding myofiber (Bruusgaard et al., 2012). The emerging use of genetically modified lines of mice with conditional ablation of >90% of satellite cells in mature skeletal muscles (Relaix & Zammit, 2012) provides powerful tools to more fully evaluate the long-term consequences of a lack of satellite cells on normal function of uninjured mature mammalian muscles.

5. MUSCLE HYPERTROPHY AND ATROPHY IN INSECTS: COMPARISON WITH OTHER SPECIES

5.1. Hypertrophy

In insects, hypertrophic growth of muscles takes place during late pupal life prior to eclosion by progressive addition of myofilaments to existing myofibrils. In this manner, myofibril diameter in IFM has been observed to increase from 4 thick filaments to ~36 thick filaments in *D. melanogaster* (Reedy & Beall, 1993). In other insect taxa, muscle growth continues in the adult stage. Muscle growth is controlled by relative concentrations of juvenile hormone (JH) and ecdysteroids, which are also involved in the molting process. In the flight capable cricket *Teleogryllus oceanicus*, treatment with the JH-analog (inhibitor) methoprene during the final nymph stage interferes with the rate of both the muscle and mitochondrial growth that normally occur in the first few days of adult life (Novicki, 1989).

Mitochondrial fusion leading to enlarged mitochondria during late pupal life has recently been found to be important for hypertrophic growth of *Drosophila* flight muscles (Rai & Nongthomba, 2013). In the desert locust *Schistocerca gregaria*, methoprene treatment of the final instar nymphs results in nymphs molting into a supernumerary nymph stage rather than to the adult stage: the flight muscles in these supernumerary nymphs show reduced growth of myofibrillar components without affecting the mitochondrial growth, in contrast with the response of *T. oceanicus* (above) (Wang, Chen, & Haunerland, 1993). Mitochondria are critical for energy production and modulation of mitochondrial number (biogenesis) and activity also occurs in the metabolic adaptation of mammalian muscle. For example in mice, modulation of the mitochondrial p43 mitochondrial receptor strongly affects muscle mass and the metabolic and contractile features of myofibers (Pessemesse et al., 2011) and chronic endurance training induces sustained mitochondrial biogenesis in skeletal muscles (Li et al., 2011). In humans increased mitochondrial biogenesis (with changes in volume and activity) is involved in phenotypic adaptations to endurance exercise (Hood, Irrcher, Ljubicic, & Joseph, 2006) and there is a loss of both mitochondria and myofibrillar volume in disuse atrophy (Hoppeler & Fluck, 2003).

5.2. Neuronal influence

Insect muscle growth is also regulated by innervation, in addition to hormones. Both the *Drosophila* IFMs and their neuronal innervation remodel simultaneously (Fernandes & Keshishian, 1998; Fernandes & VijayRaghavan, 1993). Denervation during IFM development leads to delayed muscle formation or degeneration (Fernandes & Keshishian, 2005) and denervation of flight muscles in the cricket *T. oceanicus* before the last molt also reduced muscle growth as well as that of mitochondria (Novicki, 1989). Similarly, cutting the nerves of developing leg muscles in moths emphasized that motor nerves are essential for both the accumulation of myoblasts into the correct areas of muscle development and the appropriate level of proliferation (Consoulas & Levine, 1997). The clear involvement of nerves in muscle development is also echoed in mammalian development, growth, and homeostasis; for example, during mouse embryogenesis the formation of secondary myotubes depends on innervation of the primary myotubes (Duxson & Sheard, 1995; Paul et al., 2004). The importance of innervation is emphasized for many situations of stem cell niche, tissue formation, and regeneration (Kumar & Brockes, 2012). The role of innervation in *Drosophila* in all aspects of skeletal muscle formation, growth, maintenance, and muscle loss due to neuromuscular disease and aging is an important underdeveloped area with much functional relevance to mammals. More than 20 *Drosophila* orthologs that cause neuromuscular diseases in human have been described (Reiter, Potocki, Chien, Gribskov, & Bier, 2001) and these and the role of denervation in age-related loss of skeletal muscle mass are discussed later.

5.3. Atrophy

Striking degeneration (atrophy) of adult flight muscles themselves has been observed in a number of insects in response to social or environmental cues. In particular, reproduction-associated muscle histolysis has been documented in several species of insects (Bhakthan, Borden, & Nair, 1970; Davis, Jones, & Farmer, 1989; Hocking, 1952; Johnson, 1953; López-Guillén et al., 2011). Such muscle degeneration is irreversible or complete in some (Davis et al., 1989) and incomplete in others (Bhakthan et al., 1970; Langor, 1987). Incomplete muscle degeneration that is reversible has been seen in a few reemerging Eastern larch beetle with or without regaining the flight ability (Langor, 1987) and in bark beetles (Borden & Slater, 1969). This seems to be due to striking myofiber atrophy that can subsequently be reversed by hypertrophy and may be controlled by hormonal changes. For example, once the adult bark beetle *Ips confusus* is exposed to suitable bark of a pine log (Bhakthan et al., 1970) the flight muscles undergo dramatic atrophy over about 4 days where many of the mitochondria and most of the myofilaments disappear and little structural organization remains in the myofibers. This muscle degeneration is especially striking in females and probably provides the recycling of proteins and lipids for egg yolk formation and maturation (Atkins & Farris, 1962). In males the atrophy renders them flightless, presumably to ensure their presence around egg galleries and enhance their engagement in the reproductive process (Bhakthan et al., 1970). However, postdegeneration, muscles in the reemerging bark beetle *Ips confuses* increase in volume faster in males than females (Borden & Slater, 1969).

Rapid flight muscle histolysis also occurs in queen fire ants (Jones & Davis, 1985) in response to insemination and is protease dependent, with muscle loss apparent by 6 h and completed by 24 h. Hormonal influence has been observed to dictate such degeneration and restoration of flight muscles (Stegwee, Kimmel, de Boer, & Henstra, 1963). The mechanisms of restoration of adult flight muscles are not yet understood. It appears that this striking atrophy in many insects may occur without loss of myonuclei (Unnithan & Nair, 1977) and that restoration of myofiber size mainly involves protein synthesis with formation of new myofilaments and mitochondria without any marked change in number of myonuclei (Bhakthan, Nair, & Borden, 1971). Such reversible atrophy with no loss of myonuclei appears similar to the situations for disuse atrophy of mouse muscles (discussed earlier).

Evolutionary similarities across a wide range of species are further emphasized at the mechanistic and signaling level by the molt-induced claw muscle atrophy in large-clawed crustaceans (lobsters and crabs) to facilitate withdrawal of the appendage at ecdysis: this process is coordinated by ecdysteroid hormones. This atrophy, that can reduce muscle mass by almost 80% in crabs, involves the activation of both calcium-dependent (calpains) and ubiquitin (Ub)/proteasome-dependent proteolytic systems that break down proteins to reduce muscle mass (Koenders, Yu, Chang, & Mykles, 2002). Other studies in crabs show a fourfold reduction in mass accompanied by remodeling of the contractile apparatus followed by an 11-fold increase in myofibrillar protein synthesis by the end of the premolt period: this restoration of muscle mass involves the classic protein synthesis pathways regulated by mTOR and myostatin (MacLea et al., 2012).

6. MUSCLE DEGENERATION VERSUS NECROSIS/REGENERATION IN MICE AND FLIES

As discussed earlier for a range of situation in mice, flies, and other species, degeneration may involve reversible muscle atrophy (wasting) in response to situations like transient starvation or disuse, and during molting and in reproductive situations in crustacea and insects, that is manifested by shrinking of the myofiber often without loss of myonuclei. This may be relatively easy to reverse by increasing the net protein content and return to original size. To avoid confusion, it is suggested that it may be helpful to refer to this as restoration, rather than regeneration. We propose that the term regeneration is used exclusively for situations where myofiber necrosis has occurred and inflammation and myogenesis are required to replace a significant area of damaged muscle tissue.

Long-term severe atrophy can produce complications and may be more difficult to reverse. In mice and men, severe atrophy (wasting) can also occur in situations where high levels of inflammatory cytokines are present, as in cachexia associated with cancer and acquired immune deficiency syndrome

(AIDS). Denervation also results in muscle atrophy and the consequences of long-term denervation may be hard or impossible to reverse. Some forms of degeneration due to disease can result in altered myofiber metabolism and function with disappearance of myofibrils. In other situations of disease or physical or chemical injury resulting from accidents, there is major disruption of the sarcolemma and the myofiber undergoes necrosis with destruction of most tissue components and myonuclei. In many instances, the necrosis may affect only a segment of the myofiber, with undamaged segments at either side being sealed off by newly formed plasmalemma (Papadimitriou, Robertson, Mitchell, & Grounds, 1990). Such myofiber necrosis initiates classic regeneration. In brief, in mice the overall process of regeneration is similar to that for other tissues, with initial resealing of the damaged area and infiltration of inflammatory cells to phagocytose the dead tissue and to produce cytokines and enzymes that activate the migration and proliferation of precursor cells, as well as modify the extracellular matrix. The activated myoblasts accumulate at the site of damage, myoblast proliferation starts around 18–24 h, peaks around 3–4 days, and rapidly declines to cease by about 7–10 days; the precise timing depends on the scale and type of injury (McGeachie & Grounds, 1987). Some myoblasts start to differentiate and fuse together to form myotubes from about 3.5 days and myotube formation is almost completed by 7–10 days (Robertson, Grounds, Mitchell, & Papadimitriou, 1990). After this time, myotubes fuse with the resealed ends of the myofibers (Robertson, Papadimitriou, & Grounds, 1993), any damaged innervation is restored and the myofibers increase in size with regeneration probably being largely completed by 3 weeks after necrosis. This complex series of cellular and molecular events associated with classic regeneration is clearly a vastly different process to the relatively simple restoration of atrophic myofibers by hypertrophy: hence the need to be careful with language to clearly distinguish between these two situations.

6.1. Satellite cells and regeneration in mammals

The satellite cell is essential for postnatal muscle growth and quiescent satellite cells remain as the likely major source of myoblasts for regeneration of mammalian muscles (reviewed in Grounds & Relaix, 2010; Relaix & Zammit, 2012). Satellite cells require Pax7 for normal function, as demonstrated in Pax7 null mice where myogenesis was normal *in utero*, but postnatal muscles failed to elongate and grow (Seale et al., 2000): while satellite cells were present at birth it appears that Pax7 is required for maintenance and survival of satellite cells. *Pax7* expression is maintained in virtually all quiescent satellite cells in adult mouse muscle and in many other species (reviewed in Relaix & Zammit, 2012). While it has been shown that many other cell types are capable of forming myoblasts in tissue culture and experimentally in mice, the contribution of such cells to normal myogenesis *in vivo* and regeneration of adult mammalian muscles appears to be minimal and the functional significance of these other potential sources of myoblasts remains unclear (Grounds & Relaix, 2010; Relaix & Zammit, 2012). One possibility is that they may form new myofibers *de novo* in the interstitial space (Tamaki, Uchiyama, & Akatsuka, 2010) as has been reported to occur in some situations.

6.2. Are *Drosophila* adult muscle precursors similar to mammalian satellite cells?

Drosophila AMPs (that are present in embryonic-larval stages) share similarities with vertebrate satellite cells both with respect to cellular origin and the maintenance of a quiescent undifferentiated state. However, AMPs are not located beneath the basement membrane of mature myofibers (the classic location for satellite cells in vertebrates) and they do not appear to require expression of Pax genes, whereas Pax7 expression appears essential for activity of postnatal satellite cells in limb muscles of mammals. In *Drosophila*, Numb, the inhibitor of Notch, is localized asymmetrically in the progenitor cell. After cell division the Numb negative progeny continues to express Twist and becomes the AMP (Ruiz Gomez & Bate, 1997). Similar asymmetric distribution of Numb has been observed for muscle satellite cells in vertebrates (Conboy & Rando, 2002). There is much interest in the concept of asymmetric divisions for renewing a stem cell component of the satellite cell population (Rocheteau, Gayraud-Morel, Siegl-Cachedenier, Blasco, & Tajbakhsh, 2012) although the precise *in vivo* significance is difficult to determine (Lansdorp, Falconer, Tao, Brind'Amour, & Naumann, 2012).

There is further evidence to support myogenic stem behavior of AMPs from transplantation experiments where donor myoblasts (derived from a proliferated pool of AMPs isolated from *Drosophila* larval wing disc) could make most of the muscles depending on the location (Lawrence & Brower, 1982), highlighting the importance of ectodermal signaling in genetic programming and plasticity of myoblasts. Thus, studying the origin, specification, and maintenance of AMPs through the embryonic-larval stages of *Drosophila* could provide a better understanding of muscle stem biology to complement knowledge gleaned from vertebrate systems.

Due to the absence of satellite cells in the adult fly muscles, it might be predicted that the muscles could not regenerate in response to severe damage that results in necrosis. From an evolutionary perspective, this might be due to the small size and relatively very short life span (~80 days) where there may be no compelling need for a mechanism to repair damaged adult *Drosophila* (and other insect) muscles. However, remodeling of myofibrillar structures certainly occurs in response to some forms of muscle damage in adult *Drosophila.* Physical injury of adult flies in the thoracic region resulted in upregulated expression of several muscle structural genes specific to IFMs, including *Act88F*, the only actin in IFM, *mlc* and *TpnC4* (Apidianakis et al., 2007). Similarly, mutants with a hypercontraction phenotype in IFM, where muscle loss occurs without evidence of regeneration, also resulted in upregulation in the structural genes *Act88F*, *mlp60A*, *mlp84B,* and *titin* (Montana & Littleton, 2006). However, not all the structural genes were upregulated, indicating an inefficient response for tissue remodeling. In addition, adult mutants null for *Act88F* and *Mhc* also show upregulation of other structural genes (A.M., D.T. and U.N., unpublished data), further demonstrating the remodeling capacity of *Drosophila* IFMs.

7. MUSCLE DEGENERATION IN NEUROMUSCULAR DISEASES

In human, denervation of muscle fibers leads to progressive loss of muscle mass and function. For example, in Werdnig-Hoffmann disease (acute infantile spinal muscular atrophy) apoptotic degradation with formation of apoptotic bodies was reported in many myofibers (Tews, 2002). Also in one of the most extensively studied denervation disorders, spinal muscular atrophy, loss in muscle mass is observed (Tews, 2002), very much reminiscent of the degenerating DLMs when denervated (Fernandes & Keshishian, 1998). The juvenile onset of Parkinson's diseases or Autosomal Recessive-Juvenile Parkinsonism (AR-JP) (*parkin*) is a neurodegenerative disease (Kitada et al., 1998): *Drosophila* with mutations in the related gene *parkin* show reduced life span, locomotor defects (apoptotic cell death of muscle subsets), and male sterility. Such pathological phenotypes are also observed in ectopic expression of the Parkinson's disease gene coding for α-synuclein, which is not represented in the *Drosophila* genome (Feany & Bender, 2000). The human gene *Spastin* that causes autosomal dominant hereditary paraplegia is expressed in both neurons and muscles and regulates neuromuscular synaptic activities and there are *Drosophila* homologs (Sherwood, Sun, Xue, Zhang, & Zinn, 2004). In *Drosophila*, hypomorphic mutations in the *Spastin* gene show defective flight and escapers of the protein null mutants fail to fly, suggesting a neuromuscular defect. Although the membrane-associated dystrophin, that is responsible for the lethal human disease Duchenne Muscular Dystrophy, is not present in *Drosophila*, sarcoglycans which form a trans-membrane complex with dystrophin are present and mutations in δ-sarcoglycan lead to shortened sarcomeres and disorganized M-lines (Allikian et al., 2007). These examples of phenotype overlap between species for gene expression disturbances related to a wide range of neuromuscular disorders further emphasize the value of such cross-species investigations.

8. SARCOPENIA: AGE-RELATED DEGENERATION WITH LOSS OF SKELETAL MUSCLE MASS AND FUNCTION

In adults, when skeletal muscles cease growing in length and diameter, myofibers remain at a relatively constant size (unless diameter is increased by hypertrophic exercise) and are stable until later in life when the mass and function decline: this age-related degeneration is called sarcopenia. Loss of muscle strength is one of the first hallmarks of aging whether in flies, nematodes, mice, or humans (Augustin & Partridge, 2009; Zheng et al., 2005). In humans, this loss of muscle mass and function starts after about 50 years (0.5–1% and 1.5–3% loss, respectively, per year), is reported as 14% at 65–69 years of age and >50% by 80 years (Janssen, 2010) (Fig. 9.3) and loss of muscle function may occur before loss of muscle mass: sarcopenia is associated with frailty, disability, and mortality (Sayer et al., 2013). Despite the fact that regular exercise greatly benefits muscle function with aging (Sayer et al., 2013), even highly trained athletes experience sarcopenia. The mechanisms responsible for the relentless progression of sarcopenia and the attenuated ability of aging muscle to respond to programs of conditioning and rehabilitation are largely unknown.

For mammals a wide range of factors are implicated in sarcopenia (Janssen, 2010; Sayer et al., 2013), including changes related to mitochondria (Calvani et al., 2012; Johnson, Robinson, & Nair, 2013) and metabolism within muscles (Altun et al., 2010; Rennie et al., 2009), with an impaired response to protein feeding known as anabolic resistance in older muscles (Burd, Wall, & van Loon, 2012). Increasing evidence suggests age-related impairment of protein turnover, including autophagy (Rubinsztein, Marino, & Kroemer, 2011) that can result in accumulations of many damaged proteins leading to protein aggregation with adverse effects on cell function. One form of protein aggregation called lipofuscin is elevated in aged skeletal muscles (Tohma, Hepworth, Shavlakadze, Grounds, & Arthur, 2011) of mice and increased lipofuscin is also documented in the brain, and to a lesser extent, in thorax muscles, of aging honey bees (Martin, Korall, & Förster, 1989). Lipofuscin is classically used as a marker for aging in many tissues and is associated with elevated oxidative stress that causes irreversible damage to macromolecules. Oxidative stress is widely implicated in muscle wasting and aging and can affect many aspects of cellular function (Arthur, Grounds, & Shavlakadze, 2008).

While death of myofibers via apoptosis is reported in aged muscles, the impact of apoptosis in multinucleated myofibers is more complex than for mononucleated cells (Salucci et al., 2013) and reports of apoptosis within muscles need to carefully distinguish between death of nuclei in interstitial cells, rather than myonuclei (Bruusgaard et al., 2012). Denervation of myofibers (Aagaard, Suetta, Caserotti, Magnusson, & Kjaer, 2010; Edstrom et al., 2007) with altered morphology and loss of NMJs (Chai, Vukovic, Dunlop, Grounds, & Shavlakadze, 2011) will clearly affect muscle function and is associated with sarcopenia. Other changes include alterations to the composition of extracellular matrix surrounding myofibers that can result in stiffness (Grounds, 2008). Some of these age-related changes can be ameliorated (to some extent) by exercise. It is of interest to identify biomarkers to identify the cellular events responsible for the onset and progression of sarcopenia on animal models, and ideally to apply these to clinical situations, especially to help measure the efficacy of interventions to reduce sarcopenia in older humans.

Sarcopenia has been widely studied in rats and to a lesser extent in mice. In contrast with human muscles that are composed predominantly of slow myofibers, rodent muscles are mainly fast: such differences between species need to be considered when extending observation from animal models to humans. Furthermore, there is much longer time for secondary consequences to become pronounced in humans where sarcopenia progressively manifests over >30 years, whereas the duration is far shorter in mice >1 year (from 18 to 30 months), with the normal life span of mice being a mere 3 years or less. This issue of duration of aging is far more limited in *Drosophila*, with an adult life span of only about 80 days, yet they also manifest sarcopenia. The challenge is to try and determine if there is a common molecular basis across the species for the initiation and progression of sarcopenia.

In mice, sarcopenia is evident as loss of mass of quadriceps muscle by 24 months, and pronounced by 29 months, of age (Shavlakadze, McGeachie, & Grounds, 2010). For species comparison a 15-month-old mouse is roughly equivalent to a 50-year human, and a 27-month-old mouse corresponds to a human aged >75 years. Denervation of the neuromuscular junctions of myofibers, visualized on whole muscles, has recently been documented in old mice by 24 months of age (Chai et al., 2011; Jang & Van Remmen, 2010; Valdez et al., 2010). Microarray and protein analysis studies in the Grounds laboratory of muscles from mice ages 3, 12, 24, and 29 months substantiate the observation of striking denervation of myofibers in mice that occurs sometime between 15 and 24 months of age (Barns et al., unpublished data). Rats show a similar time course of aging and sarcopenia: a recent comprehensive microarray and proteomic analysis of aging rat muscles specifically identified major changes that significantly correlated with sarcopenia, as neuromuscular junction remodeling and denervation of myofibers, plus altered mitochondrial energy metabolism (Ibebunjo et al., 2013).

In Drosophila, degenerative changes take place in aging thoracic muscles or IFMs (Takahashi, Philpott, & Miquel, 1970), that result in reduced flight ability, wing-beat frequency (Miller et al., 2008), and locomotors ability (Piazza, Gosangi, Devilla, Arking, & Wessells, 2009). Examination of IFMs in aged flies, by 49 days (life span of *Drosophila* is 60–80 days) revealed reduced sarcomere length, reduced total fiber area, increased *in vivo* interfilament distance, and altered abundance of muscle proteins (Miller et al., 2008). Also Z-band thickening and disappearance of M-line is evident in aged IFMs (Miller et al., 2008; Takahashi et al., 1970).

As the maximum metabolic rate of IFM is approximately 10 times greater than the metabolic rate of human skeletal muscles (Crossley, 1978), the energy requirement of IFMs is extremely high. So, as for vertebrates, mitochondrial biogenesis and degradation should manifest themselves in a balanced manner. This appears to be the case for flies since mitochondrial changes are associated with apoptosis and a steady increase in cell death as a result of increased expression of apoptotic genes in IFM of flies that had covered 20–60% of their life span (Zheng et al., 2005). Apoptosis was prevalent in thorax under hypoxic and induced-oxidative conditions (but not starvation), suggesting an interaction between lifelong accumulation of oxidative damage and cell death (Zheng et al., 2005). Intracellular oxidative stress is caused by ROS (reactive oxygen species: free radical derivative of oxygen) that are by-products of mitochondrial respiration (also produced by endoplasmic reticulum, peroxisomes, and lysosomes). Mitochondrial proteins are thus susceptible to such kind of stress and hence culminate in apoptotic pathways. Such mitochondrial defects and apoptosis would lead to structural and functional defects.

In old flies, it has been shown that the rate of protein synthesis is strikingly decreased in thorax muscles by 96% (compared with 15% and 33% in head and abdomen, respectively) in 35-day-old flies compared to one-day-old flies (Webster, Beachell, & Webster, 1980), leading to net protein loss. Furthermore, progressive accumulation of protein aggregates in IFM of 5-week-old flies is associated with impaired muscle function (Demontis & Perrimon, 2010). The authors found that the forkhead transcription factor FOXO and its downstream target 4E-BP are required for removing damaged proteins from muscles and hence contribute to normal muscle proteostasis. The reduction in age-related locomotory (Jones & Grotewiel, 2011; Piazza et al., 2009) and heart activities can be decreased by exercise training in young flies (Piazza et al., 2009), thus flies also serve as experimental model to assess the benefits of exercise.

8.1. Muscle as a potential tissue to enhance longevity

Since sarcopenia and aging take place in parallel, the potential participation of skeletal muscle, the largest organ system in organisms, in longevity is being explored. Recent genetic studies in the mouse, the nematode, and *Drosophila* have provided significant insights into the pathways regulating longevity. Some of these pathways include the evolutionarily conserved insulin/insulin-like growth factor signaling (IIS) pathway, the Jun kinase (JNK) pathway, and the sirtuin pathway (reviewed in Gan & Mucke, 2008; Kenyon, 2005; Lionaki, Markaki, & Tavernarakis, 2012; Salminen & Kaarniranta, 2011). Tissue-specific overexpression of factors from longevity pathways can increase life span; for example, overexpression of *foxo* in the adult *Drosophila* fat body (Giannakou et al., 2004; Hwangbo, Gershman, Tu, Palmer, & Tatar, 2004) and thoracic muscles (Demontis & Perrimon, 2010) extended life span. This suggests participation of specific tissue(s) in regulating the longevity of an organism. Genome wide transcriptional profiling in a tissue-specific manner was thus carried out at six *Drosophila* adult stages (3, 15, 20, 30, 45, and 60 days) (Zhan et al., 2007) and at two stages (3 and 40 days) in another study (Girardot, Lasbleiz, Monnier, & Tricoire, 2006). Age-related genes (4–9% of all the genes in *Drosophila*) were found to be highest in thoracic muscles and the expression of genes involved in mitochondrial and proteasomal pathways was significantly altered in these tissues (Girardot et al., 2006; Zhan et al., 2007). Moreover, transcriptional profiling of human (Zahn et al., 2006) and rat (Ibebunjo et al., 2013) muscles also delivered a similar alteration in expression of genes encoding mitochondrial components with aging. The role of skeletal muscle in longevity and prevention of age-related diseases is discussed for flies and mammals in a recent review (Demontis, Piccirillo, Goldberg, & Perrimon, 2013): this emphasized that exercise influences release of myokines by skeletal muscles with systemic benefits for other tissues in terms of their function and metabolism related to aging. These diverse studies imply that skeletal muscle is an important organ for longevity, and hence maintenance of healthy muscle for an extended time may not only increase the efficiency of metabolic activities, but also the life span.

9. CONCLUSIONS

Many of the comparisons outlined earlier are summarized in Table 9.1. This collaborative review between two groups dedicated to studies of skeletal muscle biology in both flies and mice provides novel insights into the extent of events that are similar between these vastly different species. Such commonality serves to emphasize key issues that are of fundamental importance to skeletal muscle biology, including myogenesis, homeostasis, atrophy, and hypertrophy. There is also similarity between many events involved in various neuromuscular diseases and in normal age-related sarcopenia and these areas (like atrophy/hypertrophy) are ripe for further collaborative exploration across these species. The wealth of tools now available combined with the speed and relative simplicity of experimental genetic manipulation in flies (with a life span of only 60–80 days) makes them highly attractive for proof-of-principle studies, especially for investigating and validating issues related to aging; compared with very slow and expensive equivalent studies in old mice that take >2 years. These new insights into skeletal muscle biology between flies and mice fuel exciting possibilities for more extensive collaborations across both species to enrich our knowledge of factors that control the health of myofibers.

Table 9.1 Comparison of main features of skeletal muscles for mouse (representing mammals) and fly (*Drosophila*, representing insects), with a focus on adult limb muscles and indirect flight muscles (IFM), respectively

Characteristics	Mouse skeletal muscle	*Drosophila* DLM (part of IFM)
1. Muscle morphogenesis		
Template requirement	Secondary myotubes require primary myotube	DLM require larval oblique muscles (LOM)
Nerve innervations	Forms on primary myotube	Remodeling of larval neurons that innervate developing DLM
Genetic networks	Similar	Similar
2. Myofibrillar morphology	Similar	Similar
3. Muscle organization into higher orders	Similar	Similar
4. Regeneration		
	Muscles can undergo hypertrophy or repair necrotic myofibers	Larval muscles are remodeled by fusion of AMPs to form DLM. Regeneration after necrosis has not been tested.
Satellite cells	Muscle stem cells present in postnatal/adult muscle	AMPs present in embryonic/larval wing imaginal disc. Not detected in adults.
Quiescence	Quiescent unless required	Quiescent in embryonic/larval stages

These issues are discussed in the text.

ACKNOWLEDGMENTS

We would like to thank Prof. Gayathri N. from NIMHANS, Bangalore for electron micrographs of human skeletal muscles. Research in the Nongthomba lab is funded by the Department of Science and Technology, India and the Indian Institute of Science. Over many years, the research in the Grounds' group has been funded by the National Health and Medical Research Council of Australia.

GLOSSARY OF TERMS

Myofiber An individual multinucleated muscle cell, also called a skeletal muscle fiber: groups of these form a skeletal muscle.

Myoblast A skeletal muscle precursor cell. The developmental origin of these can be variable.

Satellite cell A muscle precursor cell located on the surface of a myofiber between the sarcolemma and the basement membrane. In many mature muscles, satellite cells are normally quiescent and when activated are called myoblasts.

Muscle hypertrophy The increase in size of a muscle mass or an individual myofiber by increased protein content.

Muscle hyperplasia A process by which muscle size increases by addition of myonuclei, either to an individual myofiber or by an increase in number of myofibers. The process may overlap with hypertrophy.

Muscle homeostasis The process of maintaining normal cellular and functional stability of a muscle.

Muscle atrophy The loss of muscle mass or decreased size of individual myofibers: occurs in response to a wide range of conditions and can be transient and reversible.

Muscle degeneration A broad term to cover processes by which muscles lose their structural integrity, mass, and sometimes associated tissues like neurons.

Muscle necrosis Damage to part of a myofiber that results in breakdown of the sarcolemma and fragmentation of the sarcoplasm; it is associated with inflammation and phagocytosis of the necrotic tissue, myogenesis, and replacement of the damaged segment of myofiber.

Muscle regeneration A process of new muscle formation (myogenesis), usually in mature muscles in response to necrosis.

Sarcopenia The progressive age-related loss of muscle mass and function.

REFERENCES

Aagaard, P., Suetta, C., Caserotti, P., Magnusson, S. P., & Kjaer, M. (2010). Role of the nervous system in sarcopenia and muscle atrophy with aging: Strength training as a countermeasure. *Scandinavian Journal of Medicine & Science in Sports*, *20*, 49–64.

Allikian, M. J., Bhabha, G., Dospoy, P., Heydemann, A., Ryder, P., Earley, J. U., et al. (2007). Reduced life span with heart and muscle dysfunction in *Drosophila* sarcoglycan mutants. *Human Molecular Genetics*, *16*, 2933–2943.

Altun, M., Besche, H. C., Overkleeft, H. S., Piccirillo, R., Edelmann, M. J., Kessler, B. M., et al. (2010). Muscle wasting in aged, sarcopenic rats is associated with enhanced activity of the ubiquitin proteasome pathway. *The Journal of Biological Chemistry*, *285*, 39597–39608.

Anant, S., Roy, S., & VijayRaghavan, K. (1998). Twist and Notch negatively regulate adult muscle differentiation in *Drosophila*. *Development*, *125*, 1361–1369.

Apidianakis, Y., Mindrinos, M. N., Xiao, W., Tegos, G. P., Papisov, M. I., Hamblin, M. R., et al. (2007). Involvement of skeletal muscle gene regulatory network in susceptibility to wound infection following trauma. *PLoS One*, *2*, e1356.

Arthur, P. G., Grounds, M. D., & Shavlakadze, T. (2008). Oxidative stress as a therapeutic target during muscle wasting: Considering the complex interactions. *Current Opinion in Clinical Nutrition and Metabolic Care*, *11*, 408–416.

Atkins, M. D., & Farris, S. H. (1962). A contribution to the knowledge of flight muscle changes in the Scolytidae (Coleoptera). *The Canadian Entomologist*, *94*, 25–32.

Augustin, H., & Partridge, L. (2009). Invertebrate models of age-related muscle degeneration. *Biochimica et Biophysica Acta*, *1790*, 1084–1094.

Bate, M. (1990). The embryonic development of larval muscles in *Drosophila*. *Development*, *110*, 791–804.

Bate, M. (1993). The mesoderm and its derivatives. In M. Bate & A. M. Arias (Eds.), *The development of Drosophila melanogaster* (pp. 1013–1090). New York: Cold Spring Harbor Laboratory Press.

Bate, M., Rushton, E., & Currie, D. A. (1991). Cells with persistent twist expression are the embryonic precursors of adult muscles in *Drosophila*. *Development*, *113*, 79–89.

Beilharz, M. W., Lareu, R. R., Garrett, K. L., Grounds, M. D., & Fletcher, S. (1992). Quantitation of muscle precursor cell activity in skeletal muscle by Northern analysis of MyoD and myogenin expression: Application to dystrophic (mdx) mouse muscle. *Molecular and Cellular Neurosciences*, *3*, 326–331.

Bernard, F., Lalouette, A., Gullaud, M., Jeantet, A. Y., Cossard, R., Zider, A., et al. (2003). Control of apterous by vestigial drives indirect flight muscle development in *Drosophila*. *Developmental Biology*, *260*, 391–403.

Bhakthan, N. M., Borden, J. H., & Nair, K. K. (1970). Fine structure of degenerating and regenerating flight muscles in a bark beetle, *Ips confusus*. I. Degeneration. *Journal of Cell Science*, *6*, 807–819.

Bhakthan, N. M., Nair, K. K., & Borden, J. H. (1971). Fine structure of degenerating and regenerating flight muscles in a bark beetle, *Ips confusus*. II. Regeneration. *Canadian Journal of Zoology*, *49*, 85–89.

Bismuth, K., & Relaix, F. (2010). Genetic regulation of skeletal muscle development. *Experimental Cell Research*, *316*, 3081–3086.

Blaisdell, F. W. (2002). The pathophysiology of skeletal muscle ischemia and the reperfusion syndrome: A review. *Cardiovascular surgery*, *10*, 620–630.

Borden, J. H., & Slater, C. E. (1969). Flight muscle volume change in Ips confusus (Coleoptera: Scolytidae). *Canadian Journal of Zoology*, *47*, 29–32.

Bruusgaard, J. C., Egner, I. M., Larsen, T. K., Dupre-Aucouturier, S., Desplanches, D., & Gundersen, K. (2012). No change in myonuclear number during muscle unloading and reloading. *Journal of Applied Physiology*, *113*, 290–296.

Burd, N. A., Wall, B. T., & van Loon, L. J. (2012). The curious case of anabolic resistance: Old wives' tales or new fables? *Journal of Applied Physiology*, *112*, 1233–1235.

Calvani, R., Joseph, A. M., Adhihetty, P. J., Miccheli, A., Bossola, M., Leeuwenburgh, C., et al. (2012). Mitochondrial pathways in sarcopenia of aging and disuse muscle atrophy. *Biological Chemistry*, *394*, 393–414.

Chai, R. J., Vukovic, J., Dunlop, S., Grounds, M. D., & Shavlakadze, T. (2011). Striking denervation of neuromuscular junctions without lumbar motoneuron loss in geriatric mouse muscle. *PLoS One*, *6*, e28090.

Conboy, I. M., & Rando, T. A. (2002). The regulation of Notch signaling controls satellite cell activation and cell fate determination in postnatal myogenesis. *Developmental Cell*, *3*, 397–409.

Consoulas, C., & Levine, R. B. (1997). Accumulation and proliferation of adult leg muscle precursors in Manduca are dependent on innervation. *Journal of Neurobiology*, *32*, 531–553.

Consoulas, C., Restifo, L. L., & Levine, R. B. (2002). Dendritic remodeling and growth of motoneurons during metamorphosis of *Drosophila melanogaster*. *The Journal of Neuroscience*, *22*, 4906–4917.

Crossley, A. C. (1978). The morphology and development of the *Drosophila* muscular system. In M. Ashburner & T. R. F. Wright (Eds.), *Genetics and biology of* Drosophila (pp. 499–560). New York: Academic Press.

Currie, D. A., & Bate, M. (1991). The development of adult abdominal muscles in *Drosophila*: Myoblasts express twist and are associated with nerves. *Development*, *113*, 91–102.

Daczewska, M., Picchio, L., Jagla, T., Figeac, N., & Jagla, K. (2010). Muscle development and regeneration in normal and pathological conditions: Learning from *Drosophila*. *Current Pharmaceutical Design*, *16*, 929–941.

Davis, T. A., & Fiorotto, M. L. (2009). Regulation of muscle growth in neonates. *Current Opinion in Clinical Nutrition and Metabolic Care, 12*, 78–85.
Davis, W. L., Jones, R. G., & Farmer, G. R. (1989). Insect hemolymph factor promotes muscle histolysis in Solenopsis. *The Anatomical Record, 224*, 473–478.
Delhaas, T., Van der Meer, S. F., Schaart, G., Degens, H., & Drost, M. R. (2013). Steep increase in myonuclear domain size during infancy. *Anatomical Record (Hoboken), 296*, 192–197.
Demontis, F., & Perrimon, N. (2010). FOXO/4E-BP signaling in *Drosophila* muscles regulates organism-wide proteostasis during aging. *Cell, 143*, 813–825.
Demontis, F., Piccirillo, R., Goldberg, A. L., & Perrimon, N. (2013). The influence of skeletal muscle on systemic aging and lifespan. *Aging Cell, 12*(6), 943–949.
DeSimone, S., Coelho, C., Roy, S., VijayRaghavan, K., & White, K. (1996). ERECT WING, the *Drosophila* member of a family of DNA binding proteins is required in imaginal myoblasts for flight muscle development. *Development, 122*, 31–39.
Duxson, M. J., & Sheard, P. W. (1995). Formation of new myotubes occurs exclusively at the multiple innervation zones of an embryonic large muscle. *Developmental Dynamics, 204*, 391–405.
Edstrom, E., Altun, M., Bergman, E., Johnson, H., Kullberg, S., Ramirez-Leon, V., et al. (2007). Factors contributing to neuromuscular impairment and sarcopenia during aging. *Physiology & Behavior, 92*, 129–135.
Feany, M. B., & Bender, W. W. (2000). A *Drosophila* model of Parkinson's disease. *Nature, 404*, 394–398.
Fernandes, J., Bate, M., & Vijayraghavan, K. (1991). Development of the indirect flight muscles of *Drosophila. Development, 113*, 67–77.
Fernandes, J. J., Celniker, S. E., & VijayRaghavan, K. (1996). Development of the indirect flight muscle attachment sites in *Drosophila*: Role of the PS integrins and the stripe gene. *Developmental Biology, 176*, 166–184.
Fernandes, J. J., & Keshishian, H. (1998). Nerve-muscle interactions during flight muscle development in *Drosophila. Development, 125*, 1769–1779.
Fernandes, J. J., & Keshishian, H. (2005). Motoneurons regulate myoblast proliferation and patterning in *Drosophila. Developmental Biology, 277*, 493–505.
Fernandes, J., & VijayRaghavan, K. (1993). The development of indirect flight muscle innervation in *Drosophila melanogaster. Development, 118*, 215–227.
Figeac, N., Daczewska, M., Marcelle, C., & Jagla, K. (2007). Muscle stem cells and model systems for their investigation. *Developmental Dynamics, 236*, 3332–3342.
Gan, L., & Mucke, L. (2008). Paths of convergence: Sirtuins in aging and neurodegeneration. *Neuron, 58*, 10–14.
Giannakou, M. E., Goss, M., Junger, M. A., Hafen, E., Leevers, S. J., & Partridge, L. (2004). Long-lived *Drosophila* with overexpressed dFOXO in adult fat body. *Science, 305*, 361.
Girardot, F., Lasbleiz, C., Monnier, V., & Tricoire, H. (2006). Specific age-related signatures in *Drosophila* body parts transcriptome. *BMC Genomics, 7*, 69.
Grounds, M. D. (2008). Complexity of extracellular matrix and skeletal muscle regeneration. In S. Schiaffino & T. Partridge (Eds.), *Skeletal muscle repair and regeneration* (pp. 269–302). The Netherlands: Springer.
Grounds, M. D., & Relaix, F. (2010). Myogenic Precursor Cells, in Section I- Scientific basis of muscle disease. In D. Hilton-Jones, R. C. Griggs, K. Bushby, & G. Karpati (Eds.), *Disorders of voluntary muscles* (pp. 20–36). Cambridge University Press.
Grounds, M. D., & Shavlakadze, T. (2011). Impact of growth on properties of sarcolemma of skeletal myofibres: Clinical and scientific implications. *Bioessays, 33*, 458–468.
Grounds, M. D., Sorokin, L., & White, J. (2005). Strength at the extracellular matrix-muscle interface. *Scandinavian Journal of Medicine & Science in Sports, 15*, 381–391.
Hocking, B. (1952). Autolysis of flight muscles in a mosquito. *Nature, 169*, 1101.

Hood, D. A., Irrcher, I., Ljubicic, V., & Joseph, A. M. (2006). Coordination of metabolic plasticity in skeletal muscle. *The Journal of Experimental Biology, 209*, 2265–2275.

Hoppeler, H., & Fluck, M. (2003). Plasticity of skeletal muscle mitochondria: Structure and function. *Medicine and Science in Sports and Exercise, 35*, 95–104.

Hwangbo, D. S., Gershman, B., Tu, M. P., Palmer, M., & Tatar, M. (2004). *Drosophila* dFOXO controls lifespan and regulates insulin signalling in brain and fat body. *Nature, 429*, 562–566.

Ibebunjo, C., Chick, J. M., Kendall, T., Eash, J. K., Li, C., Zhang, Y., et al. (2013). Genomic and proteomic profiling reveals reduced mitochondrial function and disruption of the neuromuscular junction driving rat sarcopenia. *Molecular and Cellular Biology, 33*, 194–212.

Jackson, J. R., Mula, J., Kirby, T. J., Fry, C. S., Lee, J. D., Ubele, M. F., et al. (2012). Satellite cell depletion does not inhibit adult skeletal muscle regrowth following unloading-induced atrophy. *American Journal of Physiology. Cell Physiology, 303*, C854–C861.

Jang, Y. C., & Van Remmen, H. (2010). Age-associated alterations of the neuromuscular junction. *Experimental Gerontology, 46*, 193–198.

Janssen, I. (2010). Evolution of sarcopenia research. *Applied Physiology, Nutrition, and Metabolism, 35*, 707–712.

Johnson, B. (1953). Flight muscle autolysis and reproduction in aphids. *Nature, 172*, 813.

Johnson, M. L., Robinson, M. M., & Nair, K. S. (2013). Skeletal muscle aging and the mitochondrion. *Trends in Endocrinology and Metabolism, 24*, 247–256.

Jones, R. G., & Davis, W. L. (1985). Leupeptin, a protease inhibitor, blocks insemination-induced flight muscle histolysis in the fire ant *Solenopsis*. *Tissue & Cell, 17*, 111–116.

Jones, M. A., & Grotewiel, M. (2011). *Drosophila* as a model for age-related impairment in locomotor and other behaviors. *Experimental Gerontology, 46*, 320–325.

Kenyon, C. (2005). The plasticity of aging: Insights from long-lived mutants. *Cell, 120*, 449–460.

Kitada, T., Asakawa, S., Hattori, N., Matsumine, H., Yamamura, Y., Minoshima, S., et al. (1998). Mutations in the parkin gene cause autosomal recessive juvenile parkinsonism. *Nature, 392*, 605–608.

Koenders, A., Yu, X., Chang, E. S., & Mykles, D. L. (2002). Ubiquitin and actin expression in claw muscles of land crab, Gecarcinus lateralis, and American lobster, Homarus americanus: Differential expression of ubiquitin in two slow muscle fiber types during molt-induced atrophy. *The Journal of Experimental Zoology, 292*, 618–632.

Kumar, A., & Brockes, J. P. (2012). Nerve dependence in tissue, organ, and appendage regeneration. *Trends in Neurosciences, 35*, 691–699.

Labbe, R., Lindsay, T., & Walker, P. M. (1987). The extent and distribution of skeletal muscle necrosis after graded periods of complete ischemia. *Journal of Vascular Surgery, 6*, 152–157.

Langor, D. W. (1987). Flight muscle changes in the eastern larch beetle, *Dendroctonus simplex* LeConte (Coleoptera: Scolytidae). *The Coleopterists Bulletin, 41*, 351–357.

Lansdorp, P. M., Falconer, E., Tao, J., Brind'Amour, J., & Naumann, U. (2012). Epigenetic differences between sister chromatids? *Annals of the New York Academy of Sciences, 1266*, 1–6.

Lawrence, P. A., & Brower, D. L. (1982). Myoblasts from *Drosophila* wing disks can contribute to developing muscles throughout the fly. *Nature, 295*, 55–57.

Lee, S. J., Huynh, T. V., Lee, Y. S., Sebald, S. M., Wilcox-Adelman, S. A., Iwamori, N., et al. (2012). Role of satellite cells versus myofibers in muscle hypertrophy induced by inhibition of the myostatin/activin signaling pathway. *Proceedings of the National Academy of Sciences of the United States of America, 109*, E2353–E2360.

Li, L., Muhlfeld, C., Niemann, B., Pan, R., Li, R., Hilfiker-Kleiner, D., et al. (2011). Mitochondrial biogenesis and PGC-1alpha deacetylation by chronic treadmill exercise:

Differential response in cardiac and skeletal muscle. *Basic Research in Cardiology, 106,* 1221–1234.

Lionaki, E., Markaki, M., & Tavernarakis, N. (2013). Autophagy and ageing: Insights from invertebrate model organisms. *Ageing Research Reviews, 12*(1), 413–428.

Liu, J. X., Hoglund, A. S., Karlsson, P., Lindblad, J., Qaisar, R., Aare, S., et al. (2009). Myonuclear domain size and myosin isoform expression in muscle fibres from mammals representing a 100,000-fold difference in body size. *Experimental Physiology, 94,* 117–129.

López-Guillén, G., Carrasco, J. V., Cruz-López, L., Barrera, J. F., Malo, E. A., & Rojas, J. C. (2011). Morphology and structural changes in flight muscles of *Hypothenemus hampei* (Coleoptera: Curculionidae) females. *Environmental Entomology, 40,* 441–448.

MacLea, K. S., Abuhagr, A. M., Pitts, N. L., Covi, J. A., Bader, B. D., Chang, E. S., et al. (2012). Rheb, an activator of target of rapamycin, in the blackback land crab, *Gecarcinus lateralis*: Cloning and effects of molting and unweighting on expression in skeletal muscle. *The Journal of Experimental Biology, 215,* 590–604.

Martin, H., Korall, H., & Förster, B. (1989). Magnetic field effects on activity and ageing in honeybees. *Journal of Comparative Physiology, 164,* 423–431.

McCarthy, J. J., Mula, J., Miyazaki, M., Erfani, R., Garrison, K., Farooqui, A. B., et al. (2011). Effective fiber hypertrophy in satellite cell-depleted skeletal muscle. *Development, 138,* 3657–3666.

McGeachie, J. K., & Grounds, M. D. (1987). Initiation and duration of muscle precursor replication after mild and severe injury to skeletal muscle of mice. An autoradiographic study. *Cell and Tissue Research, 248,* 125–130.

Miller, A. (1950). The internal anatomy and histology of the imago of *Drosophila melanogaster.* In M. Demerec (Ed.), *Biology of* Drosophila (pp. 420–534). New York: Cold Spring Harbor Press.

Miller, M. S., Lekkas, P., Braddock, J. M., Farman, G. P., Ballif, B. A., Irving, T. C., et al. (2008). Aging enhances indirect flight muscle fiber performance yet decreases flight ability in *Drosophila. Biophysical Journal, 95,* 2391–2401.

Montana, E. S., & Littleton, J. T. (2006). Expression profiling of a hypercontraction-induced myopathy in *Drosophila* suggests a compensatory cytoskeletal remodeling response. *The Journal of Biological Chemistry, 281,* 8100–8109.

Novicki, A. (1989). Control of growth and ultrastructural maturation of a cricket flight muscle. *The Journal of Experimental Zoology, 250,* 263–272.

Papadimitriou, J. M., Robertson, T. A., Mitchell, C. A., & Grounds, M. D. (1990). The process of new plasmalemma formation in focally injured skeletal muscle fibers. *Journal of Structural Biology, 103,* 124–134.

Paul, A. C., Sheard, P. W., & Duxson, M. J. (2004). Development of a mammalian series-fibered muscle. *The Anatomical Record. Part A, Discoveries in Molecular, Cellular, and Evolutionary Biology, 278,* 571–578.

Pessemesse, L., Schlernitzauer, A., Sar, C., Levin, J., Grandemange, S., Seyer, P., et al. (2011). Depletion of the p43 mitochondrial T3 receptor in mice affects skeletal muscle development and activity. *The FASEB Journal, 26,* 748–756.

Piazza, N., Gosangi, B., Devilla, S., Arking, R., & Wessells, R. (2009). Exercise-training in young *Drosophila melanogaster* reduces age-related decline in mobility and cardiac performance. *PLoS One, 4,* e5886.

Pines, M., Das, R., Ellis, S. J., Morin, A., Czerniecki, S., Yuan, L., et al. (2012). Mechanical force regulates integrin turnover in *Drosophila* in vivo. *Nature Cell Biology, 14,* 935–943.

Rai, M., & Nongthomba, U. (2013). Effect of myonuclear number and mitochondrial fusion on *Drosophila* indirect flight muscle organization and size. *Experimental Cell Research, 319*(17), 2566–2577.

Reedy, M. C., & Beall, C. (1993). Ultrastructure of developing flight muscle in *Drosophila*. I. Assembly of myofibrils. *Developmental Biology, 160*, 443–465.

Reiter, L. T., Potocki, L., Chien, S., Gribskov, M., & Bier, E. (2001). A systematic analysis of human disease-associated gene sequences in *Drosophila melanogaster*. *Genome Research, 11*, 1114–1125.

Relaix, F., & Buckingham, M. (1999). From insect eye to vertebrate muscle: Redeployment of a regulatory network. *Genes & Development, 13*, 3171–3178.

Relaix, F., & Zammit, P. S. (2012). Satellite cells are essential for skeletal muscle regeneration: The cell on the edge returns centre stage. *Development, 139*, 2845–2856.

Rennie, M. J., Selby, A., Atherton, P., Smith, K., Kumar, V., Glover, E. L., et al. (2009). Facts, noise and wishful thinking: Muscle protein turnover in aging and human disuse atrophy. *Scandinavian Journal of Medicine & Science in Sports, 20*, 5–9.

Robertson, T. A., Grounds, M. D., Mitchell, C. A., & Papadimitriou, J. M. (1990). Fusion between myogenic cells in vivo: An ultrastructural study in regenerating murine skeletal muscle. *Journal of Structural Biology, 105*, 170–182.

Robertson, T. A., Papadimitriou, J. M., & Grounds, M. D. (1993). Fusion of myogenic cells to the newly sealed region of damaged myofibres in skeletal muscle regeneration. *Neuropathology and Applied Neurobiology, 19*, 350–358.

Rocheteau, P., Gayraud-Morel, B., Siegl-Cachedenier, I., Blasco, M. A., & Tajbakhsh, S. (2012). A subpopulation of adult skeletal muscle stem cells retains all template DNA strands after cell division. *Cell, 148*, 112–125.

Roy, S., & VijayRaghavan, K. (1998). Patterning muscles using organizers: Larval muscle templates and adult myoblasts actively interact to pattern the dorsal longitudinal flight muscles of *Drosophila*. *The Journal of Cell Biology, 141*, 1135–1145.

Rubinsztein, D. C., Marino, G., & Kroemer, G. (2011). Autophagy and aging. *Cell, 146*, 682–695.

Ruiz Gomez, M., & Bate, M. (1997). Segregation of myogenic lineages in *Drosophila* requires numb. *Development, 124*, 4857–4866.

Salminen, A., & Kaarniranta, K. (2011). AMP-activated protein kinase (AMPK) controls the aging process via an integrated signaling network. *Ageing Research Reviews, 11*, 230–241.

Salucci, S., Burattini, S., Baldassarri, V., Battistelli, M., Canonico, B., Valmori, A., et al. (2013). The peculiar apoptotic behavior of skeletal muscle cells. *Histology and Histopathology, 28*, 1073–1087.

Sayer, A. A., Robinson, S. M., Patel, H. P., Shavlakadze, T., Cooper, C., & Grounds, M. D. (2013). New horizons in the pathogenesis, diagnosis and management of sarcopenia. *Age and Ageing, 42*, 145–150.

Schiaffino, S., & Mammucari, C. (2011). Regulation of skeletal muscle growth by the IGF1-Akt/PKB pathway: Insights from genetic models. *Skeletal Muscle, 1*, 4.

Schnorrer, F., Schonbauer, C., Langer, C. C., Dietzl, G., Novatchkova, M., Schernhuber, K., et al. (2010). Systematic genetic analysis of muscle morphogenesis and function in *Drosophila*. *Nature, 464*, 287–291.

Seale, P., Sabourin, L. A., Girgis-Gabardo, A., Mansouri, A., Gruss, P., & Rudnicki, M. A. (2000). Pax7 is required for the specification of myogenic satellite cells. *Cell, 102*, 777–786.

Shavlakadze, T., Chai, J., Maley, K., Cozens, G., Grounds, G., Winn, N., et al. (2010). A growth stimulus is needed for IGF-1 to induce skeletal muscle hypertrophy in vivo. *Journal of Cell Science, 123*, 960–971.

Shavlakadze, T., & Grounds, M. (2006). Of bears, frogs, meat, mice and men: Complexity of factors affecting skeletal muscle mass and fat. *Bioessays, 28*, 994–1009.

Shavlakadze, T., McGeachie, J., & Grounds, M. D. (2010). Delayed but excellent myogenic stem cell response of regenerating geriatric skeletal muscles in mice. *Biogerontology, 11*, 363–376.

Shavlakadze, T., Soffe, Z., Anwari, T., Cozens, G., & Grounds, M. D. (2013). Short-term feed deprivation rapidly induces the protein degradation pathway in skeletal muscles of young mice. *The Journal of Nutrition, 143*, 403–409.

Sherwood, N. T., Sun, Q., Xue, M., Zhang, B., & Zinn, K. (2004). *Drosophila* spastin regulates synaptic microtubule networks and is required for normal motor function. *PLoS Biology, 2*, e429.

Sink, H. (2006). *Muscle Development in* Drosophila. USA: Landes Bioscience.

Smit, W. A., & Velzing, E. H. (1986). Fusion of myocytes with larval muscle remnants during flight muscle development in the Colorado beetle. *Tissue & Cell, 18*, 469–478.

Soler, C., & Taylor, M. V. (2009). The Him gene inhibits the development of *Drosophila* flight muscles during metamorphosis. *Mechanisms of Development, 126*, 595–603.

Sparrow, J., Hughes, S. M., & Segalat, L. (2008). Other model organisms for sarcomeric muscle diseases. *Advances in Experimental Medicine and Biology, 642*, 192–206.

Stegwee, D., Kimmel, E. C., de Boer, J. A., & Henstra, S. (1963). Hormonal control of reversible degeneration of flight muscle in the colorado potato beetle, *Leptinotarsa decemlineata* say (Coleoptera). *The Journal of Cell Biology, 19*, 519–527.

Subramanian, A., Wayburn, B., Bunch, T., & Volk, T. (2007). Thrombospondin-mediated adhesion is essential for the formation of the myotendinous junction in *Drosophila. Development, 134*, 1269–1278.

Takahashi, A., Philpott, D. E., & Miquel, J. (1970). Electron microscope studies on aging *Drosophila melanogaster.* 3. Flight muscle. *Journal of Gerontology, 25*, 222–228.

Tamaki, T., Uchiyama, Y., & Akatsuka, A. (2010). Plasticity and physiological role of stem cells derived from skeletal muscle interstitium: Contribution to muscle fiber hyperplasia and therapeutic use. *Current Pharmaceutical Design, 16*, 956–967.

Taylor, M. V. (2006). Comparison of muscle development in *Drosophila* and vertebrates. In H. Sink (Ed.), *Muscle Development in* Drosophila (pp. 169–203). USA: Landes Bioscience.

Tews, D. S. (2002). Apoptosis and muscle fibre loss in neuromuscular disorders. *Neuromuscular Disorders, 12*, 613–622.

Tissot, M., & Stocker, R. F. (2000). Metamorphosis in *Drosophila* and other insects: The fate of neurons throughout the stages. *Progress in Neurobiology, 62*, 89–111.

Tohma, H., Hepworth, A. R., Shavlakadze, T., Grounds, M. D., & Arthur, P. G. (2011). Quantification of ceroid and lipofuscin in skeletal muscle. *The Journal of Histochemistry and Cytochemistry, 59*, 769–779.

Unnithan, G. C., & Nair, K. K. (1977). Ultrastructure of juvenile hormone-induced degenerating flight muscles in a bark beetle, *Ips paraconfusus. Cell and Tissue Research, 185*, 481–490.

Vachon, P. H., Xu, H., Liu, L., Loechel, F., Hayashi, Y., Arahata, K., et al. (1997). Integrins (alpha7beta1) in muscle function and survival. Disrupted expression in merosin-deficient congenital muscular dystrophy. *The Journal of Clinical Investigation, 100*, 1870–1881.

Valdez, G., Tapia, J. C., Kang, H., Clemenson, G. D., Jr., Gage, F. H., Lichtman, J. W., et al. (2010). Attenuation of age-related changes in mouse neuromuscular synapses by caloric restriction and exercise. *Proceedings of the National Academy of Sciences of the United States of America, 107*, 14863–14868.

Van der Meer, S. F. T., Jaspers, R. T., & Degens, H. (2011). Is the myonuclear domain size fixed? *Journal of Musculoskeletal & Neuronal Interactions, 11*, 286–297.

Wang, Z., Chen, X., & Haunerland, N. H. (1993). Flight muscle development in juvenile and adult forms of the desert locust, *Schistocerca gregaria. Journal of Insect Physiology, 39*, 325–333.

Webster, G. C., Beachell, V. T., & Webster, S. L. (1980). Differential decrease in protein synthesis by microsomes from aging *Drosophila melanogaster. Experimental Gerontology, 15*, 495–497.

White, R. B., Bierinx, A. S., Gnocchi, V. F., & Zammit, P. S. (2010). Dynamics of muscle fibre growth during postnatal mouse development. *BMC Developmental Biology*, *10*, 21.

Yablonka-Reuveni, Z. (2011). The skeletal muscle satellite cell: Still young and fascinating at 50. *The Journal of Histochemistry and Cytochemistry*, *59*, 1041–1059.

Zahn, J. M., Sonu, R., Vogel, H., Crane, E., Mazan-Mamczarz, K., Rabkin, R., et al. (2006). Transcriptional profiling of aging in human muscle reveals a common aging signature. *PLoS Genetics*, *2*(7), e115.

Zhan, M., Yamaza, H., Sun, Y., Sinclair, J., Li, H., & Zou, S. (2007). Temporal and spatial transcriptional profiles of aging in *Drosophila melanogaster*. *Genome Research*, *17*, 1236–1243.

Zheng, J., Edelman, S. W., Tharmarajah, G., Walker, D. W., Pletcher, S. D., & Seroude, L. (2005). Differential patterns of apoptosis in response to aging in *Drosophila*. *Proceedings of the National Academy of Sciences of the United States of America*, *102*, 12083–12088.

CHAPTER TEN

Mathematical Modeling of Regenerative Processes

Osvaldo Chara*, Elly M. Tanaka†, Lutz Brusch*,1

*Center for Information Services and High Performance Computing (ZIH), Technische Universität Dresden, Dresden, Germany

†Center for Regenerative Therapies Dresden (CRTD), Dresden, Germany

[1]Corresponding author: e-mail address: lutz.brusch@tu-dresden.de

Contents

Abstract

In many animals, regenerative processes can replace lost body parts. Organ and tissue regeneration consequently also hold great medical promise. The regulation of regenerative processes is achieved through concerted actions of multiple organizational levels of the organism, from diffusing molecules and cellular gene expression patterns up to tissue mechanics. Our intuition is usually not adapted well to this degree of complexity and the quantitative aspects of the regulation of regenerative processes remain poorly understood. One way out of this dilemma lies in the combination of

Current Topics in Developmental Biology, Volume 108
ISSN 0070-2153
http://dx.doi.org/10.1016/B978-0-12-391498-9.00011-5

experimentation and mathematical modeling within an iterative process of model development/refinement, model predictions for novel experimental conditions, quantitative experiments testing these predictions, and subsequent model refinement. This interdisciplinary approach has already provided key insights into smaller scale processes during embryonic development and a so-far limited number of more complex regeneration processes. This review discusses selected theoretical and interdisciplinary studies and is structured along the three phases of regeneration: (1) initiation of a regeneration response, (2) tissue patterning during regenerate growth, (3) arresting the regeneration response. Moreover, we highlight the opportunities provided by extensions of mathematical models from developmental processes toward the study of related regenerative processes.

1. INTRODUCTION

The regeneration of lost body parts is a spectacular biological phenomenon (Goss, 1969; Stocum, 2012; for recent reviews, see Birnbaum & Sánchez Alvarado, 2008; Tanaka & Reddien, 2011). In general, the tissue remaining after amputation or autotomy induces one of two alternative responses: scar formation versus regeneration (Fig. 10.1A). Scar formation on the wound site prevents the regenerative response but is not at all understood in evolutionary terms (Bely & Nyberg, 2009). We can speculate that scar formation is a compromise solution that the evolution of complex organisms had to face, in analogy to a frustrated system (Binder, 2008). Most tissues in higher organisms follow this fate of scar formation when damaged.

On the regeneration-competent side, the fresh water polyp *Hydra* initiates a successful and robust regenerative response after head amputation (Browne, 1909; Galliot, 2012; Trembley, 1744) and even from small cell aggregates (Gierer et al., 1972). The more complex planaria (e.g., *Schmidtea mediterranea*) regenerate the entire body plan from almost any tiny body fragment (Gurley, Rink, & Sánchez Alvarado, 2008; Morgan, 1904). A regenerative response can also be found in some vertebrates (Brockes & Kumar, 2005; Murawala, Tanaka, & Currie, 2012; Tanaka, 2003). The zebrafish (*Danio rerio*) regenerates the fin and parts of the eye, heart, and brain (Poss, 2010). The salamander *Amblystoma mexicanum*, as its most spectacular feat, regenerates the entire limb after upper-limb amputation (Goss, 1969; Nacu & Tanaka, 2011). More complex organisms like mammals still regenerate blood, bone, skin, and liver. Organ and tissue regeneration consequently also hold great medical promise. However, the understanding of design principles

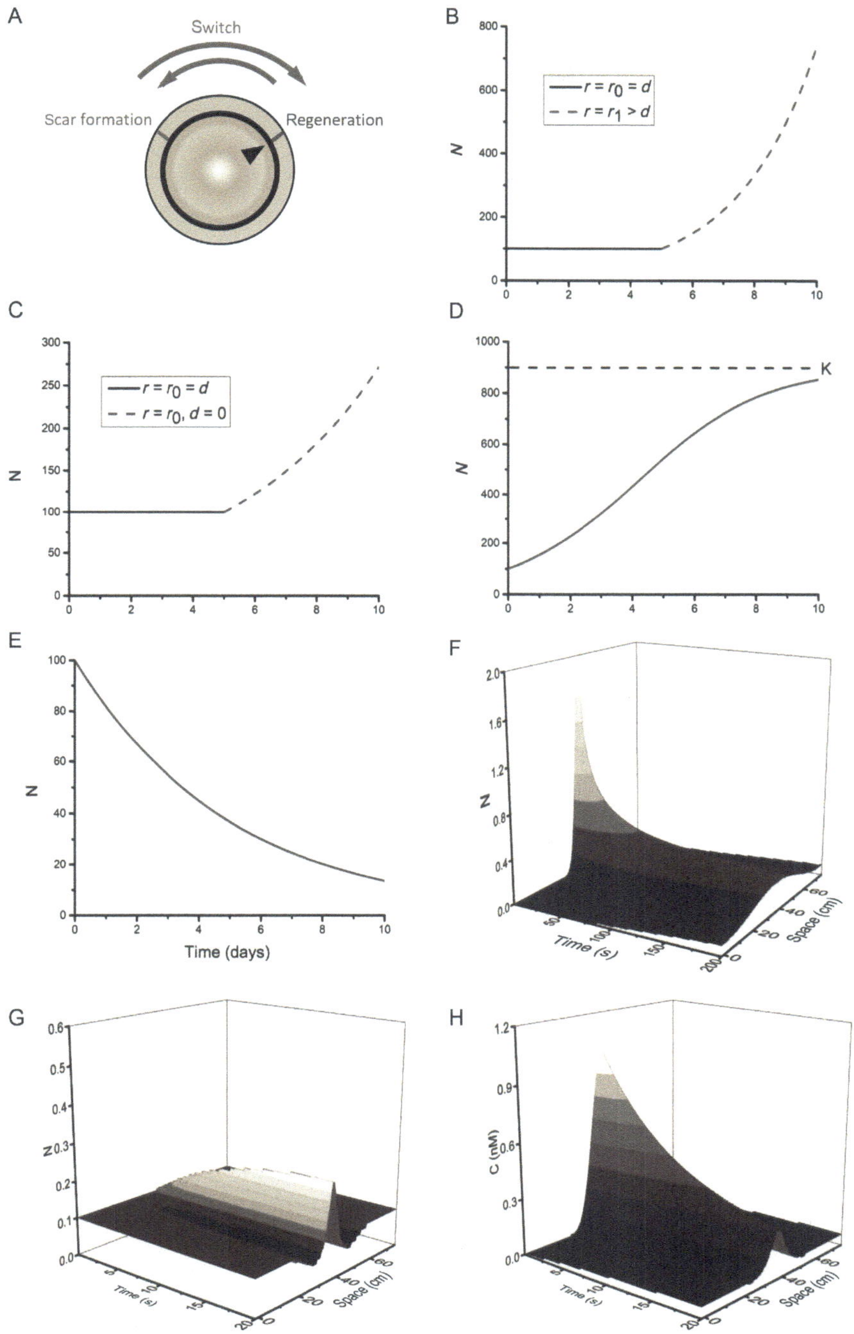

Figure 10.1 *Mathematical models of cell proliferation, apoptosis, and cell migration.* (A) A hypothetical switch illustrating that after amputation, some tissues regenerate whereas others in the same organism, or at an older age or in a closely related species, (*Continued*)

remains poorly understood even in the regeneration-competent model organisms. Such understanding is an absolute prerequisite for future medical applications.

Regenerative processes occur broadly throughout the animal kingdom but even closely related species like different planarian species possess very different regenerative capacities and many lower organisms are not capable of a robust regenerative response (Bely & Nyberg, 2009; Liu et al., 2013; Tanaka, 2003). Moreover, not all tissues of the salamander regenerate as its limb does despite the same genetic toolbox available throughout the individual. Often the regenerative capacities gradually fade out from embryonic stages toward adulthood (Brockes & Kumar, 2005). Since the same cause, amputation, triggers two different fates (scar formation vs. regeneration) in closely related species, in different tissues of the same individual and embryonic versus adult tissue of the same organism, this decision can be viewed as a switch on a largely invariant hardware (Fig. 10.1A). This switch can be tuned by quantitative parameter changes within the regulatory system of the whole tissue. Therein, complex signaling networks including feedback loops play an important role and bridge across spatial scales from signaling molecules to tissue mechanics.

Figure 10.1—Cont'd only close the wound with a scar. Hence the same regulatory systems may be tuned differently to arrive at opposite decisions. (B–E) Epimorphic responses modeled by the continuum approach. (B) Switch from steady-state cell density to exponential growth at time $t_a = 5$ days (mimicking the response to amputation) by keeping the differentiation rate $d = constant$ and increasing the proliferation rate r from $r_0 = 0.2$ to $r_1 = 0.5$ day^{-1}, corresponding to a shortening of the cell cycle from $T_{c0} = 3.5$ to $T_{c1} = 1.4$ days (see Eqs. 1–5). (C) Same as (B) but without changing the proliferation rate r and instead stopping differentiation ($d = 0$). Again, the "tissue" responds with exponential growth (Eq. 4). (D) It is assumed that amputation at time 0 induces a logistic growth with carrying capacity of the population $K = 900$ cells, as indicated (see Eq. 9). (E) Apoptosis modeled as a linear decay process with a decay rate $a = 0.02$ day^{-1} yields exponentially decaying cell density (see Eqs. 10–11). In (B–E) the initial number of cells N_0 is chosen to be 100. (F–H) Mechanisms related to morphallaxis. (F) Random cell migration modeled as cell density diffusion in a one-dimension space (see Eq. 13–14). (G) Directed cell migration driven by chemotaxis. The N cells can diffuse in a one-dimensional space and are guided by the gradient of the concentration $C(x,t)$ of a biochemical signal. (H) The signal C is produced by the cells, diffuses freely, and undergoes linear degradation (see Eqs. 16–19). The initial state is a homogeneous distribution of cells (G) and a peak of signal in the center of the domain (H). Zero-flux boundary conditions have been applied in (F, G, H). (For color version of this figure, the reader is referred to the online version of this chapter.)

These regulatory and quantitative aspects of regenerative processes remain poorly understood but we can now harness a complementary perspective. The integration of experiments with mathematical models enables us to gain mechanistic understanding of complex systems (Epstein, 2008; Fletcher, 2011; Murray, 2002, 2003). The scientific process of model development integrates and orders the wealth of experimental evidence, knowledge, and data with the goal to generate a comprehensive mechanistic understanding of complex processes. Hence the mathematical model is a tool, not the goal itself, in an iterative and interdisciplinary process. Mathematical modeling also entails the critical assessment of relevant versus less relevant system variables and regulatory interactions. Each mathematical model component is precisely defined and can be monitored during analytical or numerical model analysis, and each component can be modified or perturbed in any desired manner. Simulating a mathematical model corresponds to *in silico* experimentation and unfolds predictive power.

An example for the successful integration of experimentation and mathematical modeling is the seminal work by Hodgkin and Huxley (Hodgkin & Huxley, 1952). By measuring the axonal membrane depolarization under different conditions and linking these different observations with the help of a single mathematical model, they inferred the essential gating properties of ion channels before these molecules themselves had been identified. This ground-breaking work was then extended by many others, notably Noble and coworkers developing a comprehensive model of the heart including tissue mechanics and detailed kinetics of the ion channels in several cell types (Noble, 1962, 2002). During this incremental model development, there coexisted multiple contradicting hypotheses and quantitative comparison of model predictions to experiments resolved many of these disputes.

Hence mathematical modeling is also a powerful tool for rigorously testing and excluding hypotheses. Moreover, mathematical modeling can identify some experimental observations as the inevitable consequence of other pieces of evidence. This approach can also reveal the presently unobservable that is too fast, too slow, too small, too big, too few, or too many for today's measurement technologies. For developmental and cellular processes, there are already many examples where deeper understanding was enabled through mathematical modeling (examples include Brusch, Lorenz, Or-Guil, Bär, & Kummer, 2002; de Back, Zhou, & Brusch, 2013; del Conte-Zerial et al., 2008; Foret et al., 2012; Gierer & Meinhardt, 1972; Käfer, Hayashi, Marée, Carthew, & Graner, 2007; Merks, Glazier, Brodsky, Goligorksy, & Newman, 2006; Nakakuki et al., 2010; Schröter et al., 2012; Starruß et al., 2012; for recent reviews, see Kondo & Miura, 2010; Lewis, 2008; Meinhardt, 2012; Oates, Gorfinkiel, Gonzalez-Gaitan, & Heisenberg, 2009; Tomlin & Axelrod, 2007). Comparatively fewer mathematical models have been developed for regenerative processes of animals so far. These mathematical models and the insights they provided form the core of this review.

This review is structured along the three phases of regeneration: (1) initiation of a regenerative response, with a systematic comparison of mechanisms that expand the regenerate; (2) tissue patterning during regenerate growth, with a focus on basic pattern formation mechanisms and their implications in organizer regeneration; (3) arresting the regeneration response when the lost tissue portion has been replaced, requiring a tissue-size sensing mechanism that works for initial tissue loss of unpredictable size and at unpredictable locations. We discuss selected results from mathematical models of regeneration processes and highlight the opportunities of extensions of models of developmental processes toward the study of related regeneration processes. We then list open problems for which we expect, during the coming years, substantial progress through the interdisciplinary application of mathematical models.

2. MODELS FOR THE INITIATION OF A REGENERATIVE RESPONSE

2.1. Epimorphosis versus morphallaxis

Successful tissue regeneration implies that the injured region, as the most basic of many aspects, restores its cell number. From a naïve perspective, there are three possible and general mechanisms to accomplish that: (1) by inducing proliferation of the preexisting cells close to the amputation plane, (2) by inducing migration of cells coming from noninjured areas, or (3) by reducing the loss of proliferating cells through differentiation and apoptosis close to the amputation plane. Essentially, the first mechanism is known as *epimorphosis* while the second one, if only differentiated noncycling cells would respond, is called *morphallaxis* (Agata, Saito, & Nakajima, 2007). If cycling cells migrate into the regenerate then we speak of epimorphosis again. It is reasonable to assume that proliferation and apoptosis balance each other under homeostatic conditions, that is, before the amputation. Since cell death inhibition leads to a net cell number increase, the third mechanism above could be classified as another case of epimorphosis. Interestingly, this last mechanism is apparently not exploited by nature. On the contrary, it was demonstrated that apoptosis is actually enhanced after head amputation in *Hydra* and it is necessary to regenerate the missing head after mid-gastric bisection (Chera et al., 2009). Analogously, in mice apoptosis would be activated to promote wound healing and tissue regeneration via the so-called phoenix rising pathway (Li et al., 2010).

To mimic the regenerative response, a realistic mathematical model has to include the possibility of involving either epimorphosis or morphallaxis. In particular, cell proliferation and migration as well as cell differentiation and apoptosis should be considered.

2.2. Modeling cell proliferation, differentiation, apoptosis, and migration

The process of epimorphosis can be attributed to a number of stem cells whose proliferation is activated. A simple mathematical model describing stem cell proliferation and differentiation is

$$\frac{\mathrm{d}N}{\mathrm{d}t} = rN - \mathrm{d}N, \tag{10.1}$$

where $N(t)$ is the density of proliferating cells within some volume at time t, r is the proliferation rate (defined as the rate of total cell number increase divided by the number of proliferating cells), and d is the differentiation rate (defined as the rate of differentiated cell number increase divided by the number of proliferating cells). The following initial condition is chosen:

$$N(t=0) = N_0. \tag{10.2}$$

The proliferation rate r can in a first attempt be approximated as a constant r_0 in homeostatic conditions that increases immediately after amputation (at time t_a) and remains constant afterwards:

$$\begin{aligned} r &= r_0 = d \quad \text{if}\, t < t_a \\ r &= r_1 > d \quad \text{if} \geq t_a, \end{aligned} \tag{10.3}$$

where $r_1 > r_0$. The differential equation (Eq. 10.1) can be easily solved to obtain the expected homeostasis before amputation and exponential growth expression after amputation (Fig. 10.1B):

$$\begin{aligned} N(t) &= N_0 e^{(r_0 - d)t} = N_0 \quad \text{if}\, t < t_a \\ N(t) &= N_0 e^{(r_1 - d)(t - t_a)} \quad \text{if} \geq t_a. \end{aligned} \tag{10.4}$$

An alternative for increasing the cell number without speeding up the cell cycle, that is, $r = r_0 = r_1$, is to stop to differentiate cells ($d = 0$ after amputation) and let newly formed stem cells proliferate further. Then again, the homeostatic state with constant cell number turns into an exponential growth phase where $N = N_0 e^{r(t - t_a)}$ (Fig. 10.1C). These two models describe the behavior of a number of cells which all divide exactly at the same rate r. From the exponential cell number expression, the cell cycle length T_C is straightforwardly extracted:

$$T_C = \frac{\ln(2)}{r} \tag{10.5}$$

This quantity can be experimentally estimated by means of Bromodeoxyuridine (BrdU) labeling. This methodology involves a synthetic nucleoside analogue of thymidine that can be incorporated into replicating cells during S phase of the cell cycle. The cell cycle length is extracted from the experiments of BrdU labeling by means of a mathematical model proposed more than 20 years ago (Nowakowski, Lewin, & Miller, 1989). The standard model assumes that all the cells divide with the same rate; that is, they constitute a homogenous growing population. The model has been applied to characterize cell proliferation during nervous system development (Gonsalvez et al., 2013) as well as adult neurogenesis (Ponti et al., 2013) to mention just two recent examples.

The model described by Eq. (10.1) implies that all the cells proliferate with the same proliferation rate r. It could be the case that the tissue is composed by different cell subpopulations differing precisely in this parameter and in the initial conditions. For instance, tissue amputation could activate the proliferation of n different cell subpopulations:

$$N = N_1 + N_2 + \cdots + N_n \tag{10.6}$$

And each of them could proliferate ruled by the following equations:

$$\frac{dN_i}{dt} = r_i N_i \tag{10.7}$$

$$N_i(t = 0) = N_{i0} \tag{10.8}$$

Not all the cell subpopulations would equally contribute to the overall proliferation kinetics induced by the amputation, this being controlled by the proliferation rates and the initial conditions.

According to the model, once the amputation is performed, each cell subpopulation starts to grow exponentially. But this assumes that there is no impediment related to the space or to the availability of resources. If this would be the case, a logistic growth model can be applied for each subpopulation:

$$\frac{dN_i}{dt} = r_i\left(1 - \frac{N_i}{K_i}\right)N_i \tag{10.9}$$

where the constants K_i represent the carrying capacity of the medium for each subpopulation i. Once each subpopulation size approaches its carrying capacity, the growth velocity shrinks (Fig. 10.1D). This model was successfully applied in many fields including ecology (Sibly, Barker, Denham, Hone, & Pagel, 2005) as well as microbiology (Brusca, Irastorza, Cattoni, Ozu, & Chara, 2013) among many others.

The processes of stem cell proliferation and differentiation were considered in mathematical models of the hematopoietic system (Marciniak-Czochra, Stiehl, Ho, Jäger, & Wagner, 2009; Nakata, Getto, Marciniak-Czochra, & Alarcón, 2012; van der Wath, Wilson, Laurenti, Trumpp, & Liò, 2009). In these models, there is a term working as a "sink" in the stem cell compartment which is equal to a corresponding term conceived as a "source" in the multipotent progenitor cells. The "sink" corresponding to the stem cell differentiation is mathematically analogous to a process of cell death. Indeed, the process of apoptosis of differentiated cells can also be modeled as a first-order decay process:

$$\frac{dN_i}{dt} = -a_i N_i, \tag{10.10}$$

where a_i is the apoptosis rate of cell subpopulation i. The solution of this equation is again straightforward (Fig. 10.1E):

$$N_i(t) = N_{i0}e^{-a_i t} \tag{10.11}$$

Considering proliferation (with logistic growth) and apoptosis together yields:

$$\frac{dN_i}{dt} = r_i\left(1 - \frac{N_i}{K_i}\right)N_i - a_i N_i \tag{10.12}$$

If the density of proliferating cells is much smaller than their carrying capacity then the logistic expression converges to Eq. (10.7) in which the parameters r_i of this expression correspond to $r_i - a_i$ of Eq. (10.12). Therefore, proliferation, differentiation, and apoptosis can all be accounted for by the same simple expression in the continuum formalism for the cell density.

The previous ordinary differential equations constitute deterministic models of stem cell proliferation and differentiation as well as apoptosis of differentiated cells. N, the cell density, represents the *average* density of stem cells and progenitors which may change or not in time and/or space. When spatiotemporal experimental information of stem cells and progenitors is available, stochastic modeling had demonstrated to be an interesting alternative. For instance, by assuming subpopulations of stem cells stochastically dividing by symmetric or asymmetric cell division, Simons and coworkers could explain tissue homeostasis of intestinal crypt (Snippert et al., 2010), esophageal epithelium repair (Doupé et al., 2012), as well as the clonal distribution of retina (He et al., 2012).

As mentioned earlier, amputation could induce epimorphosis but also morphallaxis. The involved process of random cell migration (for directed migration, see below) in a continuous formalism can be modeled as diffusion of cell density N (Murray, 2002):

$$\frac{\mathrm{d}N}{\mathrm{d}t} = D\frac{\partial^2 N}{\partial x^2}, \tag{10.13}$$

where the parameter D is the diffusion coefficient of the moving cells and is related to the mean squared displacement of cell trajectories. The model predicts that cells that are initially restricted to a localized spatial region will spread over the whole space. In the continuous framework, the initial condition could be modeled, for instance, by a Gaussian distribution:

$$N(t=0,x) = A\mathrm{e}^{(x-\mu)^2/\sigma^2} \tag{10.14}$$

The relevant parameters of the initial condition are μ and σ, the location and the width of the spatial distribution with approximately 95% of the cell population being localized between $\mu - 2\sigma$ and $\mu + 2\sigma$. The spatial distribution of N, which initially started as a narrow distribution with a ratio μ/σ higher than one disseminates all over the space (Fig. 10.1F) and concomitantly decreases this ratio. In the same vein than what was discussed for proliferating cells, this model can be generalized by assuming more than one subpopulation of motile cells, each one characterized by a diffusion coefficient D_i:

$$\frac{\mathrm{d}N_i}{\mathrm{d}t} = D_i \frac{\partial^2 N_i}{\partial x^2} \tag{10.15}$$

In the framework of regeneration, this model can be applied to explain how the reduction in the cell number in the wound induces a movement of cells coming from the remaining cells in the neighboring (and more populated) regions.

Another interesting possibility is to assume that cells migrate by chemotaxis, for example, guided by the gradient of a chemical factor (Murray, 2002):

$$\frac{\mathrm{d}N}{\mathrm{d}t} = D_N \frac{\partial^2 N}{\partial x^2} - \alpha \frac{\partial}{\partial x}\left(N \frac{\partial C}{\partial x}\right) \tag{10.16}$$

$$\frac{\mathrm{d}C}{\mathrm{d}t} = hN - kC + D_C \frac{\partial^2 C}{\partial x^2} \tag{10.17}$$

The model considers a signaling molecule of concentration $C(x)$ which is produced by the cells N(x). The cells are able to diffuse, as in the previous model, and the parameter controlling this process is the cell diffusion coefficient D_N (first term of the right-hand side of Eq. 10.16). The signal is produced at a rate which is proportional to the number of cells N, is linearly degraded and follows diffusion (first, second, and third term of the right-hand side of Eq. (10.17), respectively). The interesting feature of the model can be found in the second term of the right-hand side of Eq. (10.16). This term encodes the chemotactic behavior: the cells distribute over space driven by the spatial gradient of C and controlled by the parameter α. Therefore, the "flux" of chemoattracted cells is maximal where the spatial gradient also shows a maximum. By starting with the following initial conditions:

$$C(t=0,\, x) = A\mathrm{e}^{(x-\mu)^2/\sigma^2} \tag{10.18}$$

$$N(t=0,\, x) = N_0, \tag{10.19}$$

the cells "follow" the gradient of C. Hence, the cells orient forming a spatial pattern although they started homogeneously distributed. Figure 10.1G and H shows how the initial signal distribution widens due to diffusion but then approaches a stationary profile as the attracted cells provide a localized source of further signal. The neighborhood of the cell aggregate is depleted of cells and cell density remains unchanged at distances larger than $\sqrt{D_C/k}$ where the signal has decayed. The homogeneous state is stable for large enough D_N.

As mentioned, the initiation of the regenerative response involves cell proliferation, differentiation, apoptosis, and cell migration as the key cellular processes. All of them were included in the models containing systems of ordinary or partial differential equations shown in this section. This continuous approach also plays an important role in the so-called reaction–diffusion models (RDMs) of organizer regeneration discussed in the next section.

3. MODELS FOR TISSUE PATTERNING DURING REGENERATE GROWTH

Nonmathematical as well as mathematical models were proposed to understand the process of tissue patterning during regenerate growth. Among the first class, the polar coordinate model (PCM) was a very influential model proposed to understand the process of regeneration (and specially the role of intercalation) in insect imaginal discs, insect legs, and amphibian limbs (Bryant, French, & Bryant, 1981; French, Bryant, & Bryant, 1976; reviewed in Nacu & Tanaka, 2011). As a step from the first into the second class, Meinhardt developed the mechanistic boundary model that yields the PCM-rules en passant (Meinhardt, 1983). Mathematical models of pattern formation have traditionally received much attention (Gierer & Meinhardt, 1972; Kondo & Miura, 2010; Meinhardt, 1982; Murray, 2003; Turing, 1952). Whereas the fundamental principles appear rather well understood today, we still lack a one-to-one correspondence between essential components in the theory of pattern formation and specific sets of molecules and genes. The latter is slowly emerging from studies of particular model organisms as *Hydra* and planaria as well as mammalian organs. Another well-studied system of embryonic development, the *Drosophila* wing imaginal disc, is now also moving into the focus of regeneration studies (Martín, Herrera, & Morata, 2009). Salamander limbs have attracted significant modeling efforts (reviewed in Newman et al., 2008; Nacu & Tanaka, 2011). In this section, after emphasizing the role of spatial scales for molecular positioning systems, the contributions of the so-called RDMs of organizer regeneration are discussed, with special emphasis on the Gierer–Meinhardt model in the context of *Hydra* regeneration (Gierer & Meinhardt, 1972; Meinhardt, 1982). While these continuum approaches have proven successful as descriptions for density patterns of cells and molecules, cell-based models allow to incorporate and infer additional cellular properties including cell polarity, cell shape, and cell mechanics (Deutsch & Dormann, 2005). We end this section showing how such a cell-based model has been used to elucidate the mechanisms responsible for liver regeneration in mice after intoxication with a prototypical inducer of pericentral liver damage (Höhme et al., 2010, 2007).

3.1. Embryological versus regeneration spatial scales, Crick's diffusion estimation

Based on the works of Wolpert (Wolpert, 1969) and Child (Child, 1941), Crick highlighted the role of diffusion in embryogenesis by putting forward the Source–Sink model (Crick, 1970). He considered a simple one-dimensional model constituted by a linear array of n spherical cells, each of diameter d (Fig. 10.2). Additional important model assumptions are:

(1) The first cell on the left works as "source" of a certain biochemical signal while the cell, with number n, on the right constitutes a "sink" of this signal.

(2) The diffusion coefficient of the signal of low molecular weight (such as ATP) in an aqueous medium would be $D \sim 2.7 \times 10^{-7}\ \mathrm{cm}^2\ \mathrm{s}^{-1}$.

(3) The time t needed to set up a signal gradient would be about 3 h (close to the time suggested by Wolpert).

Crick calculated n, the number of cells constituting the line, as follows:

$$n = \frac{\sqrt{2tD}}{d} \tag{10.20}$$

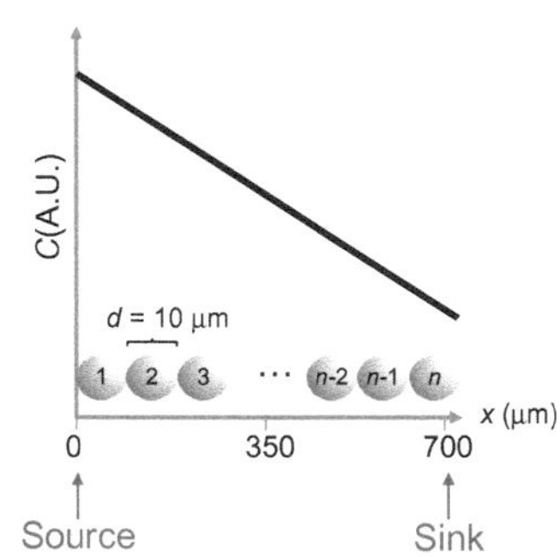

Figure 10.2 *Source–Sink model.* The model, proposed by Francis Crick in 1970, considers a putative morphogenetic signal at concentration C produced by a source located at $x = 0$ and degraded by a sink at the opposite boundary ($x = 700$). A linear array of n cells of diameter $d = 10$ μm is considered. By analyzing this model, Francis Crick showed that diffusion can play an important role in embryonic development. *The figure has been reprinted and modified, with permission, from Crick (1970).* (For color version of this figure, the reader is referred to the online version of this chapter.)

If the cell diameter d is 10 μm, the line would be covered by $n=70$ cells. More than four decades later, the idea is still appealing (Schier & Needleman, 2009). Indeed, although morphogens like Wnt have a molecular weight higher than 30 kD which could change the diffusion coefficient and, consequently, the number of cells involved (Shimizu, 2012) the simplicity of the above calculus is remarkable. Independent arguments on the necessary balance between gradient precision and robustness argue for the same number of cells spanning the gradient (Lander, Lo, Nie, & Wan, 2009). Moreover, experimental evidence for Fgf8 morphogen gradients in living zebrafish embryos supported Crick's idea qualitatively, where the "sink" in the zebrafish example corresponds to the signal-receiving/endocytosing cells (Yu et al., 2009).

The success of the Source–Sink model implies that a diffusive mechanism is *sufficient* to explain the establishment of morphogenetic gradients in embryonic development. Could this model be useful to understand the processes of regeneration? In principle, the distances and the times involved in regeneration seem to be higher than those implicated in development. Hence, although diffusion of signaling molecules could play an important role in regenerative processes, other mechanisms should as well be considered.

3.2. RDM of organizer regeneration

A fundamental and widely applied model of self-organizing and regenerating tissue patterns is the RDM. The RDM encompasses coupled positive and negative feedback loops in reaction systems or signaling systems which can dynamically self-organize spatial patterns starting from random initial distributions of the components given different ranges of their spatial influence (Gierer & Meinhardt, 1872; Kondo & Miura, 2010; Meinhardt, 1982; Meinhardt, 2008; Turing, 1952).

A typical example of the RDM with two coupled components is defined by two partial differential equations

$$\frac{\mathrm{d}u}{\mathrm{d}t} = f(u, v) + D_u \frac{\partial^2 u}{\partial x^2} \qquad (10.21)$$

$$\frac{\mathrm{d}v}{\mathrm{d}t} = g(u, v) + D_v \frac{\partial^2 v}{\partial x^2} \qquad (10.22)$$

where $u(x,t)$ and $v(x,t)$ represent the concentrations of two biochemical species, which react according to the arbitrary functions $f(u,v)$ and $g(u,v)$ of the two concentrations and which diffuse with the diffusion coefficients D_u and D_v. The two reaction terms $f(u,v)$ and $g(u,v)$ capture in a lumped or abstract manner how a possibly long reaction or signaling pathway regulates u and v in response to the state of u and v.

A version of the RDM with linear functions $f(u,v)$ and $g(u,v)$ was proposed and studied analytically and numerically by Turing (Turing, 1952). He discovered the productive role of diffusion for the self-organization of spatial patterns from random initial distributions and derived a quantitative criterion for the onset of pattern formation as a function of reaction rates and diffusion constants, Eq. 10.23. Turing focused on the instability of a homogeneous system toward periodic patterns, not on the self-organization of monotonous gradients. Also, the properties of the self-organized stationary pattern were not in the focus of this seminal work and could not be studied in this linear version of the RDM since they are determined by the nonlinearities, for example, saturation due to Michaelis–Menten or Hill kinetics or non-linear inhibitor feedback. Since his focus was not on the steady state pattern, Turing did not yet address the following question of pattern regeneration.

Twenty years later, Gierer and Meinhardt developed a nonlinear RDM and numerically studied the properties of self-organizing stationary and dynamic patterns (Gierer & Meinhardt, 1972; Meinhardt, 1982). They discovered that any pattern-forming RDM can be classified as either activator–inhibitor (AI) or activator–substrate-depletion (AS) type. In the AI type, one component has to autocatalytically enhance its own production and is called activator whereas a second component called inhibitor is produced in response to activator presence and suppresses further activator production. In case the inhibitor could not dilute due to transport away from the activator location then a small initial activator elevation would be damped back to the homogeneous state. In the opposite case, the activator will grow at that location until dilution and nonlinear reaction saturation will stop further growth. In the AS type, the autocatalytic component consumes the substrate. Concurrent lack of substrate will damp the activator elevation back to the homogeneous state if the substrate in the neighborhood would not be mobile enough to enter the activator location. Hence, both types are closely related, possess nested positive and negative feedback loops, and depend critically on the diffusion constant of the second component. Meinhardt has successfully applied the RDM to several regeneration scenarios including *Hydra* (see Section 3.3 for a particular example) and planarian regeneration and could explain a number of complex phenotypes (Meinhardt, 1982).

Turing's quantitative instability criterion

$$\frac{D_v}{D_u} > \frac{\left(\sqrt{f_u g_v - f_v g_u} + \sqrt{-f_v g_u}\right)^2}{(f_u)^2} \qquad (10.23)$$

predicts exactly how much faster the inhibitor or substrate has to diffuse than the activator for the activator elevation to persist. The abbreviations f_u, g_u, f_v, and g_v refer to the rates, that is, first derivatives, with which the functions $f(u,v)$ and $g(u,v)$ change when u and v are changed, respectively. These four quantities have to be evaluated for the values u and v that correspond to the unstable homogeneous stationary state. Thus these four quantities become functions of the model parameters and hence a parameter shift can induce or suppress the instability of the homogeneous state toward a heterogeneous pattern. Letting the system evolve from the unstable state, the neighborhood of the emerging activator elevation then receives so much inhibitor, or depletes its substrate, that another activator elevation gets its chance only at large enough distance from the first. The minimal distance between activator elevations, or pattern wavelength for periodic arrangements, was predicted by Turing:

$$l = 2\pi\left(\frac{f_u}{2D_u} + \frac{g_v}{2D_v}\right)^{-1/2} \tag{10.24}$$

This characteristic length scale l of the self-organizing pattern is solely determined by the properties of the two components and independent of the geometry, size, and boundary conditions of the spatial domain in which the RDM is contained. Those domain properties will however determine how many, if any, activator elevations can be accommodated within that domain.

A variant of the RDM with *bistable* local kinetics (introduced in Meinhardt, 1972) and conservation of the total amount of one component within the domain has been analyzed by Edelstein-Keshet and coworkers (Mori, Jilkine, & Edelstein-Keshet, 2008). Therein a superthreshold perturbation of the homogeneous state is required to excite the coexisting bistable state. A front will extend the spatial region of the excited state until the consumption of the conserved component halts the front, termed wave pinning (Mori et al., 2008). This pattern-forming mechanism is fundamentally different from the linear instability mechanism and could underlie those systems that require a superthreshold stimulus for regeneration. Pattern formation in such multistable systems could be triggered by additional directed transport processes, including cell flow and advection of molecules. Examples of these coupled mechanochemical systems have been reviewed in Howard, Grill, and Bois (2011).

The concept of the RDM has helped to understand the formation and spatial localization of organizers, that is, localized signaling centers, during embryonic development (Kondo & Miura, 2010; Meinhardt, 1982; Murray, 2003). Starting from an unpatterned cell mass, the temporal dynamics of a RDM can establish an organizer in a field of otherwise low activity. As discussed next, these predictions describe aspects of *Hydra* regeneration from fragments of the body column (Meinhardt, 1982) or from dissociated and reaggregated cells (Hobmayer et al., 2000; Technau et al., 2000).

3.3. Gierer-Meinhardt model of hydra regeneration

As mentioned in the introduction, after the initiation of the regenerative response, tissue patterning processes during regenerate growth are generated. The RDM developed by Gierer and Meinhardt (Gierer & Meinhardt, 1972; Meinhardt, 2008) can describe the process of pattern formation needed in regeneration. In particular, this model describes many features observed in the regenerating *Hydra* by involving two biochemical signals A and H, both able to diffuse over a one-dimensional space (representing the head–foot axis of *Hydra*). The first signal, the activator A, is upregulating its own production while it is inhibited by the second signal H (from the German word *Hemmstoff*, for inhibitor). The model differential equations look like:

$$\frac{\mathrm{d}A}{\mathrm{d}t} = \frac{\rho A^2}{H} - \mu_A A + D_A \frac{\partial^2 A}{\partial x^2} + \rho_A \tag{10.25}$$

$$\frac{\mathrm{d}H}{\mathrm{d}t} = \rho A^2 - \mu_H H + D_H \frac{\partial^2 H}{\partial x^2} \tag{10.26}$$

The model describes a process by which both, the activator and the inhibitor are upregulated by a nonlinear autocatalytic production term A^2 (first term of the right-hand side of Eqs. 10.25 and 10.26) and linearly degraded with rate constants μ_A and μ_H (second term of the right-hand side of Eqs. 10.25 and 10.26). Additionally, both signals are able to diffuse with diffusion coefficients D_A and D_H (third term of the right-hand side of Eqs. 10.25 and 10.26) and the inhibitor is downregulating the activator production (first term of the right-hand side of Eq. 10.25). The self-enhancement of the activator is not linear in order to overcome the linear degradation. This condition is compatible with an activator working only in a dimeric configuration. The factor ρ is the *source density* or *competence* and describes the cells' ability to promote the self-enhancement of the activator.

Starting from a homogenous initial condition (i.e., both the activator and the inhibitor are homogenously distributed in the space) a stable spatial pattern is formed if the inhibitor distributes faster and has a more rapid turnover than the activator ($D_H > D_A$ and $\mu_H > \mu_A$, respectively). Parameter ρ_A (Eq. 10.25) is one of the regeneration keys and is responsible for the production of activator which is independent of the activator itself. When *Hydra* amputation is mimicked by removing the activator from a region of the space, this parameter allows that new activator is produced there (Fig. 10.3A–C).

Experimental evidence indicates that *Hydra* fragments from all positions are capable to regenerate a new polyp. The model can reproduce this phenomenon by assuming that the source density is enhanced by the activator, while undergoing a linear degradation, diffusion, and also showing an activator-independent production, as it is encoded in the following expression:

$$\frac{\mathrm{d}\rho}{\mathrm{d}t} = \gamma_\rho A - \mu_\rho \rho + D_\rho \frac{\partial^2 \rho}{\partial x^2} + \rho_\rho \qquad (10.27)$$

If the inhibitor has a small time constant (allowing the reappearance of the new organizer immediately after the original organizer removal) while the source density has a larger one, then, the organizing region inhibits the formation of other competing organizing regions at short times while inducing a new organizing region at that position closest to an amputated organizing region.

The specific RDM of Eqs. (10.25) and (10.26) is also capable of establishing an organizer *de novo* from unpatterned cell aggregates as described in the previous section and as it has been observed in experiments with dissociated cells of *Hydra* (Hobmayer et al., 2000; Technau et al., 2000). This self-organization process can be illustrated by simulating Eqs. (10.25) and (10.26) on a spatial domain with the same shape as the reported cell aggregate. For suitably chosen parameter values four organizers forming *de novo* can be observed, as in the mentioned experiments (Fig. 10.3D–F, and Supplementary Material, http://dx.doi.org/10.1016/B978-0-12-391498-9.00011-5). In this simulation, the particular shape of the aggregate differs from a circle, which is isotropic, and hence the shape triggered the break of angular symmetry such that the organizers could place themselves at large enough distance to each other. If coupled to cell migration upward the gradients of *u* or *v*, then the RDM would also let the experimentally observed aggregate shape emerge from an initially circular aggregate shape. Such bidirectional coupling of a pattern forming RDM with shape dynamics has been introduced as morphodynamics (Salazar-Ciudad, Jernvall, &

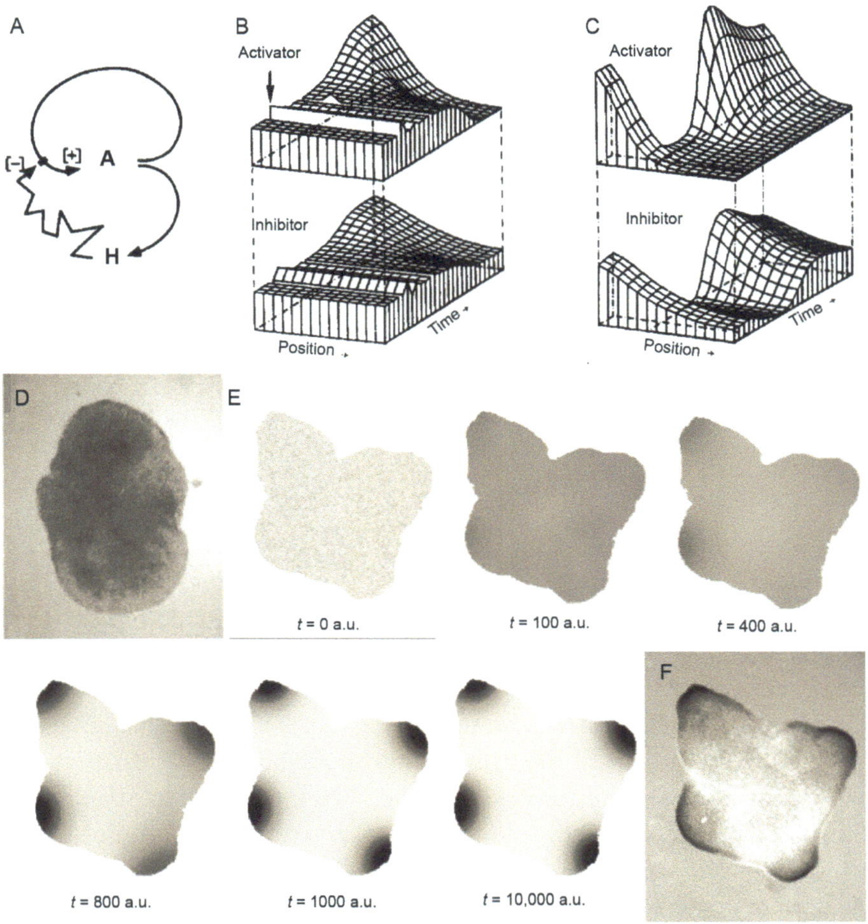

Figure 10.3 *Activator–Inhibitor model.* (A) Model diagram of activator (*A*) and inhibitor (*H*) components linked by positive and negative regulatory interactions (see Eqs. 10.25–10.26). (B) Space–time profiles of simulated activator and inhibitor dynamics under normal developmental conditions. Starting from homogenous concentrations of both components, the one-dimensional "tissue" self-organizes a pattern in which the activator is concentrated at the left border. (C) Simulation results for the regeneration condition with identical parameter values as in (B) but removing the activator peak after the initial three time steps. The "tissue" regenerates the original pattern (Meinhardt, 1982). (D) Experimental confirmation of the homogeneous initial distribution of the head marker HyTcf shown by hybridization 24 h after plating a *Hydra* cell aggregate. (E) Time course of the activator–inhibitor model of (A–C) simulated on a two-dimensional spatial domain which was trimmed to the shape of the experimental *Hydra* cell reaggregate reported in Hobmayer et al. (2000) and shown in (F). The spatial distribution of the activator variable is represented by a color scale with maximum 10 and darker color denoting larger values. Starting from homogeneous initial conditions with (*Continued*)

Newman, 2003). For a perfectly circular aggregate, tiny density fluctuations would get amplified and break the angular symmetry spontaneously. It has also been reported that spheroids of reaggregated *Hydra* cells may undergo osmosis-driven shape oscillations that provide a symmetry break (Kücken, Soriano, Pullarkat, Ott, & Nicola, 2008). As a model prediction, a small guidance cue like a fraction of an organizer could be used to place the new organizer at a desired position.

The various RDMs are able to explain processes of tissue pattern formation and self-organization, reproducing many phenomena observed during embryonic development and regeneration. While these RDMs describe density patterns of cells and molecules, cell-based models have allowed the inference of cell properties such as the orientation of the division axis. The next section demonstrates this complementary approach.

3.4. Cell-based model of liver regeneration

The extraordinary regeneration capacity of the mammalian liver had already inspired the fate of Prometheus in Greek Myth. Upon intoxication, dying hepatocytes are replaced by new hepatocytes, largely through proliferation of the remaining hepatocytes (Höhme et al., 2007). One open question is which cues guide the activation of proliferation and transport of newly created hepatocytes such that dead hepatocytes can be replaced without distorting the intact parts of the tissue architecture and the vascular network in particular. Drasdo and coworkers provided a proof of principle that mathematical modeling in combination with microscopy data from confocal laser scans, image segmentation, and three-dimensional tissue reconstruction can be used to address this question (Höhme et al., 2010).

Figure 10.3—Cont'd small added noise, all cells (here lattice nodes) first upregulate the activator and tiny asymmetries amplify until all weaker activator regions become suppressed by the stronger ones' inhibitor. The strong activator regions repel each other and thereby tend to form in or move into corners of the irregular aggregate shape. The similarity between distributions at 10^3 and 10^4 arbitrary time units (a.u.) indicates that a stable state has been reached. An animation of the full simulation with activator (inhibitor) distribution in the left (right) panel is available online as supplementary material, http://dx.doi.org/10.1016/B978-0-12-391498-9.00011-5. The multi-scale modeling and simulation software Morpheus was used as it can account for irregular, image-derived spatial domain shapes (Starruß, de Back, Brusch & Deutsch, 2014). See also Meinhardt (2012) for a similar simulation in a different geometry. (F) Snapshot of HyTcf hybridization at 48 h after plating of the *Hydra* cell aggregate shows the same pattern as in the simulation. *Panels A–C have been reprinted, with permission, from the classic book Meinhardt (1982). Panel D: Image reprinted with permission from Hobmayer et al. (2000). Panel F: Image reprinted with permission from Hobmayer et al. (2000).* (For color version of this figure, the reader is referred to the online version of this chapter.)

Their microscopy snapshots captured hepatocyte rearrangements at lobule scale during liver regeneration in mice after CCl_4 intoxication. CCl_4 is a prototypical chemical inducing liver damage around central veins. The building blocks of this mathematical model are the image-derived blood vessel network and spheres representing single hepatocytes. Both building blocks are deformable according to force balances that are evaluated at each time step of the simulation. These and the following model components also need to be parameterized and previously published data has instructed most of the parameter choices, for details see (Höhme et al., 2010). Each sphere has the options to migrate, to grow, to divide, or to die and these are executed according to rules that incorporated the configuration within a neighborhood, long-range signals as well as a stochastic component. Defining these rules requires mechanistic insight into the regulation of cell-scale processes during liver regeneration but this mechanistic insight is currently lacking. Several alternative mechanisms were proposed including (1) homogeneous activation of hepatocyte proliferation and hepatocyte displacement due to cell–cell repulsion, (2) directed cell migration toward the necrotic lesion possibly guided by chemoattractants secreted by dying hepatocytes, and (3) the alignment of cell divisions in the direction of the closest blood vessel.

In order to test these competing hypotheses, all of them were implemented as separate models and their emergent tissue-scale behavior simulated. The authors compared the time courses of hepatocyte density, lesion closure, and hepatocyte-vessel contact area between each of the models and the experimental data. After exploring the parameter space of each model, these three emergent properties and the simulated overall tissue architecture were found to be in agreement with the experimental data only for the model corresponding to the aforementioned mechanism number (3) (Fig. 10.4A–E). No other tested mechanism was able to replace the role of daughter cell alignment with blood vessels in these simulations. This model prediction was subsequently tested using independent experimental data. Altogether, this study demonstrated how multiscale mathematical modeling can be used to infer cell-scale properties from tissue-scale experimental observables.

4. MODELS FOR ARRESTING THE REGENERATIVE RESPONSE

4.1. Central versus decentral growth control

During embryonic development and tissue regeneration alike, the individual-specific size of an organ collectively results from parallel,

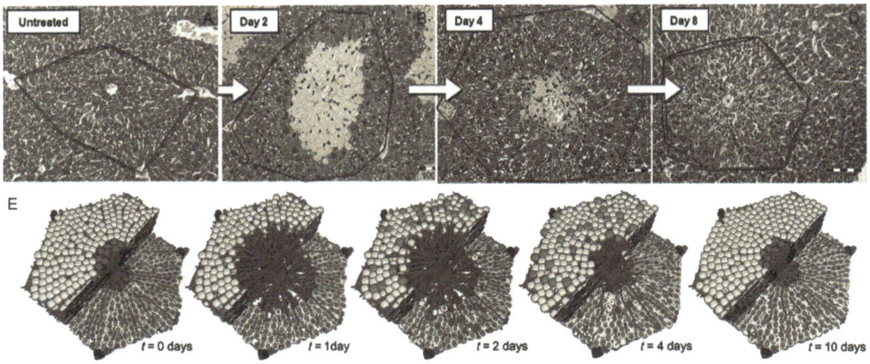

Figure 10.4 *Liver regeneration model.* (A–D) Mouse liver lobules with CCl4 induced liver damage in the central region, visible as gray domain in (B, C). Light microscopy images for (A) Control, (B) 2 days, (C) 4 days, and (D) 8 days after administration of CCl4 illustrate how the liver cell mass regenerates. Lines in (A–D) denote the borders of a single liver lobule in cross section. (E) Simulation of liver regeneration due to liver cell proliferation within a digital liver lobule. *All images reprinted with permission from Höhme et al. (2010).* (See color plate.)

decentralized decisions of single cells. Thus, organ size regulation represents a multiscale problem. The single cell's size and shape distributions together with cell number determine organ proportion and macroscopic extension. Thereby, information simply accumulates from the small to the large scale. But how is the information about the instantaneous organ size transferred back to the small scale of single cells where it controls cell proliferation such that the regenerative growth process is arrested once the organ's target size is reached? This question is equally pressing for developmental as well as for regeneration research. Different modes of information transfer and corresponding cellular sensing mechanisms have been studied as part of developmental biology.

(i) Bacterial biofilms employ quorum sensing to evaluate cell density (Ben-Jacob, Cohen, & Levine, 2000). A similar integrative role is played by molecules termed "chalones" in tissues (Lander, Gokoffski, Wan, Nie, & Calof, 2009).

(ii) *Hydra* employs long-range signals to pattern its body column and to bud off daughter polyps when a critical size is exceeded. Such behavior has been reproduced by RDMs (Gierer & Meinhardt, 1972; Meinhardt, 1982). Hence, growth is not arrested but excess tissue will be separated from the completely regenerated tissue.

(iii) The mammalian liver which operates in relation to the organismal metabolic needs seems to be controlled centrally by circulating growth stimuli that reflect organismal demand.

(iv) Embryonic appendage development employs an intrinsic size regulation mechanism until central hormone levels entrain it to organismal size in a later proportional growth phase. The organ-intrinsic growth regulation phenomenon is the most relevant to tissue regeneration in higher organisms and is still controversial today.

We review the competing hypotheses on organ-intrinsic growth regulation and the corresponding models in more detail below. We restrict the survey to those mechanisms implicated in *Drosophila* wing imaginal disc growth, a primary experimental paradigm of tissue size regulation. We emphasize the *central* versus *decentral* nature of each mechanism for organ-intrinsic growth regulation. Since regenerative processes have to cope with a range of diverse starting conditions, any central growth control mechanism needs to deliver the right dose of growth stimulus for the required size increase dependent on the size of the responding tissue and on the geometric relation between stimulus source and responding tissue. Whereas this dosage may be hard-coded in the organizer behavior for a stereotypic developmental process, it would need to be "recomputed" specifically for each instance of damage or amputation. The advantage of a decentral growth control mechanism such as intercalation is that damage sensing and growth response can be linked locally. The mathematical models reviewed in this section, notably the cell-based models, are defined by extended rule sets and parameter sets which we do not describe in full detail here, please see the cited references for these details.

An elegant test of the alternative hypotheses for growth regulation is provided by controlled elimination of groups of cells from the disc. Morata and coworkers have induced apoptosis of cells in the disc either randomly or within clones of transformed cells and observed that the discs regenerate perfectly both from massive cell loss as well as from loss of localized clones (Herrera, Martin, & Morata, 2013; Perez-Garijo, Shlevkov, & Morata, 2009; Repiso, Bergantiños, & Serras, 2013). The apoptotic clones were multiple and induced at random locations within the disc but many repetitions of these experiments with many random combinations of apoptotic clones have consistently yielded the growth arrest at the correct size. From these results we extrapolate to a hypothetical experiment with a single apoptotic clone at a desired location. It is unlikely that this single clone would yield a size defect since all combinations of clone locations resulted in correctly regenerated discs and it is hard to imagine a way in which all the single-clone errors always cancelled out.

For each of the three hypothetical growth regulation mechanisms discussed below, we will use a *Gedanken experiment* with induced cell loss in a compact region at desired location (e.g., the white sector in Fig. 10.5B) and then derive the model prediction regarding the regenerated disc size. Those models that will predict the recovery of the original disc size are then more likely explanations whereas others that contradict the experimental evidence may apply only to early developmental stages. This argument can be corroborated by quantitative analysis of the mathematical models but this shall be the subject of a separate work.

4.2. Mechanical feedback models

Mechanical forces have been proposed to spread the information about overall tissue geometry within the tissue (Shraiman, 2005). Growth against the peripheral layer of cells will increase the physical pressure within the tissue and can eventually arrest growth at suprathreshold pressures. Amputation would relieve this pressure through small spatial relaxation of the new tissue margin and thereby reactivate proliferation throughout the entire remaining tissue. This closed negative feedback loop of growth-dependent pressure increases and pressure-hindered growth represents a *decentral* mechanism and is called mechanical feedback model (Aegerter-Wilmsen, Aegerter, Hafen, & Basler, 2007; Hufnagel, Teleman, Rouault, Cohen, & Shraiman, 2007). So far, this model has been studied in the context of *Drosophila* wing imaginal disc growth during development. Two slightly different variants of the model have been implemented as a continuum model (Aegerter-Wilmsen et al., 2007) and a cell-based model (Hufnagel et al., 2007), respectively.

The disc is an epithelium with the central part of the wing pouch approximately confined to a planar subspace such that it cannot fold into the third dimension. In the model, proliferation is maintained by and contributes to a particular physical stress distribution that compresses inner regions and tangentially stretches outer cell layers (Fig. 10.5). Coupled to a concentration gradient of diffusible growth stimuli, a homogeneous proliferation rate emerges in model simulations as a new property at the tissue scale without the need to define a unique proliferation rate at the cell scale (Aegerter-Wilmsen et al., 2007; Hufnagel et al., 2007). The final disc size is determined by a balance of heterogeneous growth stimulus, compression, and stretching forces (Fig. 10.5). Growth is arrested since a closed outer belt of cells under stretch (outer/blue areas in Fig. 10.5) encapsulates the inner part of the disc (Aegerter-Wilmsen et al., 2007; Hufnagel et al., 2007). The liver regeneration model discussed in Section 3 is another instance of the mechanical feedback model since also there proliferation is arrested once the open space of the lesion has been filled with cells and pressure within the tissue arrests further proliferation.

We now consider the above-described Gedanken experiment in analogy to actual disc regeneration experiments (Herrera et al., 2013; Perez-Garijo et al., 2009). If the homeostatic stress distribution is perturbed by cell loss in a sector or a narrow ray near the periphery of the disc (white box in Fig. 10.5B), then the formerly closed outer belt will break open even for small localized damage in that region and the compression of the formerly encapsulated inner tissue will be relieved. This mechanical relaxation response has been observed through changes of the force-dependent optical birefringence of the tissue (Nienhaus, Aegerter-Wilmsen, & Aegerter, 2009). In our Gedanken experiment, proliferation will fill the void in the tissue but it is very unlikely that the newly formed cells in that region would restore the highly stretched configuration of the closed outer belt which is necessary to compress the inner part of the tissue and consequently arrest proliferation. The new cells would rather fill the void in a relaxed

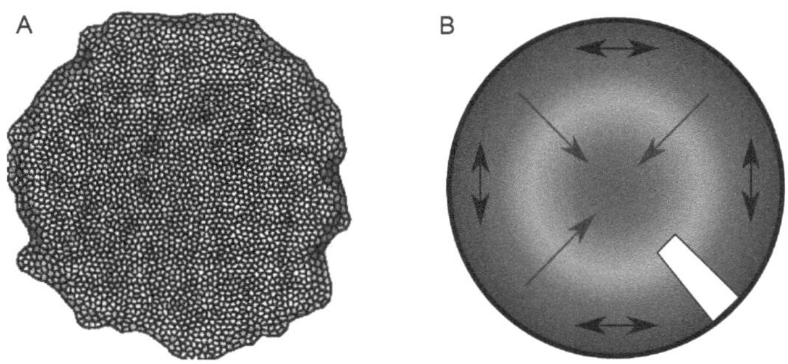

Figure 10.5 *Mechanical feedback model.* (A–B) One of the proposed mechanisms for tissue size regulation combines a *central* growth factor source (not shown) which drives proliferation in the center of the wing disc with a *decentral* mechanical feedback mechanism through proliferation-inhibiting compression forces (radial/red arrows in B for direction and darkness/intensity of red color in the center of A and B for amplitude of forces) and proliferation-promoting stretch forces (tangential/blue arrows in B for direction and darkness/intensity of blue color in the periphery of A and B for amplitude of forces). (A) Snapshot of a cell-based simulation starting from a small and relaxed cell aggregate shows the buildup of the force pattern (Hufnagel et al., 2007). Ongoing growth in the center stretches outer cell layers tangentially which consequently compresses the inner cell mass such that the net proliferation rate decreases sharply at the target size. (B) Sketch of the force pattern and location of the hypothetical tissue loss that is discussed in a Gedanken experiment (see text). *Panel A was reprinted with permission from Hufnagel et al. (2007).* (See color plate.)

configuration and growth will continue, thereby starting to stretch the new outer belt, until a new balance at a *larger* disc size can eventually arrest growth. This would not explain the reported disc regeneration with normal size (Herrera et al., 2013; Perez-Garijo et al., 2009). The overgrowth, as wrongly predicted by mechanical feedback models, could be corrected through a more active role of the *central* growth stimulus gradient, as discussed in Aegerter-Wilmsen et al. (2012) and the next section.

4.3. RDMs of growth regulation

The interplay between diffusible growth promoting morphogen gradients (including Decapentaplegic, Dpp) as well as diffusible growth repressing morphogen gradients (including Brinker, Brk) has been studied in great detail (Schwank, Restrepo, & Basler, 2008; Serrano & O'Farrell, 1997; Wartlick, Mumcu, Kicheva, et al., 2011). This mechanism can again be represented by reaction diffusion equations (Meinhardt, 1982). The local balance of the instantaneous concentrations of the antagonistic stimuli could determine the proliferation response during embryonic development (Schwank et al., 2008). A different option is to process the temporal history of concentration changes and let any cell divide when the *relative* stimulus increment since the last division exceeds a threshold, for example, 40–50% as measured for the wing disc (Wartlick, Mumcu, Kicheva, et al., 2011). Then the predicted cell cycle time

$$T_c(t) \sim \left(\frac{1}{C(x,t)}\frac{\partial C(x,t)}{\partial t}\right)^{-1} \tag{10.28}$$

becomes uniform for the observed *scaling* stimulus gradient (Wartlick, Mumcu, Kicheva, et al., 2011). The time dependence of the local stimulus $C(x,t)$ results from a combination of time-dependent stimulus release from the central organizer $C_0(t)$, retardation due to signal propagation between organizer and receiving cell, and displacement of the receiving cell within a stimulus gradient by proliferation of cells between the receiving cell and the organizer. Since the latter proliferation is ruled by Eq. 10.28 and the scaling gradient has the property

$$\frac{1}{C(x,t)}\frac{\partial C(x,t)}{\partial t} = \frac{1}{C_0(t)}\frac{\partial C_0(t)}{\partial t} \tag{10.29}$$

then, after inserting Eq. (10.29) into Eq. (10.28), $T_C(t)$ is determined by the time course $C_0(t)$ of stimulus release from the central organizer.

Both above options constitute *central* mechanisms ruled by the dynamics of the stimulus source(s) and these can well explain the developmental data sets. But in the Gedanken experiment in analogy to real experiments (Herrera et al., 2013; Perez-Garijo et al., 2009), a clone of apoptotic cells far away from the organizer would have to transfer information on the loss of cells within the clone back to the organizer such that an additional impulse $\partial C_0(t)/\partial t$ of growth stimulus can be released. This cannot be excluded but appears to be a difficult requirement for all central growth control mechanisms. The next difficult question would then be how the organizer could integrate damage signals from different wounds throughout the disc.

4.4. Slope sensor models

Using as proliferation trigger the spatial slope of a concentration gradient instead of its local values has been proposed as a simple and robust mechanism for relaying the information of the current tissue size to each cell (Day & Lawrence, 2000; García-Bellido, 2009; García-Bellido and García-Bellido, 1998; Lawrence & Struhl, 1996; Rogulja & Irvine, 2005). If the overall expression profiles interpolate between fixed extreme values at the organ boundaries then the local slope would reflect organ size (Day & Lawrence, 2000; Lawrence & Struhl, 1996). If individual cells locally sense a suprathreshold slope then they might proliferate and intercalate new expression levels such that a shallower slope results (García-Bellido, 2009; García-Bellido and García-Bellido, 1998). This decentral mechanism would arrest developmental disc growth at the target size that has been evolutionarily encoded as the same threshold value in each cell. Each cell may compare this threshold value to a function of its local slope signal and the absolute expression level. This way, either linear or exponential expression gradients, thanks to the constant relative slope of the latter, could serve as input to the slope sensor.

Over the past few years, experimental evidence on the role of Fat and Hippo signaling pathways in disc growth shows that cells indeed sense differences of Dachsous (Ds) and Four-jointed (Fj) levels as compared to their neighboring cells (Reddy & Irvine, 2008; Staley & Irvine, 2012). Both Ds and Fj are expressed as concentric gradients with extremes at the disc margin and center, respectively, but opposite orientation (Rogulja, Rauskolb, & Irvine, 2008). Fat and Ds are large trans-membrane proteins that bind each other whereas Fj is a Golgi-localized kinase that can modulate Fat and Ds activities (Ishikawa, Takeuchi, Haltiwanger, & Irvine, 2008). A suprathreshold difference of Ds or Fj among neighboring cells inhibits the tumor suppressor Hippo through Fat and thereby activates proliferation (Cho et al., 2006; Rogulja & Irvine, 2005).

In the Gedanken experiment in analogy to real experiments (Herrera et al., 2013; Perez-Garijo et al., 2009), apoptosis of cells anywhere with intermediate expression levels would induce a large slope for cells at the wound margin and correspondingly trigger proliferation until the expression gap has been filled by cells expressing the intermediate levels. Such a system will regenerate the original tissue size as long as the central extreme expression level is retained. This interpolation of expression levels, representing positional values, has been tested in grafting experiments with cricket legs (Bando et al., 2009). If a tissue graft was reinserted in reversed orientation then two large tissue portions of the size of the graft were regenerated until a zig-zag profile of positional values could fulfill the slope threshold criterion everywhere again, as expected from mathematical models (Meinhardt, 1982). Hence, the slope sensor model is an attractive candidate to be considered further.

4.5. Proposed integrated model

The reaction–diffusion and mechanical feedback models can explain many aspects of the developmental processes of *de novo* pattern formation in initially uniform fields of cells (Aegerter-Wilmsen et al., 2012; Meinhardt, 1982; Wartlick, Mumcu, Jülicher, & Gonzalez-Gaitan, 2011). Once established, these patterns can be translated into gradients of cell-bound properties, for example, the expression level of plasma membrane components, as proposed (Lewis & Wolpert, 1976; Slack, 1980) and found experimentally (Rogulja, Rauskolb, & Irvine, 2008). These cell-bound properties can be stably maintained over long time at low cost since they do not diffuse away. Upon cell division, daughter cells could initially establish their own cell-bound properties by averaging over their neighbors' properties and then maintain this once established property (García-Bellido, 2009). Therewith, a growing tissue would maintain a gradient of cell-bound properties. The slope of this gradient would gradually decrease as a result of daughter cells shifting previously neighboring cells with fixed properties further apart. Dachsous expression in the fly imaginal discs and Prod1 (CD59) in the newt limb appear to encode such large-scale gradients of cell-bound properties. A slope-sensing mechanism would then inform each cell any time about the macroscopic tissue extension and proliferation could cease as the slope approaches a target value. This way, multiple mechanisms would be required subsequently and mutagenesis studies of overall growth defects would reveal components of seemingly competing mechanisms.

5. OPEN PROBLEMS

When compared to developmental processes during embryogenesis, two fundamental differences become evident: developmental processes operate (1) on small embryonic length scales and involving hundreds of cells and (2) from defined starting conditions whereas regenerative processes bridge to larger adult length scales involving millions of cells and function from a range of diverse starting conditions. Hence a central question is: Can developmental processes be reinstated and adapted or are there entirely new regenerative processes to be discovered?

So far, the modeling of regenerative processes has often focused on explaining the principles that govern growth control and patterning, basically assuming the responses of only a single cell type. While this may suffice for a system such as the *Drosophila* wing disc, regeneration in other systems including *Hydra* and the salamander limb clearly involve the participation of multiple cell types that retain a separate identity during regeneration (Hobmayer et al., 2012; Kragl et al., 2009). For modeling the dynamics of these coupled populations, it will be important for experimentalists to determine whether the different cell types show common properties, or distinct profiles of cell proliferation, cell death, and migration. From a molecular point of view, it will be important to determine whether one common injury associated signal coordinates the responses of different cell types or multiple signals act in parallel. Finally, the interaction and coordination among the different cell types will also come into play during tissue restoration that will eventually need to be incorporated into models. From a modeling perspective, the integration of cell-based models with continuum approaches (via differential equations) will be necessary to truly integrate the large-scale spatial control of cellular responses and the concrete cellular behaviors that occur.

Interestingly, regeneration occurs over many different spatial scales. For example, small pieces of planaria regenerate to small individuals, while larger pieces regenerate to larger animals. In salamanders, regeneration occurs in small larvae or adults that can be 10 times bigger than the larvae. Therefore, the mechanisms controlling proliferation and patterning must be robust over different spatial scales. This consideration of size will be an important aspect of evaluating different models and will require experiments in animals of different sizes to help determine how scaling is occurring. In considering scaling, it is likely that regenerating structures have a natural range of scale at which patterning can be completed. Then there should be a second phase of matching the regenerate with overall body size. When this second phase starts and how it is controlled will be fascinating future challenges for biologists and mathematicians alike.

6. CONCLUSIONS AND PERSPECTIVES

Many studies of regenerative processes have provided qualitative mechanistic models or pathway maps. Translating these qualitative models into mathematical models will allow their internal consistency to be verified, to clarify hidden assumptions, and to define the specific conditions and parameter ranges for which the anticipated emergent behavior indeed results. The current blossoming of quantitative experimentation in the regeneration field is providing the foundation to extend the interdisciplinary approach from developmental to regenerative processes and to benefit from mathematical modeling more widely. Mathematical modeling has already contributed to a better understanding of regenerative processes with examples from lower as well as higher organisms and from the three phases of regeneration. (1) Key processes of the initiation phase of a regeneration response can be described by comparatively simple models and the task is now to model the signaling pathways that tune the parameter values of the simple models, thereby extending them. (2) Tissue patterning during regenerate growth is a very attractive and very active research area that benefits from classic and abstract models which now have to be linked to the discovered molecular components. (3) The arrest of the regeneration response through tissue size regulation mechanisms still poses fundamental challenges. Central and decentral regulatory mechanisms appear to operate in sequence such that the latter regulate regenerative processes. However, mathematical modeling needs to accelerate in this area. We hence like to advocate a closer integration of experimentation and mathematical modeling based on an iterative process of model development, model predictions, quantitative experiments, and model refinement.

ACKNOWLEDGMENTS

We are grateful to Salvador Herrera, Fabian Rost, Michael Kücken, Andreas Deutsch, Hans Meinhardt, Kenneth Irvine, Brigitte Galliot, and Gines Morata for fruitful discussions. The authors acknowledge support by the Human Frontier Science Program (HFSP, grant RGP0016/2010), the cluster of excellence cfaed and the German Ministry for Education and Research (BMBF, Grants 0316169A and 0315734).

REFERENCES

Aegerter-Wilmsen, T., Aegerter, C. M., Hafen, E., & Basler, K. (2007). Model for the regulation of size in the wing imaginal disc of Drosophila. *Mechanisms of Development, 124*, 318–326.

Aegerter-Wilmsen, T., Heimlicher, M. B., Smith, A. C., Barbier de Reuille, P., Smith, R. S., Aegerter, C. M., et al. (2012). Integrating force-sensing and signaling pathways in a model for the regulation of wing imaginal disc size. *Development, 139*, 3221–3231.

Agata, K., Saito, Y., & Nakajima, E. (2007). Unifying principles of regeneration I: Epimorphosis versus morphallaxis. *Development, Growth & Differentiation, 49*, 73–78.

Bando, T., Mito, T., Maeda, Y., Nakamura, T., Ito, F., Watanabe, T., et al. (2009). Regulation of leg size and shape by the Dachsous/Fat signalling pathway during regeneration. *Development, 136*, 2235–2245.

Bely, A. E., & Nyberg, K. G. (2009). Evolution of animal regeneration: Re-emergence of a field. *Trends in Ecology & Evolution, 25*, 161–170.

Ben-Jacob, E., Cohen, I., & Levine, H. (2000). Cooperative self-organization of microorganisms. *Advances in Physics, 49*, 395–554.

Binder, P. M. (2008). Frustration in complexity. *Science, 320*, 322–323.

Birnbaum, K. D., & Sánchez Alvarado, A. (2008). Slicing across kingdoms: Regeneration in plants and animals. *Cell, 132*, 697–710.

Brockes, J. P., & Kumar, A. (2005). Appendage regeneration in adult vertebrates and implications for regenerative medicine. *Science, 310*, 1919–1923.

Browne, E. N. (1909). The production of new hydranths in Hydra by the insertion of small grafts. *The Journal of Experimental Zoology*, 7, 1–13.

Brusca, M. I., Irastorza, R. M., Cattoni, D. I., Ozu, M., & Chara, O. (2013). Mechanisms of interaction between Candida albicans and Streptococcus mutans: An experimental and mathematical modelling study. *Acta Odontologica Scandinavica, 71*, 416–423.

Brusch, L., Lorenz, W., Or-Guil, M., Bär, M., & Kummer, U. (2002). Fold-Hopf bursting in a model for calcium signal transduction. *Zeitschrift für Physikalische Chemie, 216*, 487–497.

Bryant, S. V., French, V., & Bryant, P. J. (1981). Distal regeneration and symmetry. *Science, 212*, 993–1002.

Chera, S., Ghila, L., Dobretz, K., Wenger, Y., Bauer, C., Buzgariu, W., et al. (2009). Apoptotic cells provide an unexpected source of Wnt3 signaling to drive hydra head regeneration. *Developmental Cell, 17*, 279–289.

Child, C. M. (1941). *Patterns and problems of development.* Chicago: University of Chicago Press.

Cho, E., Feng, Y., Rauskolb, C., Maitra, S., Fehon, R., & Irvine, K. D. (2006). Delineation of a Fat tumor suppressor pathway. *Nature Genetics, 38*, 1142–1150.

Crick, F. (1970). Diffusion in embryogenesis. *Nature, 225*, 420–422.

Day, S. J., & Lawrence, P. A. (2000). Measuring dimensions: The regulation of size and shape. *Development, 127*, 2977–2987.

de Back, W., Zhou, J. X., & Brusch, L. (2013). On the role of lateral stabilization during early patterning in the pancreas. *Journal of the Royal Society Interface, 10*, 20120766.

del Conte-Zerial, P., Brusch, L., Rink, J., Collinet, C., Kalaidzidis, Y., Zerial, M., et al. (2008). Membrane identity and GTPase cascades regulated by toggle and cut-out switches. *Molecular Systems Biology*, *4*, 206.

Deutsch, A., & Dormann, S. (2005). *Cellular automaton modeling of biological pattern formation: Characterization, applications, and analysis*. Boston: Birkhäuser.

Doupé, D. P., Alcolea, M. P., Roshan, A., Zhang, G., Klein, A. M., Simons, B. D., et al. (2012). A single progenitor population switches behavior to maintain and repair esophageal epithelium. *Science*, *337*, 1091–1093.

Epstein, J. M. (2008). Why model? *Journal of Artificial Societies and Social Simulation*, *11*, 12.

Fletcher, D. A. (2011). To model or not to model? *Molecular Biology of the Cell*, *22*, 909–910.

Foret, L., Dawson, J. E., Villaseñor, R., Collinet, C., Deutsch, A., Brusch, L., et al. (2012). A general theoretical framework to infer endosomal network dynamics from quantitative image analysis. *Current Biology*, *22*, 1381–1390.

French, V., Bryant, P. J., & Bryant, S. V. (1976). Pattern regulation in epimorphic fields. *Science*, *193*, 969–981.

Galliot, B. (2012). Hydra, a fruitful model system for 270 years. *The International Journal of Developmental Biology*, *56*, 411–423.

García-Bellido, A. (2009). The cellular and genetic bases of organ size and shape in Drosophila. *The International Journal of Developmental Biology*, *53*, 1291–1303.

García-Bellido, A. C., & García-Bellido, A. (1998). Cell proliferation in the attainment of constant sizes and shapes: The Entelechia model. *The International Journal of Developmental Biology*, *42*, 353–362.

Gierer, A., Bode, H., Berking, S., Schaller, H., Trenkner, E., Hansmann, G., et al. (1972). Regeneration of hydra from reaggregated cells. *Nature New Biol*, *239*, 98–101.

Gierer, A., & Meinhardt, H. (1972). A theory of biological pattern formation. *Kybernetik*, *12*, 30–39.

Gonsalvez, D. G., Cane, K. N., Landman, K. A., Enomoto, H., Young, H. M., & Anderson, C. R. (2013). Proliferation and cell cycle dynamics in the developing stellate ganglion. *The Journal of Neuroscience*, *33*, 5969–5979.

Goss, R. J. (1969). *Principles of regeneration*. New York: Academic Press.

Gurley, K. A., Rink, J. C., & Sánchez Alvarado, A. (2008). Beta-catenin defines head versus tail identity during planarian regeneration and homeostasis. *Science*, *319*, 323–327.

He, J., Zhang, G., Almeida, A. D., Cayouette, M., Simons, B. D., & Harris, W. A. (2012). How variable clones build an invariant retina. *Neuron*, *75*, 786–798.

Herrera, S. C., Martin, R., & Morata, G. (2013). Tissue homeostasis in the wing disc of drosophila melanogaster: Immediate response to massive damage during development. *PLoS genetics*, *9*, e1003446.

Hobmayer, B., Jenewein, M., Eder, D., Eder, M. K., Glasauer, S., Gufler, S., et al. (2012). Stemness in Hydra—A current perspective. *The International Journal of Developmental Biology*, *56*, 509–517.

Hobmayer, B., Rentzsch, F., Kuhn, K., Happel, C. M., von Laue, C. C., Snyder, P., et al. (2000). Wnt signalling molecules act in axis formation in the diploblastic metazoan hydra. *Nature*, *407*, 186–189.

Hodgkin, A., & Huxley, A. F. (1952). A quantitative description of membrane current and its application to conduction and excitation in nerve. *The Journal of Physiology*, *117*, 500–544.

Höhme, S., Brulport, M., Bauer, A., Bedawy, E., Schormann, W., Hermes, M., et al. (2010). Prediction and validation of cell alignment along microvessels as order principle to restore tissue architecture in liver regeneration. *PNAS*, *107*, 10371–10376.

Höhme, S., Hengstler, J. G., Brulport, M., Schäer, M., Bauer, A., Gebhardt, R., et al. (2007). Mathematical modelling of liver regeneration after intoxication with ccl4. *Chemico-Biological Interactions*, *168*, 74–93.

Howard, J., Grill, S. W., & Bois, J. (2011). Turing's next steps: The mechanochemical basis of morphogenesis. *Nature Reviews Molecular Cell Biology*, *12*, 400–406.
Hufnagel, L., Teleman, A. A., Rouault, H., Cohen, S. M., & Shraiman, B. I. (2007). On the mechanism of wing size determination in fly development. *PNAS*, *104*, 3835–3840.
Ishikawa, H. O., Takeuchi, H., Haltiwanger, R. S., & Irvine, K. D. (2008). Four-jointed is a Golgi kinase that phosphorylates a subset of cadherin domains. *Science*, *321*, 401–404.
Käfer, J., Hayashi, T., Marée, A. F. M., Carthew, R. W., & Graner, F. (2007). Cell adhesion and cortex contractility determine cell patterning in the drosophilaretina. *PNAS*, *104*, 18549–18554.
Kondo, S., & Miura, T. (2010). Reaction–diffusion model as a framework for understanding biological pattern formation. *Science*, *329*, 1616–1620.
Kragl, M., Knapp, D., Nacu, E., Khattak, S., Maden, M., Epperlein, H. H., et al. (2009). Cells keep a memory of their tissue origin during axolotl limb regeneration. *Nature*, *460*, 60–65.
Kücken, M., Soriano, J., Pullarkat, P. A., Ott, A., & Nicola, E. M. (2008). An osmoregulatory basis for shape oscillations in regenerating hydra. *Biophysical Journal*, *95*, 978–985.
Lander, A. D., Gokoffski, K. K., Wan, F. Y., Nie, Q., & Calof, A. L. (2009). Cell lineages and the logic of proliferative control. *PLoS Biology*, *7*, e15.
Lander, A. D., Lo, W. C., Nie, Q., & Wan, F. Y. (2009). The measure of success: Constraints, objectives, and tradeoffs in morphogen-mediated patterning. *Cold Spring Harbor Perspectives in Biology*, *1*, a002022.
Lawrence, P. A., & Struhl, G. (1996). Morphogens, compartments, and pattern: Lessons from drosophila? *Cell*, *85*, 951–961.
Lewis, J. (2008). From signals to patterns: Space, time, and mathematics in developmental biology. *Science*, *322*, 399–403.
Lewis, J., & Wolpert, L. (1976). The principle of non-equivalence in development. *Journal of Theoretical Biology*, *62*, 479–490.
Li, F., Huang, Q., Chen, J., Peng, Y., Roop, D. R., Bedford, J. S., et al. (2010). Apoptotic cells activate the "phoenix rising" pathway to promote wound healing and tissue regeneration. *Science Signaling*, *3*, ra13.
Liu, S.-Y., Selck, C., Friedrich, B., Lutz, R., Vila-Farré, M., Dahl, A., et al. (2013). Reactivating head regrowth in a regeneration-deficient planarian species. *Nature*, *500*, 81–84.
Marciniak-Czochra, A., Stiehl, T., Ho, A. D., Jäger, W., & Wagner, W. (2009). Modeling asymmetric cell division in hematopietic stem cells—Regulation of self-renewal is essential for efficient repopulation. *Stem Cell Dev*, *18*, 377–385.
Martín, F. A., Herrera, S. C., & Morata, G. (2009). Cell competition, growth and size control in the Drosophila wing imaginal disc. *Development*, *136*, 3747–3756.
Meinhardt, H. (1982). *Models of biological pattern formation*. London: Academic Press.
Meinhardt, H. (1983). Cell determination boundaries as organizing regions for secondary embryonic fields. *Developmental Biology*, *96*, 375–385.
Meinhardt, H. (2008). Models of biological pattern formation: From elementary steps to the organization of embryonic axes. *Current Topics in Developmental Biology*, *81*, 1–63.
Meinhardt, H. (2012). Modeling pattern formation in hydra: A route to understanding essential steps in development. *The International Journal of Developmental Biology*, *56*, 447–462.
Merks, R. M. H., Glazier, J. A., Brodsky, S. V., Goligorksy, M. S., & Newman, S. A. (2006). Cell elongation is key to in silico replication of in vitro vasculogenesis and subsequent remodeling. *Developmental Biology*, *289*, 44–54.
Morgan, T. H. (1904). Notes on regeneration. *Biol Bulletin*, *6*, 159–172.
Mori, Y., Jilkine, A., & Edelstein-Keshet, L. (2008). Wave-pinning and cell polarity from a bistable reaction–diffusion system. *Biophysical Journal*, *94*, 3684–3697.

Murawala, P., Tanaka, E. M., & Currie, J. D. (2012). Regeneration: The ultimate example of wound healing. *Seminars in Cell & Developmental Biology, 23*, 954–962.

Murray, J. D. (2002). *Mathematical biology: I. An introduction* (3rd ed.). New York: Springer.

Murray, J. D. (2003). *Mathematical biology: II. Spatial models and biomedical applications* (3rd ed.). New York: Springer.

Nacu, E., & Tanaka, E. M. (2011). Limb regeneration: A new development? *Annual Review of Cell and Developmental Biology, 27*, 409–440.

Nakakuki, T., Birtwistle, M. R., Saeki, Y., Yumoto, N., Ide, K., Nagashima, T., et al. (2010). Ligand-Specific c-Fos expression emerges from the spatiotemporal control of ErbB network dynamics. *Cell, 141*, 884–896.

Nakata, Y., Getto, P., Marciniak-Czochra, A., & Alarcón, T. (2012). Stability analysis of multi-compartment models for cell production systems. *Journal of Biological Dynamics, 6*, 2–18.

Newman, S. A., Christley, S., Glimm, T., Hentschel, H. G., Kazmierczak, B., Zhang, Y. T., et al. (2008). Multiscale models for vertebrate limb development. *Current Topics in Developmental Biology, 81*, 311–340.

Nienhaus, U., Aegerter-Wilmsen, T., & Aegerter, C. M. (2009). Determination of mechanical stress distribution in Drosophila wing discs using photoelasticity. *Mechanisms of Development, 126*, 942–949.

Noble, D. (1962). Modification of hodgkin-huxley equations applicable to purkinje fibre action and pace-maker potentials. *Journal of Physiology-London, 160*, 317–352.

Noble, D. (2002). Modeling the heart–from genes to cells to the whole organ. *Science, 295*, 1678–1682.

Nowakowski, R. S., Lewin, S. B., & Miller, M. W. (1989). Bromodeoxyuridine immunohistochemical determination of the lengths of the cell cycle and the DNA-synthetic phase for an anatomically defined population. *Journal of Neurocytology, 18*, 311–318.

Oates, A. C., Gorfinkiel, N., Gonzalez-Gaitan, M., & Heisenberg, C. P. (2009). Quantitative approaches in developmental biology. *Nature Reviews. Genetics, 10*, 517–530.

Perez-Garijo, A., Shlevkov, E., & Morata, G. (2009). The role of Dpp and Wg in compensatory proliferation and in the formation of hyperplastic overgrowths caused by apoptotic cells in the Drosophila wing disc. *Development, 136*, 1169–1177.

Ponti, G., Obernier, K., Guinto, C., Jose, L., Bonfanti, L., & Alvarez-Buylla, A. (2013). Cell cycle and lineage progression of neural progenitors in the ventricular-subventricular zones of adult mice. *PNAS, 110*, E1045–E1054.

Poss, K. D. (2010). Advances in understanding tissue regenerative capacity and mechanisms in animals. *Nature Reviews. Genetics, 11*, 710–722.

Reddy, B. V., & Irvine, K. D. (2008). The Fat and Warts signaling pathways: New insights into their regulation, mechanism and conservation. *Development, 135*, 2827–2838.

Repiso, A., Bergantiños, C., & Serras, F. (2013). Cell fate respecification and cell division orientation drive intercalary regeneration in Drosophila wing discs. *Development, 140*, 3541–3551.

Rogulja, D., & Irvine, K. D. (2005). Regulation of cell proliferation by a morphogen gradient. *Cell, 123*, 449–461.

Rogulja, D., Rauskolb, C., & Irvine, K. D. (2008). Morphogen control of wing growth through the Fat signaling pathway. *Developmental Cell, 15*, 309–321.

Salazar-Ciudad, I., Jernvall, J., & Newman, S. A. (2003). Mechanisms of pattern formation in development and evolution. *Development, 130*, 2027–2037.

Schier, A. F., & Needleman, D. (2009). Developmental biology: Rise of the source–sink model. *Nature, 461*, 480–481.

Schröter, C., Ares, S., Morelli, L. G., Isakova, A., Hens, K., Soroldoni, D., et al. (2012). Topology and dynamics of the zebrafish segmentation clock core circuit. *PLoS Biology, 10*, e1001364.

Schwank, G., Restrepo, S., & Basler, K. (2008). Growth regulation by Dpp: An essential role for Brinker and a non-essential role for graded signaling levels. *Development, 135*, 4003–4013.

Serrano, N., & O'Farrell, P. H. (1997). Limb morphogenesis: Connections between patterning and growth. *Current Biology*, 7, R186–R195.

Shimizu, H. (2012). Transplantation analysis of developmental mechanisms in Hydra. *The International Journal of Developmental Biology*, *56*, 463–472.

Shraiman, B. I. (2005). Mechanical feedback as a possible regulator of tissue growth. *PNAS*, *102*, 3318–3323.

Sibly, R. M., Barker, D., Denham, M. C., Hone, J., & Pagel, M. (2005). On the regulation of populations of mammals, birds, fish, and insects. *Science*, *309*, 607–610.

Slack, J. M. W. (1980). A serial threshold theory of regeneration. *Journal of Theoretical Biology*, *82*, 105–140.

Snippert, H. J., van der Flier, L. G., Sato, T., van Es, J. H., van den Born, M., Kroon-Veenboer, C., et al. (2010). Intestinal crypt homeostasis results from neutral competition between symmetrically dividing Lgr5 stem cells. *Cell*, *143*, 134–144.

Staley, B. K., & Irvine, K. D. (2012). Hippo signaling in Drosophila: Recent advances and insights. *Developmental Dynamics*, *241*, 3–15.

Starruß, J., Peruani, F., Jakovljevic, V., Søgaard-Andersen, L., Deutsch, A., & Bär, M. (2012). Pattern-formation mechanisms in motility mutants of Myxococcus xanthus. *Interface Focus*, *2*, 774–785.

Starruß, J., de Back, W., Brusch, L., & Deutsch, A. (2014). Morpheus: A user-friendly modeling and simulation environment for multiscale and multicellular systems biology. *Bioinformatics*, http://dx.doi.org/0.1093/bioinformatics/btt772.

Stocum, D. L. (2012). *Regenerative Biology and Medicine* (2nd ed.). New York: Academic Press.

Tanaka, E. M. (2003). Regeneration: If they can do it, why can't we? *Cell*, *113*, 559–562.

Tanaka, E. M., & Reddien, P. W. (2011). The Cellular Basis for Animal Regeneration. *Developmental Cell*, *21*, 172–185.

Technau, U., Cramer von Laue, C., Rentzsch, F., Luft, S., Hobmayer, B., Bode, H. R., et al. (2000). Parameters of self-organization in Hydra aggregates. *PNAS*, *97*, 12127–12131.

Tomlin, C. J., & Axelrod, J. D. (2007). Biology by numbers: Mathematical modelling in developmental biology. *Nature Reviews. Genetics*, *8*, 331–340.

Trembley, A. (1744). *Mémoires, pour servir à l'histoire d'un genre de polypes d'eau douce, à bras en forme de cornes*. Leiden: Jean et Herman Verbeek.

Turing, A. M. (1952). The Chemical Basis of Morphogenesis. *Philosophical Transactions of the Royal Society of London. Series B, Biological Sciences*, *237*, 37–72.

van der Wath, R. C., Wilson, A., Laurenti, E., Trumpp, A., & Liò, P. (2009). Estimating dormant and active hematopoietic stem cell kinetics through extensive modeling of Bromodeoxyuridine label-retaining cell dynamics. *PLoS ONE*, *4*, e6972.

Wartlick, O., Mumcu, P., Jülicher, F., & Gonzalez-Gaitan, M. (2011). Understanding morphogenetic growth control—Lessons from flies. *Nature Reviews. Molecular Cell Biology*, *12*, 594–604.

Wartlick, O., Mumcu, P., Kicheva, A., Bittig, T., Seum, C., Jülicher, F., et al. (2011). Dynamics of Dpp Signaling and Proliferation Control. *Science*, *331*, 1154–1159.

Wolpert, L. (1969). Positional information and the spatial pattern of cellular differentiation. *Journal of Theoretical Biology*, *25*, 1–47.

Yu, S. R., Burkhardt, M., Nowak, M., Ries, J., Petrásek, Z., Scholpp, S., et al. (2009). Fgf8 morphogen gradient forms by a source-sink mechanism with freely diffusing molecules. *Nature*, *461*, 533–536.

CHAPTER ELEVEN

Re-engineering Development to Instruct Tissue Regeneration

Beatrice Tonnarelli[*,†], **Matteo Centola**[*,†], **Andrea Barbero**[*,†], **Rolf Zeller**[‡], **Ivan Martin**[*,†,1]

[*]Department of Biomedicine, University of Basel, Tissue engineering group, Hebelstrasse Basel, Switzerland
[†]Department of Surgery, University Hospital Basel, Tissue engineering group, Hebelstrasse Basel, Switzerland
[‡]Department of Biomedicine, Developmental Genetics, University of Basel, Mattenstrasse, Basel, Switzerland
[1]Corresponding author: e-mail address: ivan.martin@usb.ch

Contents

Abstract

With few exceptions, tissue regeneration strategies based on the conventional combination of cells, scaffolding materials, and soluble factors (tissue engineering) have introduced a rather limited clinical impact. While it is being recognized that the non-convincing benefits of engineered grafts require more fundamental knowledge on mechanisms of action and potency factors, the attempt to mimic and recapitulate developmental events has inspired an evolution of the paradigm. In the context of skeletal regeneration, a "developmental engineering" approach has been advocated to generate intermediate grafts (i.e., hypertrophic cartilage templates) which, as suggested by limb developmental biology, are capable of autonomous spatial and temporal evolution into fully functional bone organs. However, limited consideration has been given to the fact that the recipient site within adult organisms may not be compatible with well-established developmental processes. This can be due to the possibly restricted function of resident progenitors, to the critical mechanical and physical boundary conditions of mature organs, or to the strong role of inflammatory signals and immune cells at repair sites. We thus propose that predictable, orderly, and durable tissue regeneration should be based on a "developmental RE-engineering" paradigm, with the challenge to instruct the execution of developmental programs in the context of an adult system.

Current Topics in Developmental Biology, Volume 108
ISSN 0070-2153
http://dx.doi.org/10.1016/B978-0-12-391498-9.00005-X

ABBREVIATIONS

BMP bone morphogenetic protein
FGF fibroblast growth factor
IHH Indian hedgehog
MSC mesenchymal stem/stromal cells
PTHrP parathormone-related protein
TGFβ transforming growth factor β
VEGF vascular endothelial growth factor
WNT wingless-type protein

1. INTRODUCTION

Tissue engineering has developed as an interdisciplinary field with the aim to generate tissue analogs that are capable to support or induce regeneration in cases of severely compromised healing or tissue degeneration (Langer & Vacanti, 1993). Tissue engineering paradigms are typically based on combining (i) an appropriate source of progenitor/stem cells, (ii) a biomaterial (i.e., scaffold) to deliver the cells and/or provide a structural/instructive template for tissue development, and (iii) a specific cocktail of soluble factors to regulate tissue growth and differentiation. In spite of significant investments and collaborations of scientists from rather different fields (such as cellular and molecular biology, material science, and biomedical engineering), tissue engineered grafts have to date not extensively translated to the clinic and helped to improve human health (Hunziker et al., 2006; Jakob et al., 2012). With few notable exceptions (Ferrari, Pellegrini, Matsui, Mavilio, & De Luca, 2006), standardization of the engineering process, reproducibility, efficacy, safety, and cost-effectiveness are not yet sufficient for broad acceptance of tissue-engineered implants by healthcare systems. Difficulties in translating tissue engineering to clinic in general manner may be due to several issues, one of which has been the lack of in-depth understanding of the molecular and cellular mechanisms underlying regeneration and healing. This in turn has hampered the design of tissue-engineered constructs, which induce and/or control tissue regeneration and healing upon their implantation.

Therefore, different strategies have been developed to move tissue engineering and regenerative medicine forward and increase their biological and medical impact. In particular, it has been proposed that, rather than *ex vivo* engineering of tissue grafts, it may be more relevant to design instructive implants capable to recruit endogenous cells at the repair site and reactivate the endogenous morphoregulatory systems controlling embryonic tissue development. This strategy, referred to as "Developmental Engineering," is based on controlled recapitulation of embryonic morphogenetic events in adult tissues (Lenas, Moos, & Luyten, 2009a, 2009b). Therefore, developmental engineering strategies require a profound understanding of both developmental and tissue engineering processes.

In this review, we will discuss the potential of developmental engineering with a specific focus on bone replacement. To this aim, we first elucidate the key mechanisms controlling embryonic development of limb skeleton and endochondral ossification and the mechanisms underlying endogenous bone repair. Next, we compare classical tissue engineering with "developmental engineering" approaches by highlighting conceptual differences. Finally, we discuss the need to "RE-engineer" developmental processes in order to develop regeneration strategies that will be effective in adult, fully developed and functional tissues.

2. LIMB SKELETAL DEVELOPMENT

The long bones in vertebrate embryos form by a process called endochondral ossification, which is best studied during limb skeletal development (Fig. 11.1, Wuelling & Vortkamp, 2011). The formation of the cartilage models of the future limb skeletal elements is initiated by the aggregation and condensation of mesenchymal progenitors that express the Sox9 transcription factor (Barna & Niswander, 2007). These chondrogenic progenitors form the cartilage models, which become surrounded by a layer of fibroblastic cells that give rise to the perichondrium. The cartilage model of the future bone grows as chondrocytes are arranged in juxtaposed regions of resting, proliferating columnar, and hypertrophic chondrocytes that together form the growth plate (Day & Yang, 2008; Ehlen, Buelens, & Vortkamp, 2006; Horton & Degnin, 2009; Karsenty, Kronenberg, & Settembre, 2009). In parallel, the perichondral cells begin to differentiate into osteoblasts, which form the primary bone collar. Osteoblast progenitors invade the hypertrophic

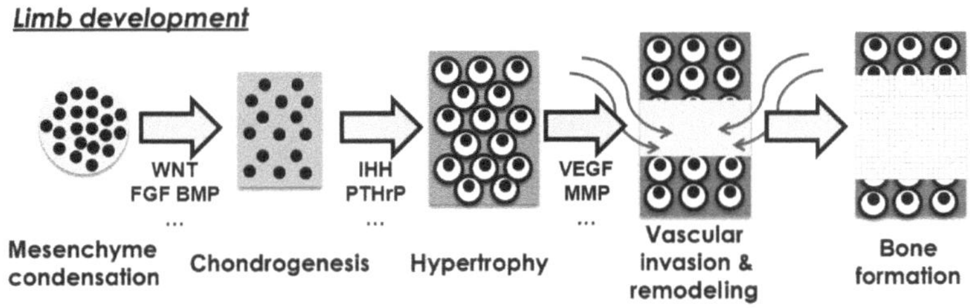

Figure 11.1 *Crucial phases occurring during limb development.* Mesenchymal cells condensate, expand, and undergo chondrogenic differentiation driven by Wnt, FGF, and BMP pathways; chondrocytic cells become hypertrophic, get spatially organized by IHH/PTHrP feedback loop, and produce Matrix metalloproteinases and VEGF that will cause matrix remodeling and vascular invasion. Osteoprogenitors, ingrowing together with invading vessels, deposit new bony matrix replacing the remodeled cartilage template. (For color version of this figure, the reader is referred to the online version of this chapter.)

region to replace the chondrocytes eliminated by cell death, which hallmarks the onset of endochondral ossification (Maes et al., 2010).

Primarily, the extensive genetic analysis of skeletal development in mouse embryos has revealed that endochondral bone development is orchestrated by rather complex signaling interactions involving many of the major embryonic signaling pathways, namely the Indian hedgehog (IHH), wingless-type protein (WNT), fibroblast growth factors (FGF), bone morphogenetic protein (BMP), and parathyroid hormone-related protein (PTHrP) pathways (Fig. 11.1, Day & Yang, 2008; Ehlen et al., 2006; Horton & Degnin, 2009; Karsenty et al., 2009). During proliferative expansion of mesenchymal progenitors, FGFs control growth and inhibit chondrogenesis in concert with WNT signaling (Mariani, Ahn, & Martin, 2008; ten Berge, Brugmann, Helms, & Nusse, 2008). In contrast, BMP and possibly transforming growth factor-β (TGFβ) signaling are required for the aggregation and differentiation of the Sox9 positive chondrogenic progenitors in mouse limb buds (Barna & Niswander, 2007; Bénazet et al., 2012; Karamboulas, Dranse, & Underhill, 2010). During progression of skeletal development, BMP signaling might be required for hypertrophic differentiation as transgene-mediated overexpression of a constitutively active BMP receptor accelerates differentiation of chondrocytes toward the hypertrophic phenotype (Kobayashi, Lyons, McMahon, & Kronenberg, 2005). The apparently context-dependent functions of BMPs in chondrocyte specification, cartilage template growth, and endochondral bone development echo the spatiotemporal relativity of BMPs during limb bud outgrowth and patterning in mouse embryos (reviewed by Zeller, Lopez-Rios, & Zuniga, 2009). Probably, the best understood regulatory interaction is the negative regulatory feedback loop between IHH (produced by the prehypertrophic chondrocytes) and PTHrP (produced by distal periarticular chondrocytes). This IHH/PTHrP feedback loop regulates the pools of proliferating and terminally differentiating hypertrophic chondrocyte and the spatial arrangement of the growth and differentiation zones (Ehlen et al., 2006; Mak, Kronenberg, Chuang, Mackem, & Yang, 2008; St-Jacques, Hammerschmidt, & McMahon, 1999; Vortkamp et al., 1996). In addition, high levels of WNT activity are required in the perichondrium to promote osteoblast differentiation during both endochondral and intramembranous ossification (Chen et al., 2007; Day, Guo, Garrett-Beal, & Yang, 2005). WNT signaling is also required for initiation of chondrocyte hypertrophy by inhibiting PTHrP signaling and promotes their final maturation (Guo, Mak, Taketo, & Yang, 2009).

These developmental studies (Fig. 11.1) have uncovered the underlying morphoregulatory systems and revealed the immense complexity and dynamics of the different signaling pathway interactions during limb bud and long bone development (reviewed by Karsenty et al., 2009; Zeller et al., 2009). Key features of the underlying morphoregulatory systems include inherent redundancy and pathway interconnectivity, which contribute to robustness. These systems consist of spatiotemporally dynamic populations of inducer (=organizer) and responder cell populations (e.g., the growth plate) that progress in a largely self-regulatory manner (see, e.g., Bénazet et al., 2009) from proliferative expansion to specification and differentiation. Therefore, a major challenge is to translate these insights from normal vertebrate development to the engineering of replacement tissues for human patients. Developmental engineering of cartilage and bone for tissue repair will need to integrate the knowledge about these complex systems into regenerative medicine for the benefit of human patients. Albeit the fact that developmental processes show astonishing evolutionary conservation of the basic underlying mechanisms, important species-specific differences exist (such as spatiotemporal organization of morphogenesis and gestation periods), which may severely hamper translation of knowledge gained in studying, for example, mouse models to human regenerative medicine (Zhu et al., 2009). As such mouse–human discrepancies are apparent in disease models, drug development, and cell-based therapies, the development and validation of "humanized" models for tissue engineering and regenerative medicine is important (Wilkie, 2003; Zhang et al., 2010).

3. ENDOGENOUS BONE REPAIR PROCESSES

In case of damage, adult bone tissue initiates a reparative process able to restore native features with no apparent scar formation. Bone healing is tightly regulated and progresses from inflammation to repair and remodeling (Caron et al., 2012; Claes, Recknagel, & Ignatius, 2012). Complete regeneration involves the coordination of multiple interconnected processes of anatomical, biomechanical, biochemical, and cellular nature (Marsell & Einhorn, 2011). Ideal conditions for successful repair should be created, starting with conservative and/or surgical stabilization of the bone fragments. Bone either heals by direct deposition of a mineralized matrix (=primary direct healing if the bone fragments are in close contact) or by development of a reparative cartilaginous callus (=secondary indirect healing). In the latter case, a hematoma clot initially fills the defect site, stabilizes the fracture, and acts as a trap for inflammatory cells. Cellular and matrix debris first attract polymorphonuclear neutrophils, whose reaction is to secret chemokines that mediate the recruitment of macrophages (innate immune reaction). Subsequently, lymphocytes invade the fracture site and trigger the adaptive immune response, resulting in activation of antibody-mediated processes and promoting the secretion of proinflammatory cytokines. These include members of the TGFβ superfamily and proangiogenic factors. *De novo* vascularization and locally produced chemotactic signals allow mesenchymal stromal progenitor cells (MSC) to invade the repair site. These differentiate into chondrocytes and produce a cartilaginous callus that is subsequently remodeled and replaced by mineralized bone tissue. Such repair processes recapitulate aspects of bone formation during embryonic development, which renders the healing process "a unique postnatal window in the biological and molecular processes that are involved in bone development" (Grimes, Jepsen, Fitch, Einhorn, & Gerstenfeld, 2011). During bone repair, the key signaling pathways that regulate the cellular interactions and matrix deposition are identical to the ones regulating embryonic bone development (i.e., IHH, PTHrP, BMPs, see Fig. 11.1). In addition, the ongoing vascularization reestablishes normoxic conditions in the core of the callus and facilitates the migration and homing of progenitors, similar to ossification centers during formation of the limb skeletal bones (Grimes et al., 2011). However, the inflammatory molecular and cellular processes, together with the evolving biomechanical environment, render postnatal bone repair more rapid and streamlined than embryonic skeletal development (Gerstenfeld, Cullinane, Barnes, Graves, & Einhorn, 2003).

Despite the robustness of the normal bone repair process, bone healing is compromised in at least 5–10% of all cases (up to 40% in patients with risk factors such as smoking and diabetes). Biomechanical and biological factors may impair endogenous bone repair in large bone defects (i.e., due to trauma, infection, or tumor resection), skeletal abnormalities, concomitant pathologies (i.e., necrosis and osteoporosis), and treatments with anti-inflammatory drugs, which may all require intervention using bone substitutes (Claes et al., 2012; Jakob et al., 2012; Welting et al., 2011). In all these conditions, untreated defects, delayed lesions, or nonunions are associated with the onset of degenerative processes, resulting in severe pain and disabilities for millions of people worldwide. Therefore, there is a significant clinical demand for regenerative treatments that support fracture healing (Jakob et al., 2012). Current strategies include (i) bone substitute materials, (ii) the delivery of osteoinductive growth factors, and/or (iii) the use of bone autografts. Natural and synthetic materials, mainly calcium phosphate or titanium based, can structurally replace bone but lack durability and regenerative capacity. The delivery of BMP (e.g., BMP-2) induces osteogenesis, but only at doses way above physiological concentrations, which renders such therapeutic strategies very expensive and potentially unsafe (Deutsch, 2010). Thus, the current clinical gold standard is to use autografts for treating bone defects in trauma and reconstructive surgery. This procedure consists in the harvest of autologous bone chips from a healthy donor site (e.g., the iliac crest) and their direct transplantation into the defect as osteogenic grafts. Unfortunately, the limited availability of autologous material together with considerable morbidity at the donor site and the difficulties in matching the topology of the graft with the injured site have limited the clinical satisfaction of bone chip grafts. The exploitation of allo- and/or xenografts has been severely compromised by potential transmission of unknown pathogens and possible immunogenicity. Thus, none of these therapeutic approaches have an established potential for long-lasting recovery (Redman, Oldfield, & Archer, 2005). These biological and clinical observations demonstrate that reparative strategies depend on endogenous healing conditions and may not be sufficient to promote complete and stable bone repair. Therefore, the development of alternative bone substitute devices and strategies are a major challenge and priority for biomedical orthopedic research.

4. CLASSICAL APPROACHES TO BONE TISSUE ENGINEERING

The need of alternative strategies underlies the multidisciplinary field of tissue engineering (Langer & Vacanti, 1993), which in the context of bone tissues has predominantly relied on the *ex vivo* generation of tissue replacements resembling native bone (Fig. 11.2A). This is generally achieved by combining autologous bone forming progenitor cells with supportive biomaterials and osteogenic growth factors. The embedding of progenitors into a scaffolding template facilitates the handling of the construct during preclinical and clinical procedures, provides initial mechanical support following engraftment, and represents the first extracellular matrix substrate. This matrix will be recognized and remodeled and serve likely as a stimulus of osteogenic and osteoinductive processes for the progenitors. The exponential development of biomaterials has promoted the design of engineered

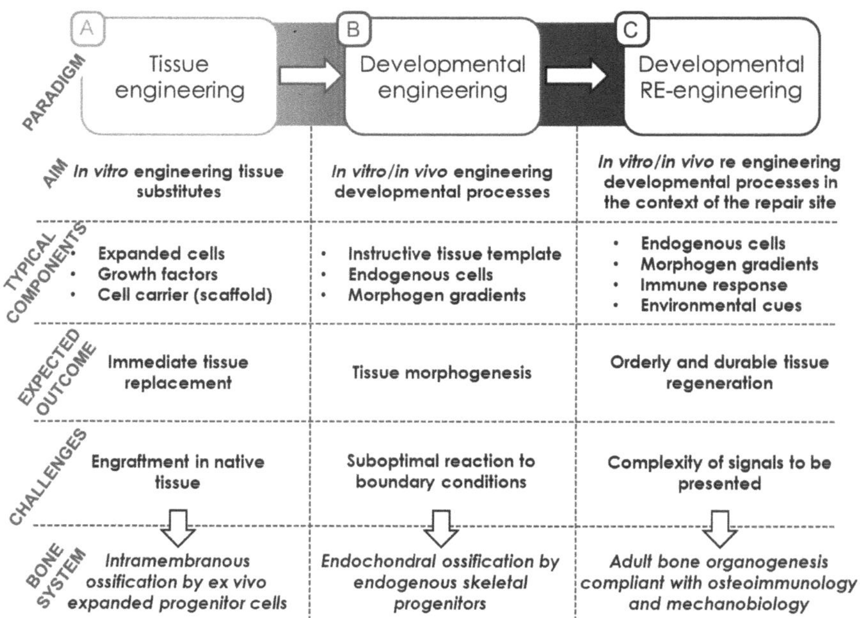

Figure 11.2 *Comparison between tissue engineering* (column A), *developmental engineering* (column B), and *developmental re-engineering* (column C) paradigms, from a conceptual framework to a possible practical implementation in the context of engineering bone substitutes/grafts. (For color version of this figure, the reader is referred to the online version of this chapter.)

bone substitutes that can be tailored in shape, size, architecture, topography, and composition (Barradas, Yuan, van Blitterswijk, & Habibovic, 2011). The cocktail of signaling molecules to which the progenitors are exposed by being cultured in differentiation media under defined environmental conditions (such as oxygen tension, mechanical stimuli, in coculture with other cell types) are continuously, but empirically, refined to improve bone formation (Braccini et al., 2005; Güven et al., 2011; Papadimitropoulos et al., 2011; Sheehy et al., 2013; Siddappa et al., 2008). In spite of these improvements, the current conventional protocols used to engineer bone analogs in a direct and efficient manner suffer from large interdonor variation in cell differentiation, massive hypoxic–necrotic core zones due to delayed vascularization following implantation, an apparent lack of tissue organization, and the need for prolonged *ex vivo* manipulation (Jakob et al., 2012). As a consequence, the survival and integration rates of grafts implanted into the defect site are inadequate, especially when scaffold sizes and engineered implants need to be critically large (Fig. 11.2A).

Recently, alternative strategies aiming to move toward more effective graft manufacturing processes have been proposed (Martin, Smith, & Wendt, 2009). Bioreactor-based culture systems such as online controlled and automated closed environments may help to standardize bioprocesses by enabling temporally controlled regimes of physical stimulation of the progenitor cells in culture. Such bioreactors should be useful to minimize product variability and lower costs in comparison to extensive manual labor. Furthermore, intraoperative procedures have been improved to save costs and time. For example, progenitors cells can be isolated, concentrated, and embedded in a supportive matrix (e.g., fibrin glue) for immediate reinjection into the defect sites to overcome *ex vivo* expansion and differentiation (Mehrkens et al., 2012). Alternatively, *in situ* regeneration could be enhanced by promoting the contribution of endogenous progenitors to instructive scaffolds, hence exploiting the potential of the body as a bioreactor (Stevens et al., 2005). Despite improvements have been implemented into graft manufacturing processes, the bottlenecks discussed before remain and delay the translation of research into bone tissue engineering products that benefit human health (Fig. 11.2A).

5. DEVELOPMENTAL ENGINEERING OF BONE

Bone tissue engineering usually relies on direct differentiation of progenitors toward the osteogenic lineage and production of mineralized matrix in attempting to accelerate tissue recovery. From an embryonic perspective, direct osteogenic commitment typically occurs in development of flat bones (intramembranous ossification), but not for long bones, which are ossified by replacing the initial cartilage templates (endochondral ossification). Engineered long bone grafts may need to recapitulate developmental process to faithfully mimic the structure and function of the native tissues and to successfully repair the defect site (Lenas et al., 2009a, 2009b). A possible strategy to achieve this goal is to engineer the developmental process rather than the mere replacement tissue. The "developmental engineering" paradigm (reviewed by Lenas et al., 2009a, 2009b) proposes to mimic the embryological processes underlying tissue growth and differentiation by exploiting the key mechanisms of ontogenesis (Fig. 11.2B). Therefore, substitutes to repair long bone defects should be generated by phenocopying the steps of normal endochondral ossification, which starts with condensation of the mesenchymal progenitors, their chondrogenesis, and hypertrophy. This is then followed by formation of a bony collar through perichondral ossification, matrix remodeling, and vascularization, while the cartilage model is resorbed and bone matrix deposited (Kronenberg, 2003). The modularity of the endochondral process should permit direct translation into controlled *ex vivo* experimental procedures. Observation and control of the ongoing process may be facilitated by the identification of defined molecular markers for each lineage and maturation stage (such as, e.g., collagen type X marking hypertrophic chondrocytes, Riminucci et al., 1998), thereby increasing the robustness of the entire engineering process. Moreover, since hypertrophic chondrocytes not only survive better under hypoxic conditions before vascular anastomosis, but also produce vascular endothelial growth factor (VEGF) to attract endothelial cells, the strategy of implantation of hypertrophic cartilage is expected to improve survival of the graft following implantation (Lenas et al., 2009a, 2009b). Last but not least, the formation of a functional long bone replacement will ultimately require homing of hematopoietic cells to the bone marrow, which appears to depend strictly on endochondral ossification (Chan et al., 2008). Thus, faithful recapitulation of morphogenesis should enable formation of well-organized and fully functional organs (Fig. 11.2B).

We and others have recently succeeded in recapitulating aspects of endochondral ossification using mouse embryonic stem cells, chick embryonic mesenchymal stem/stromal cells MSC, and clinically relevant cell sources such as human bone marrow-derived MSC (Jukes et al., 2008; Lenas et al., 2009a, 2009b; Oliveira et al., 2009; Scotti et al., 2013, 2010). In this series of studies, bone marrow-derived MSC are induced to differentiate into hypertrophic chondrocytes *in vitro*. The resulting chondrogenic templates are implanted into nude mice, where they remodel into bone under the influence of endogenous progenitors (Bianco et al., 2013). These bone-like structures are vascularized and hematopoietic niches are formed (Scotti et al., 2013, 2010). Recently, it has been shown that mouse embryonic stem cells can be induced to form chondrogenic mesodermal cells. These will differentiate into hypertrophic chondrocytes under the influence of BMP4, while treatment with growth differentiation factor-5 in conjunction with inhibiting Sonic hedgehog and BMP will result in nonhypertrophic chondrocytes with articular characteristics (Craft et al., 2013). Interestingly, some of the cell–cell signaling interactions required for embryonic skeletal development and endochondral ossification such as morphogenetic IHH signaling are activated and are functionally relevant for bone tissue morphogenesis from human MSC (Scotti et al., 2010). While these recent advances represent the first steps toward developmental engineering of cartilage and bone (Fig. 11.2B), it will be important to mark both donor and host cells to investigate their fates and respective contributions to the resulting bone-like structures. In addition, there is an urgent need to identify the instructive signals required for the *in vitro* and *in vivo* induction/remodeling processes. The identification and understanding of underlying molecular circuitry will aid the development of next-generation approaches for *in situ* bone morphogenesis, based on cell-free grafts. For example, morphogenetic key signals may be delivered from engineered and then decellularized endochondral matrices. These could confer the necessary 3D structure and deliver the required signals at the time of implantation to reactivate the embryonic chondrogenesis/endochondral ossification programs by endogenous progenitors (Badylak, Weiss, Caplan, & Macchiarini, 2012; Sadr et al., 2012).

At this stage, developmental engineering strategies (Fig. 11.2B), which would instruct orderly and robust tissue regeneration and communication with the endogenous microenvironment at the repair site, have not yet been developed (Lenas et al., 2009a, 2009b; Scotti et al., 2013, 2010). The spatiotemporal arrangement of the different cell types and the expression of key regulatory signals in an engineered chondrogenic tissue construct fail to, for example, reproduce the growth plate of developing long bones (reviewed by Karsenty et al., 2009). This is likely the consequence of a failure of the engineered bone constructs to self-organize as is the case during embryonic development (Fig. 11.2B). This may be due to the absence of spatiotemporally coordinated cell proliferation, fate assignment and differentiation, and/or the absence of tightly regulated morphogenetic signaling (reviewed by Karsenty et al., 2009; Zeller et al., 2009). Conventional culture systems impose rather rigid physical constraints, while bone development in embryos relies on highly dynamic spatiotemporal modulation of the underlying signaling systems and cell characteristics (see before). Therefore, next-generation extracellular matrix-based strategies might be taking advantage of smart biomaterials such as hydrogels (Barrère, Mahmood, de Groot, & van Blitterswijk, 2008; Sala et al., 2011). Smart biomaterials would aid mimicking of the spatiotemporal controlled release of morphogenetic factors in the attempt to create structured signaling gradients for directed cell behavior (Lienemann, Lutolf, & Ehrbar, 2012). In combination with systems approaches to predict the core morphogenetic pathways and events, this will aid the implementation of more effective and controlled developmental engineering approaches (Bénazet et al., 2009; Lefevre, Rieckhof, & Califano, 2012; Moledina et al., 2012).

6. DEVELOPMENTAL RE-ENGINEERING OF BONE TISSUE

Faithful recapitulation of endochondral ossification by developmental engineering may be not sufficient for preclinical models and clinical translation, as the *in vivo* repair processes take place in a complex adult as opposed to embryologic environment. Therefore, we here introduce and discuss the concept of "developmental RE-engineering" as a potential next-generation approach to tissue/organ repair (Fig. 11.2C). Re-engineering should be based on adult human systems and thus take into account: (i) the stage of commitment of adult stem cells, thus their possibly restricted potential of differentiation capacity and 3D self-organization; (ii) the distinct environmental cues provided by the surrounding adult tissues upon engraftment, such as mechanical stimuli, matrix stiffness, oxygen tension, molecular cues and signals; (iii) the inflammatory and immune responses at the site of damage/engraftment.

As a first step toward "developmental re-engineering," the molecular signature of the bona fide skeletal stem cells of human adult origin needs to be investigated (Bianco et al., 2013) and compared with the one of, for example, mouse limb skeletal progenitors. It is not unlikely that adult stem cells respond either in a highly heterogeneous manner or not at all to the application of specific morphogenetic signals, as they lack the necessary receptors and/or signal transduction circuits to receive and interpret the relevant morphogenetic cues. This could explain the incomplete response and differentiation of adult MSC treated with FGF and WNT ligands (Centola et al., 2013). Recently, it has become possible to differentiate mouse and human embryonic stem cells (Craft et al., 2013; Oldershaw et al., 2010) and induced pluripotent stem cells (Yokote & Yokoo, 2013; Zhu et al., 2009) into chondrocytes in an efficient and organized manner. These stem cells are valuable tools to reproduce developmental processes *ex vivo* and to gain novel insights to directed differentiation of various types of human tissues (Cohen & Melton, 2011; Zhu et al., 2009). A challenge will be to reconcile the results obtained by *in vitro* analysis of multipotent stem and progenitor cells with clinically compliant cells derived from adult patients and the requirements of these patients that suffer from, for example, age-related degenerative or pathological processes.

The environmental cues and paracrine signals that stem and progenitors cells receive during *in vitro* preconditioning and upon implantation into the defect site may also be significantly different from an embryo and might substantially influence graft development (Fig. 11.2C). As an example, tissue developing in the embryo is not at all at the same scale compared to redeveloping one in the adult, that is, regeneration of a tissue involves a much larger number of cells, leads to the reestablishment of a much larger structure, and might generate spatially different morphogenetic fields, compared to developing limb. These dimensional adjustments may require further engineering of the manufacturing process, for example, by introducing a dynamic culture regime (e.g., use perfusion culture to shorten times of molecular signaling between distant areas (Wendt, Riboldi, Cioffi, & Martin, 2009)) or by exploiting a modular design (e.g., generate larger tissue structures by bringing together constructs developed at a smaller scale).

Looking from a yet alternative perspective, there is increasing evidence that *in vitro* engineered skeletal tissues respond favorably to mechanical regimes (i.e., compression, perfusion), which may improve graft quality and reproducibility in comparison to classical and simple static cell culture approaches (Braccini et al., 2005; Candrian et al., 2008; Martin et al., 2009). It has been shown that reduced oxygen tension has a major effect on the chondrogenic differentiation potential of MSC, which indicates that experimental conditions reproducing the variation in oxygen levels during fracture healing should be used (Mumme et al., 2012). Moreover, cell–substrate and cell–cell interactions imposed by the *ex vivo* environment significantly alter cell specification and determination (Gilbert et al., 2010). In particular, biomechanical and rheological properties as well as substrate topography, stiffness, and functionalization are known to alter cell fates (McMurray et al., 2011; Gobaa et al., 2011; Lee, Knight, Campbell, & Bader, 2011; Unadkat et al., 2011). Therefore, these constraints must be carefully considered in the attempt to best mimic the endogenous environment and they must be adapted as cell differentiation progresses. The design of smart scaffolds presenting a regenerative environment to endogenous progenitors will likely become an integrated part of graft manufacturing. This, together with an in-depth mechanistic understanding of the molecular signaling interactions, will result in active biomaterials able to instruct cell behavior in a spatiotemporally controlled manner rather than being inert cell carriers (Fig. 11.2C). Such next-generation biomaterials, together with spatiotemporally controlled release of growth factors, will be key to developmental engineering (Fisher, Khademhosseini, Langer, & Peppas, 2010) and should allow translation of mechanistic findings in animal models to human regenerative medicine.

Last but not least, another important aspect to consider for "developmental re-engineering" is the fact that inflammatory signals and cells of the immune system, not active during embryonic skeletal development, play fundamental roles during fracture healing. In this context, the field of "osteoimmunology" has developed to analyze the interplay between the skeleton and the immune system, as a consequence of shared auto- and paracrine mediators, direct cell–cell interactions, and a common milieu. The interplay between skeletal progenitors, lymphocytes, and macrophages is illustrated by the fact that addition of cytokines, chemokines, and other signaling factors induced by cell–cell contact can strongly regulate cell differentiation and engraftment. For example, addition of interleukin-1β to *in vitro* cultures of engineered hypertrophic cartilage constructs mimics the inflamed environment at the fracture site and enhances vessel invasion, matrix remodeling, and bone formation (Scotti et al., 2013). Mutual and dynamic modulation patterns have been depicted: mesenchymal cell populations display immunomodulatory properties, while immune cells may impact on MSC differentiation (Arron & Choi, 2000; Takayanagi, 2012). For example, specific lymphocyte subpopulations have been found at the defect site and shown to govern bone formation by MSC, likely through the production of soluble factors as tumor necrosis factor-α and interferon-γ (Reinke et al., 2013). This indicates that the signaling interactions that control the crosstalk between MSC and cells of the immune systems must be integrated in the design of skeletal regeneration strategies (Fig. 11.2C).

7. CONCLUSIONS

We discuss the complexity of applying a "developmental re-engineering" paradigm to bone repair, replacement, and ultimately regenerative medical procedures (Fig. 11.3). Artificial bone organogenesis should aim to imitate normal embryonic skeletal development but also consider mechanobiology and osteoimmunology to promote directed expansion and differentiation of adult mesenchymal progenitors *in vitro* and long-lasting engraftment and replacement *in vivo*. The aim is not only to mimic and recapitulate embryological processes during fracture healing, but also to adopt 3D organogenesis to an adult fully immune- and mechanically-competent organism. Ultimately, the complexity of identifying and controlling spatiotemporally regulated morphogenetic signals should be tackled by integrating knowledge from multiple

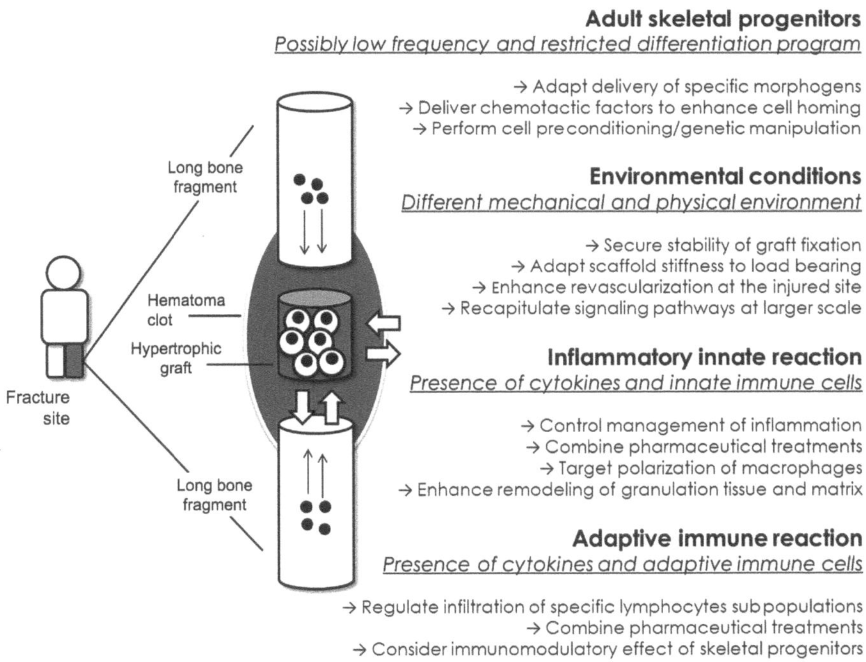

Figure 11.3 *Developmental re-engineering for adult skeletal systems. Crucial differences from the embryological fields and processes* (*Bold* characters) determine a new set of issues to be incorporated in the design of developmentally inspired grafts (*Italic and underlined*), which may be addressed by a variety of possible approaches and research directions (arrow-introduced items). (For color version of this figure, the reader is referred to the online version of this chapter.)

scientific domains and streamlined within systems biology approaches. We speculate that a "developmental re-engineering" approach will offer the potential to regenerate functional and durable tissues in a highly efficient, predictable, and controlled manner, beyond the context of skeletal elements specifically discussed in this perspective.

ACKNOWLEDGMENTS

The authors would like to acknowledge the Swiss National Science Foundation to support the work within the program "Sinergia" (SNF Grant No. CRSII3_136179/1) and the Division III (SNF Grant No. NMS1725).

GLOSSARY

Intramembranous ossification mesenchymal progenitors at the site of formation or regeneration of flat bone tissues (e.g., skull) directly differentiate into bone forming, osteoblastic cells (for in-depth review, please see Kronenberg, 2003).

Endochondral ossification mesenchymal progenitors at the site of formation or regeneration of long bones (e.g., limb) first differentiate into cartilaginous cells, generating a cartilage template, which subsequently undergoes hypertrophy and remodeling. The latter facilitates vessel invasion, arrival of osteoprogenitors and, at last, deposition of mineralized bone tissue (for in-depth review, please see Kronenberg, 2003).

Tissue engineering *ex vivo* generation of tissue substitutes to restore damaged integrity and functionality (for in-depth review, please see Langer & Vacanti, 1993).

Intraoperative procedures single surgery during which autologous progenitor cells are harvested, processed, and reinjected in the defect site, possibly in association with biomaterials (for in-depth review, please see Jakob et al., 2012).

Biomaterial 3D supportive structure with basic functions of delivery of cellular aggregates and mechanical support at a defect site, inert or active in instructing cell differentiation by means of chemical/physical properties or by the presence of functionalized growth factors, eventually delivered with spatial/temporal/environmental-dependent control (*smart* scaffold) (for in-depth review, please see Barrère et al., 2008).

Rheological properties viscoelastic features of the substrate where the cells adhere to (for in-depth review, please see Lee et al., 2011).

Topography characterization of the microscopic level of the substrate features (for in-depth review, please see Unadkat et al., 2011).

REFERENCES

Arron, J. R., & Choi, Y. (2000). Bone versus immune system. *Nature, 408*, 535–536.

Badylak, S. F., Weiss, D. J., Caplan, A., & Macchiarini, P. (2012). Engineered whole organs and complex tissue. *Lancet, 379*, 943–952.

Barna, M., & Niswander, L. (2007). Visualization of cartilage formation: Insight into cellular properties of skeletal progenitors and chondrodysplasia syndromes. *Developmental Cell, 12*, 931–941.

Barradas, A. M., Yuan, H., van Blitterswijk, C. A., & Habibovic, P. (2011). Osteoinductive biomaterials: Current knowledge of properties, experimental models and biological mechanisms. *European Cells & Materials, 21*, 407–429.

Barrère, F., Mahmood, T. A., de Groot, K., & van Blitterswijk, C. A. (2008). Advanced biomaterials for skeletal tissue regeneration: Instructive and smart functions. *Materials Science and Engineering: R: Reports, 59*, 38–71.

Bénazet, J. D., Bischofberger, M., Tiecke, E., Gonçalves, A., Martin, J. F., Zuniga, A., et al. (2009). A self-regulatory system of interlinked signaling feedback loops controls mouse limb patterning. *Science, 20*, 1050–1053.

Bénazet, J. D., Pignatti, E., Nugent, A., Unal, E., Laurent, F., & Zeller, R. (2012). Smad4 is required to induce digit ray primordia and to initiate the aggregation and differentiation of chondrogenic progenitors in mouse limb buds. *Development, 139*, 4250–4260.

Bianco, P., Cao, X., Frenette, P. S., Mao, J. J., Robey, P. G., Simmons, P. J., et al. (2013). The meaning, the sense and the significance: Translating the science of mesenchymal stem cells into medicine. *Nature Medicine, 19*, 35–42.

Braccini, A., Wendt, D., Jaquiery, C., Jakob, M., Heberer, M., Kenins, L., et al. (2005). Three-dimensional perfusion culture of human bone marrow cells and generation of osteoinductive grafts. *Stem Cells, 23*, 1066–1072.

Candrian, C., Vonwil, D., Barbero, A., Bonacina, E., Miot, S., Farhadi, J., et al. (2008). Engineered cartilage generated by nasal chondrocytes is responsive to physical forces resembling joint loading. *Arthritis and Rheumatism, 58*, 197–208.

Caron, M. M. J., Emans, P. J., Surtel, D. A. M., Cremers, A., Voncken, J. W., Welting, T. J. M., et al. (2012). Activation of NF-κB/p65 facilitates early chondrogenic differentiation during endochondral ossification. *PLoS One, 7*, e33467.

Centola, M., Tonnarelli, B., Schären, S., Glaser, N., Barbero, A., & Martin, I. (2013). Priming 3D cultures of human mesenchymal stromal cells towards cartilage formation via developmental pathways. *Stem Cells and Development, 22*, 2849–2858.

Chan, C. K., Chen, C. C., Luppen, C. A., Kim, J. B., DeBoer, A. T., Wei, K., et al. (2008). Endochondral ossification is required for haematopoietic stem-cell niche formation. *Nature, 457*, 490–494.

Chen, Y., Whetstone, H. C., Youn, A., Nadesan, P., Chow, E. C., Lin, A. C., et al. (2007). Beta-catenin signaling pathway is crucial for bone morphogenetic protein 2 to induce new bone formation. *The Journal of Biological Chemistry, 282*, 526–533.

Claes, L., Recknagel, S., & Ignatius, A. (2012). Fracture healing under healthy and inflammatory conditions. *Nature Reviews. Rheumatology, 8*, 133–143.

Cohen, D. E., & Melton, D. (2011). Turning straw into gold: Directing cell fate for regenerative medicine. *Nature Reviews. Genetics, 12*, 243–252.

Craft, A. M., Ahmed, N., Rockel, J. S., Baht, G. S., Alman, B. A., Kandel, R. A., et al. (2013). Specification of chondrocytes and cartilage tissues from embryonic stem cells. *Development, 140*, 2597–2610.

Day, T. F., Guo, X., Garrett-Beal, L., & Yang, Y. (2005). Wnt/beta-catenin signaling in mesenchymal progenitors controls osteoblast and chondrocyte differentiation during vertebrate skeletogenesis. *Developmental Cell, 8*, 739–750.

Day, T. F., & Yang, Y. (2008). Wnt and hedgehog signaling pathways in bone development. *The Journal of Bone and Joint Surgery. American Volume, 90*, 19–24.

Deutsch, H. (2010). High-dose bone morphogenetic protein-induced ectopic abdomen bone growth. *The Spine Journal, 10*, e1–e4.

Ehlen, H. W., Buelens, L. A., & Vortkamp, A. (2006). Hedgehog signaling in skeletal development. *Birth Defects Res C Embryo Today: Reviews, 78*, 267–279.

Ferrari, S., Pellegrini, G., Matsui, T., Mavilio, F., & De Luca, M. (2006). Gene therapy in combination with tissue engineering to treat epidermolysis bullosa. *Expert Opinion on Biological Therapy, 6*, 367–378.

Fisher, O. Z., Khademhosseini, A., Langer, R., & Peppas, N. A. (2010). Bioinspired materials for controlling stem cell fate. *Accounts of Chemical Research, 43*, 419–428.

Gerstenfeld, L. C., Cullinane, D. M., Barnes, G. L., Graves, D. T., & Einhorn, T. A. (2003). Fracture healing as a post-natal developmental process: Molecular, spatial, and temporal aspects of its regulation. *Journal of Cellular Biochemistry, 88*, 873–884.

Gilbert, P. M., Havenstrite, K. L., Magnusson, K. E., Sacco, A., Leonardi, N. A., Kraft, P., et al. (2010). Substrate elasticity regulates skeletal muscle stem cell self-renewal in culture. *Science, 329*, 1078–1081.

Gobaa, S., Hoehnel, S., Roccio, M., Negro, A., Kobel, S., & Lutolf, M. P. (2011). Artificial niche microarrays for probing single stem cell fate in highthroughput. *Nature Methods, 8*, 949–955.

Grimes, R., Jepsen, K. J., Fitch, J. L., Einhorn, T. A., & Gerstenfeld, L. C. (2011). The transcriptome of fracture healing defines mechanisms of coordination of skeletal and vascular development during endochondral bone formation. *Journal of Bone and Mineral Research, 26*, 2597–2609.

Guo, X., Mak, K. K., Taketo, M. M., & Yang, Y. (2009). The Wnt/beta-catenin pathway interacts differentially with PTHrP signaling to control chondrocyte hypertrophy and final maturation. *PLoS One, 4*, e6067.

Güven, S., Mehrkens, A., Saxer, F., Schaefer, D. J., Martinetti, R., Martin, I., et al. (2011). Engineering of large osteogenic grafts with rapid engraftment capacity using mesenchymal and endothelial progenitors from human adipose tissue. *Biomaterials, 32*, 5801–5809.

Horton, W. A., & Degnin, C. R. (2009). FGFs in endochondral skeletal development. *Trends in Endocrinology and Metabolism, 20*, 341–348.

Hunziker, E., Spector, M., Libera, J., Gertzman, A., Woo, S. L., Ratcliffe, A., et al. (2006). Translation from research to applications. *Tissue Engineering, 12*, 3341–3364.

Jakob, M., Saxer, F., Scotti, C., Schreiner, S., Studer, P., Scherberich, A., et al. (2012). Perspective on the evolution of cell-based bone tissue engineering strategies. *European Surgical Research, 49*, 1–7.

Jukes, J. M., Both, S. K., Leusink, A., Sterk, L. M., van Blitterswijk, C. A., & de Boer, J. (2008). Endochondral bone tissue engineering using embryonic stem cells. *Proceedings of the National Academy of Sciences of the United States of America, 105*, 6840–6845.

Karamboulas, K., Dranse, H. J., & Underhill, T. M. (2010). Regulation of BMP-dependent chondrogenesis in early limb mesenchyme by TGFbeta signals. *Journal of Cell Science, 15*, 2068–2076.

Karsenty, G., Kronenberg, H. M., & Settembre, C. (2009). Genetic control of bone formation. *Annual Review of Cell and Developmental Biology, 25*, 629–648.

Kobayashi, T., Lyons, K. M., McMahon, A. P., & Kronenberg, H. M. (2005). BMP signaling stimulates cellular differentiation at multiple steps during cartilage development. *Proceedings of the National Academy of Sciences of the United States of America, 102*, 18023–18027.

Kronenberg, H. M. (2003). Developmental regulation of the growth plate. *Nature, 423*, 332–336.

Langer, R., & Vacanti, J. P. (1993). Tissue engineering. *Science, 260*, 920–926.

Lee, D. A., Knight, M. M., Campbell, J. J., & Bader, D. L. (2011). Stem cell mechanobiology. *Journal of Cellular Biochemistry, 112*, 1–9.

Lefevre, C., Rieckhof, G., & Califano, A. (2012). Reverse-engineering human regulatory networks. *Wiley Interdisciplinary Reviews. Systems Biology and Medicine, 4*, 311–325.

Lenas, P., Moos, M., & Luyten, F. P. (2009a). Developmental engineering: A new paradigm for the design and manufacturing of cell-based products. Part I: From three-dimensional cell growth to biomimetics of in vivo development. *Tissue Engineering. Part B, Reviews, 15*, 381–394.

Lenas, P., Moos, M., & Luyten, F. P. (2009b). Developmental engineering: A new paradigm for the design and manufacturing of cell-based products. Part II: From genes to networks:

Tissue engineering from the viewpoint of systems biology and network science. *Tissue Engineering. Part B, Reviews, 15*, 395–422.
Lienemann, P. S., Lutolf, M. P., & Ehrbar, M. (2012). Biomimetic hydrogels for controlled biomolecule delivery to augment bone regeneration. *Advanced Drug Delivery Reviews, 64*, 1078–1089.
Maes, C., Kobayashi, T., Selig, M. K., Torrekens, S., Roth, S. I., Mackem, S., et al. (2010). Osteoblast precursors, but not mature osteoblasts, move into developing and fractured bones along with invading blood vessels. *Developmental Cell, 17*, 329–344.
Mak, K. K., Kronenberg, H. M., Chuang, P. T., Mackem, S., & Yang, Y. (2008). Indian hedgehog signals independently of PTHrP to promote chondrocyte hypertrophy. *Development, 135*, 1947–1956.
Mariani, F. V., Ahn, C. P., & Martin, G. R. (2008). Genetic evidence that FGFs have an instructive role in limb proximal-distal patterning. *Nature, 453*, 401–405.
Marsell, R., & Einhorn, T. A. (2011). The biology of fracture healing. *Injury, 42*, 551–555.
Martin, I., Smith, T., & Wendt, D. (2009). Bioreactor-based roadmap for the translation of tissue engineering strategies into clinical products. *Trends in Biotechnology, 27*, 495–502.
McMurray, R. J., Gadegaard, N., Tsimbouri, P. M., Burgess, K. V., McNamara, L. E., et al. (2011). Nanoscale surfaces for the long-term maintenance of mesenchymal stem cell phenotype and multipotency. *Nature Materials, 10*, 637–644.
Mehrkens, A., Saxer, F., Güven, S., Hoffmann, W., Müller, A. M., Jakob, M., et al. (2012). Intraoperative engineering of osteogenic grafts combining freshly harvested, human adipose-derived cells and physiological doses of bone morphogenetic protein-2. *European Cells & Materials, 24*, 308–319.
Moledina, F., Clarke, G., Oskooei, A., Onishi, K., Günther, A., & Zandstra, P. W. (2012). Predictive microfluidic control of regulatory ligand trajectories in individual pluripotent cells. *Proceedings of the National Academy of Sciences of the United States of America, 105*(109), 3264–3269.
Mumme, M., Scotti, C., Papadimitropoulos, A., Todorov, A., Hoffmann, W., Bocelli-Tyndall, C., et al. (2012). Interleukin-1β modulates endochondral ossification by human adult bone marrow stromal cells. *European Cells & Materials, 24*, 224–236.
Oldershaw, R. A., Baxter, M. A., Lowe, E. T., Bates, N., Grady, L. M., Soncin, F., et al. (2010). Directed differentiation of human embryonic stem cells toward chondrocytes. *Nature Biotechnology, 28*, 1187–1194.
Oliveira, S. M., Mijares, D. Q., Turner, G., Amaral, I. F., Barbosa, M. A., & Teixeira, C. C. (2009). Engineering endochondral bone: In vivo studies. *Tissue Engineering. Part A, 15*, 635–643.
Papadimitropolous, A., Scherberich, A., Güven, S., Theilgaard, N., Crooijmans, H. J., Santini, F., et al. (2011). A 3D in vitro bone organ model using human progenitor cells. *European Cells & Materials, 21*, 445–458.
Redman, S. N., Oldfield, S. F., & Archer, C. W. (2005). Current strategies for articular cartilage repair. *Journal of European Cells and Materials, 9*, 23–32.
Reinke, S., Geissler, S., Taylor, W. R., Schmidt-Bleek, K., Juelke, K., Schwachmeyer, V., et al. (2013). Terminally differentiated $CD8^+$ T cells negatively affect bone regeneration in humans. *Science Translational Medicine, 5*, 177ra36.
Riminucci, M., Bradbeer, J. N., Corsi, A., Gentili, C., Descalzi, F., Cancedda, R., et al. (1998). Vis-à-vis cells and the priming of bone formation. *Journal of Bone and Mineral Research, 13*, 1852–1861.
Sadr, N., Pippenger, B. E., Scherberich, A., Wendt, D., Mantero, S., Martin, I., et al. (2012). Enhancing the biological performance of synthetic polymeric materials by decoration with engineered, decellularized extracellular matrix. *Biomaterials, 33*, 5085–5093.

Sala, A., Hänseler, P., Ranga, A., Lutolf, M. P., Vörös, J., Ehrbar, M., et al. (2011). Engineering 3D cell instructive microenvironments by rational assembly of artificial extracellular matrices and cell patterning. *Integration Biology (Camb)*, *3*, 1102–1111.

Scotti, C., Piccinini, E., Takizawa, H., Todorov, A., Bourgine, P., Papadimitropoulos, A., et al. (2013). Engineering of a functional bone organ through endochondral ossification. *Proceedings of the National Academy of Sciences of the United States of America*, *110*, 3997–4002.

Scotti, C., Tonnarelli, B., Papadimitropoulos, A., Scherberich, A., Schaeren, S., Schauerte, A., et al. (2010). Recapitulation of endochondral bone formation using human adult mesenchymal stem cells as a paradigm for developmental engineering. *Proceedings of the National Academy of Sciences of the United States of America*, *107*, 7251–7256.

Sheehy, E. J., Vinardell, T., Buckley, C. T., & Kelly, D. J. (2013). Engineering osteochondral constructs through spatial regulation of endochondral ossification. *Acta Biomaterialia*, *9*, 5484–5492.

Siddappa, R., Martens, A., Doorn, J., Leusink, A., Olivo, C., Licht, R., et al. (2008). cAMP/PKA pathway activation in human mesenchymal stem cells in vitro results in robust bone formation in vivo. *Proceedings of the National Academy of Sciences of the United States of America*, *105*, 7281–7286.

Stevens, M. M., Marini, R. P., Schaefer, D., Aronson, J., Langer, R., & Shastri, V. P. (2005). In vivo engineering of organs: The bone bioreactor. *Proceedings of the National Academy of Sciences of the United States of America*, *102*, 11450–11455.

St-Jacques, B., Hammerschmidt, M., & McMahon, A. P. (1999). Indian hedgehog signaling regulates proliferation and differentiation of chondrocytes and is essential for bone formation. *Genes & Development*, *13*, 2072–2086.

Takayanagi, H. (2012). New developments in osteoimmunology. *Nature Reviews Rheumatology*, *8*, 684–689.

ten Berge, D., Brugmann, S. A., Helms, J. A., & Nusse, R. (2008). Wnt and FGF signals interact to coordinate growth with cell fate specification during limb development. *Development*, *135*, 3247–3257.

Unadkat, H. V., Hulsman, M., Cornelissen, K., Papenburg, B. J., Truckenmüller, R. K., Carpenter, A. E., et al. (2011). An algorithm-based topographical biomaterials library to instruct cell fate. *Proceedings of the National Academy of Sciences of the United States of America*, *108*, 16565–16570.

Vortkamp, A., Lee, K., Lanske, B., Segre, G. V., Kronenberg, H. M., & Tabin, C. J. (1996). Regulation of rate of cartilage differentiation by Indian hedgehog and PTH-related protein. *Science*, *273*, 613–622.

Welting, T. J., Caron, M. M., Emans, P. J., Janssen, M. P., Sanen, K., Coolsen, M. M., et al. (2011). Inhibition of cyclooxygenase-2 impacts chondrocyte hypertrophic differentiation during endochondral ossification. *European Cells & Materials*, *22*, 420–436.

Wendt, D., Riboldi, S. A., Cioffi, M., & Martin, I. (2009). Potential and bottlenecks of bioreactors in 3D cell culture and tissue manufacturing. *Advanced Materials*, *21*, 3352–3367.

Wilkie, A. O. (2003). Why study human limb malformations? *Journal of Anatomy*, *202*, 27–35.

Wuelling, M., & Vortkamp, A. (2011). Chondrocyte proliferation and differentiation. *Endocrine Development*, *21*, 1–11.

Yokote, S., & Yokoo, Y. (2013). Organogenesis for kidney regeneration. *Current Opinion in Organ Transplantation*, *18*, 186–190.

Zeller, R., Lopez-Rios, J., & Zuniga, A. (2009). Vertebrate limb bud development: moving towards integrative analysis of organogenesis. *Nature Rev Genet*, *10*, 845–858.

Zhang, X., Huang, C. T., Chen, J., Pankratz, M. T., Xi, J., Li, J., et al. (2010). Pax6 is a human neuroectoderm cell fate determinant. *Cell Stem Cell*, *7*, 90–100.

Zhu, H., Guo, Z., Jiang, X., Li, H., Wang, X., Yao, H., et al. (2009). A protocol for isolation and culture of mesenchymal stem cells from mouse compact bone. *Nature Protocols*, *5*, 550–560.

INDEX

Note: Page numbers followed by "*f*" indicate figures and "*t*" indicate tables.

Lin Xu and Hai Huang, Figure 1.1 Plant regeneration in nature and in tissue culture. (A) Adventitious shoots regenerating from a damaged branch of a tree (*Prunus* × *Yedoensis*). (B) Leaf of a jade plant (*Crassula ovata*) regenerating roots on the soil when detached. (C) Callus regenerated from *Arabidopsis* leaf explants on CIM containing 2.2 μM 2,4-dichlorophenoxyacetic acid (2,4-D) and 0.2 μM kinetin. (D) Shoot regeneration from *Arabidopsis* root explants on SIM containing 12.5 μM 6-(γ,γ-dimethylallylamino) purine (2-IP) and 0.57 μM indole-3-acetic acid (IAA).

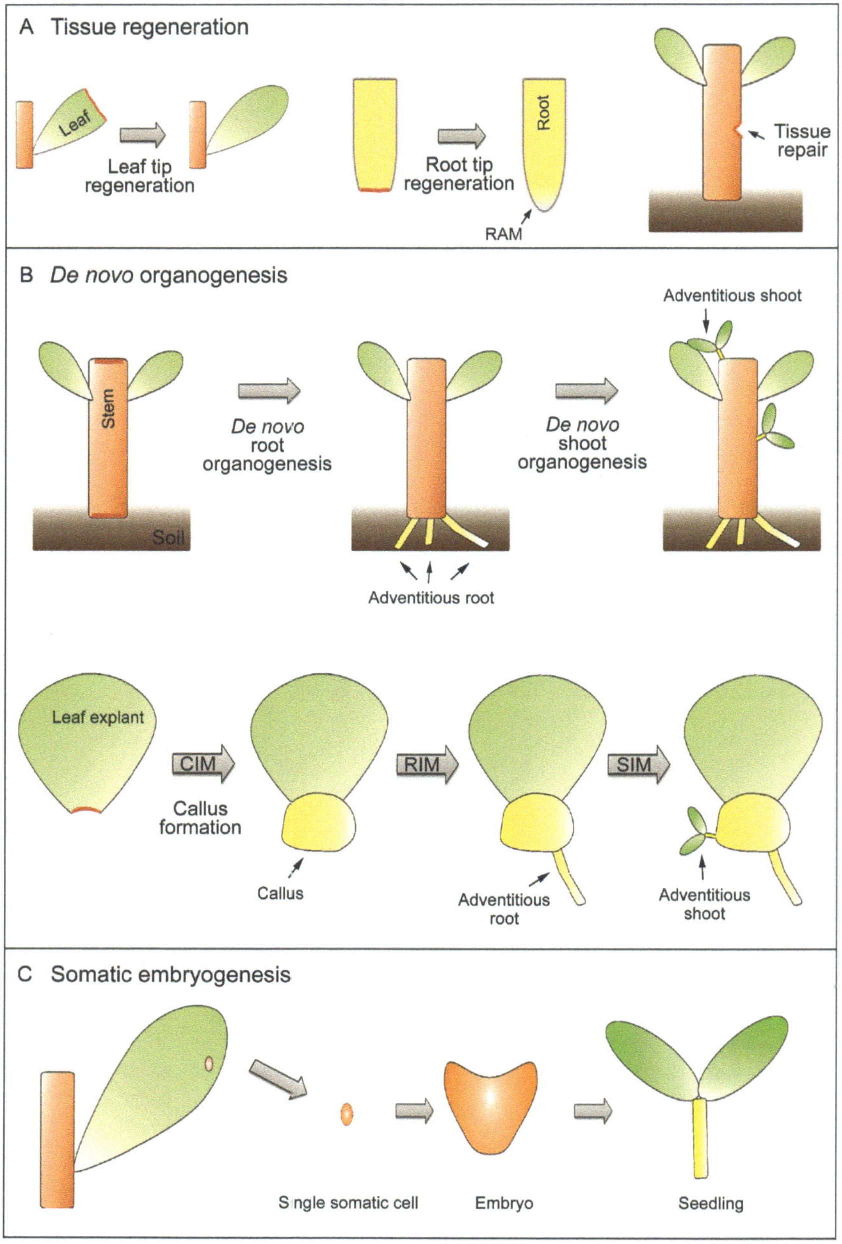

Lin Xu and Hai Huang, Figure 1.2 Different types of plant regeneration. (A) Tissue regeneration. Excised tips of very young leaves (left) and roots (middle) can re-grow via tissue regeneration. Stem tissues can also be repaired after damage (right). (B) *De novo* organogenesis. Upper panels show *de novo* root and shoot organogeneses during the propagation of plant materials via cuttings. Lower panels show a plant regenerated from leaf explants induced to form callus, adventitious roots, and adventitious shoots on CIM, RIM, and SIM, respectively, in tissue culture. (C) Somatic embryogenesis. A single somatic cell can be induced by auxin to form an embryo, which develops into a whole plant, reflecting the totipotency of plant cells. Red lines in (A) and (B) indicate wounding sites.

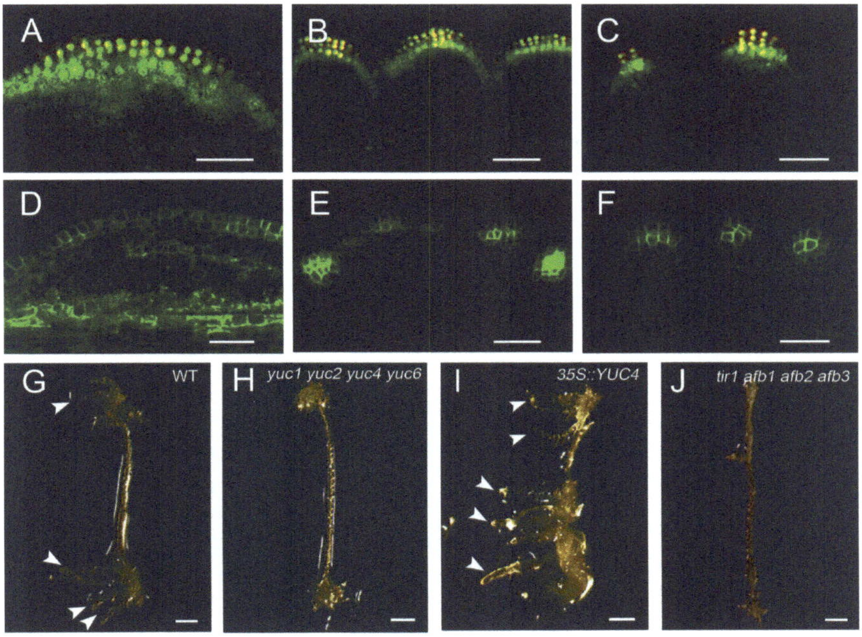

Ying Hua Su and Xian Sheng Zhang, Figure 2.3 Auxin response, polar transport, and biosynthesis in *Arabidopsis* root regeneration. *Arabidopsis* seedlings (ecotype Columbia) are grown on MS medium (Murashige & Skoog, 1962) for 10 days. Root segments (5 mm) are cut and preincubated on CIM (Che et al., 2006) for 6 days and then transferred to RIM (Che et al., 2006) for root induction. (A–C) *DR5rev::YFP* signals (yellow) and *PWOX5::GFP* signals (green) at the edges of callus incubated on CIM for 6 days (A) and on RIM for 2 days (B) and 4 days (C). (D–F) *PIN1::GFP* signals (green) at the edges of callus incubated on CIM for 6 days (D) and on RIM for 2 days (E) and 4 days (F). (G–J) Regenerated roots from callus of wild-type (WT) plants (G), *yuc1 yuc2 yuc4 yuc6* mutants (H), *35S::YUC4* (I), and *tir1 afb1 afb2 afb3* mutants (J) grown on RIM for 10 days. Arrowheads indicate regenerated roots. Scale bars = 150 μm (A–F) and 500 μm (G–J).

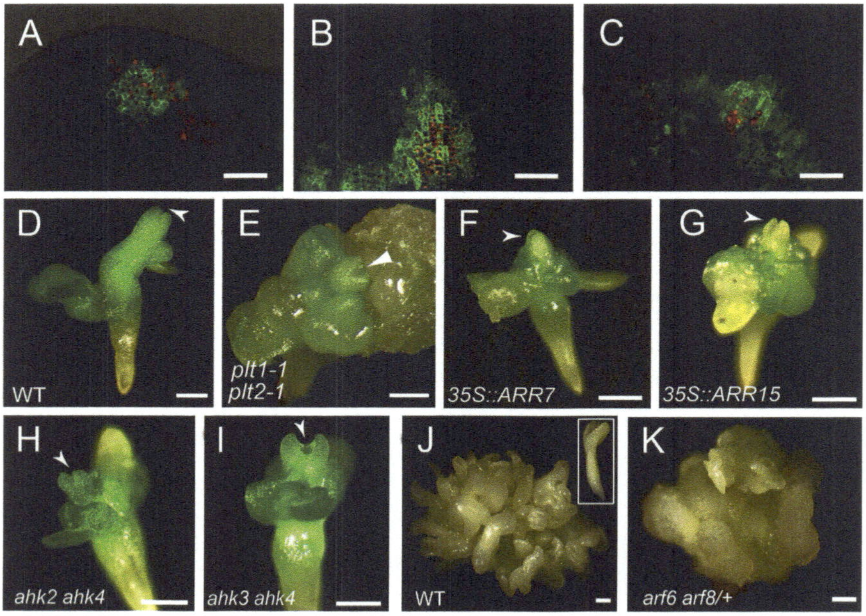

Ying Hua Su and Xian Sheng Zhang, Figure 2.4 Cytokinin and auxin response in *Arabidopsis* somatic embryogenesis. The process of *Arabidopsis* somatic embryogenesis has been reported by Su et al. (2009). Green primary somatic embryos (PSEs) can be induced from explants (zygotic embryos, ecotype Columbia) cultured on medium containing 2,4-D after 10 days. PSEs are then transferred into liquid medium containing 2,4-D (ECIM) for 14 days to form embryonic calli. The resulting calli are transferred into 2,4-D free liquid medium (SEIM) to induce secondary somatic embryos (SSEs). (A–C) *PARR7::GFP* signals (green) and *PWOX5::RFP* signals (red) at the edges of embryonic callus incubated on SEIM for 1 day (A), 2 days (B), and 3 days (C). (D–I) Phenotypes of PSE induction from explants of WT (D), *plt1 plt2* mutants (E), *35S::ARR7* (F), *35S::ARR15* (G), *ahk2 ahk4* mutants (H), and *ahk3 ahk4* mutants (I) grown on solid medium for 10 days. Arrowheads indicate PSEs. (J–K) Phenotypes of SSE induction from calli of WT (J) and *arf6 arf8/+* mutants (K) grown on SEIM for 8 days. Scale bars = 60 μm (A–C), 0.4 mm (D–I), and 1.2 mm (J–K).

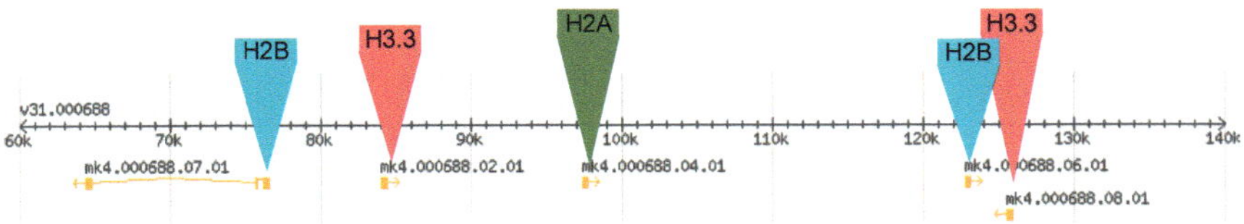

Sofia M.C. Robb and Alejandro Sánchez Alvarado, Figure 3.1 Largest identifiable *S. mediterranea* histone gene cluster. This is a screen capture from SmedGD of genomic contig v31.000688. It is the largest identifiable histone cluster and contains two H2B genes, two H3.3 genes, and one H2A gene. Only the first exon of the MAKER transcript mk4.000688.07.01 is the single exon H2B gene.

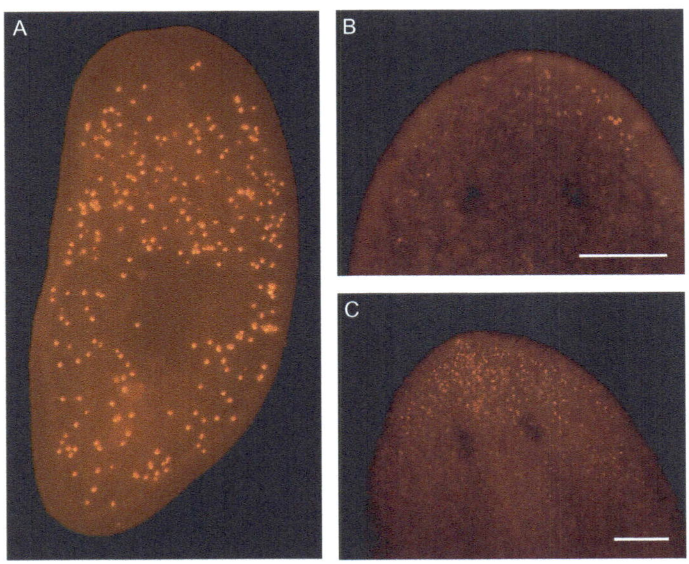

Sofia M.C. Robb and Alejandro Sánchez Alvarado, Figure 3.3 Whole-mount immunohistochemistry. (A) Animals immunostained with antibodies against (A) H3P, (B) H3K9K14-Ac, and (C) H3K9-Me3.

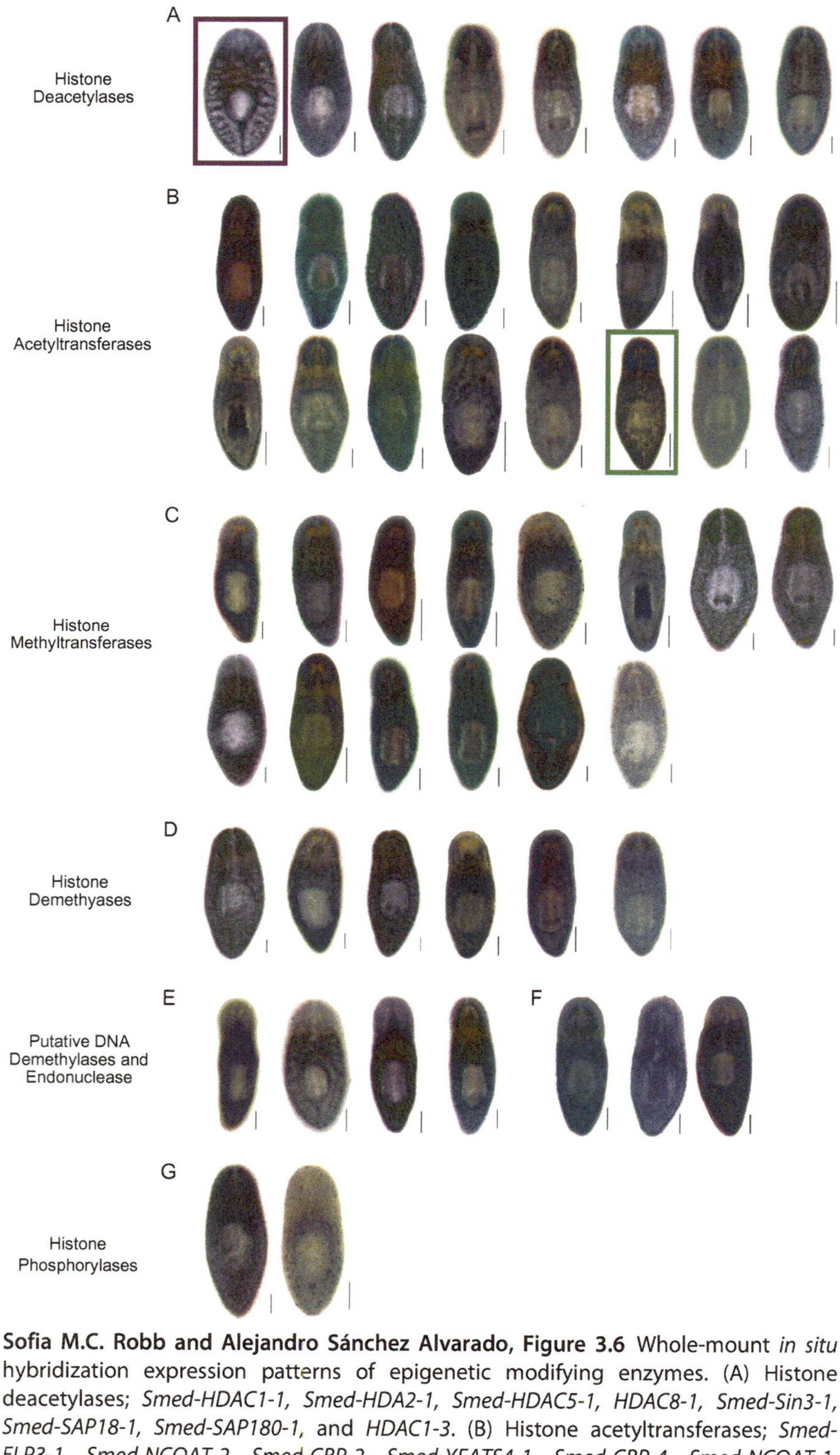

Sofia M.C. Robb and Alejandro Sánchez Alvarado, Figure 3.6 Whole-mount *in situ* hybridization expression patterns of epigenetic modifying enzymes. (A) Histone deacetylases; *Smed-HDAC1-1*, *Smed-HDA2-1*, *Smed-HDAC5-1*, *HDAC8-1*, *Smed-Sin3-1*, *Smed-SAP18-1*, *Smed-SAP180-1*, and *HDAC1-3*. (B) Histone acetyltransferases; *Smed-ELP3-1*, *Smed-NCOAT-2*, *Smed-CBP-2*, *Smed-YEATS4-1*, *Smed-CBP-4*, *Smed-NCOAT-1*,

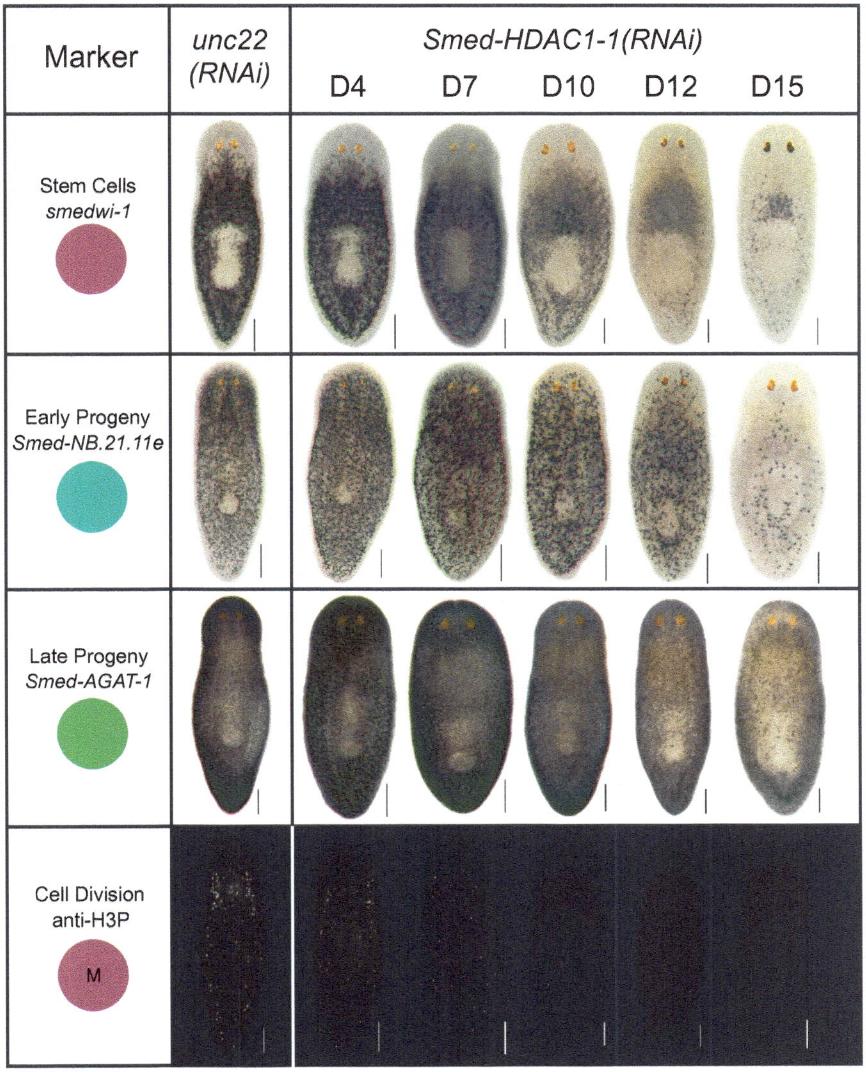

Sofia M.C. Robb and Alejandro Sánchez Alvarado, Figure 3.7 Stem cell lineage analysis of intact *Smed-HDAC1-1(RNAi)* animals. Intact *Smed-HDAC1-1(RNAi)* animals were fixed on D4, D7, D10, D12, and D15 and tested for the presence of the four markers *smedwi-1*, *Smed-NB.21.11e*, *Smed-AGAT-1*, and anti-H3P. *unc22(RNAi)* animals were used as a control.

Sofia M.C. Robb and Alejandro Sánchez Alvarado, Figure 3.6—Cont'd *Smed-MYST3-1, Smed-MYST4-2, Smed-MYST2-2, Smed-Ep300-2, Smed-MYST4-1, Smed-HAT-1, Smed-CBP-5, Smed-CBP-6, Smed-Ep300-3*, and *Smed-Tip60-1*. (C) Histone methyltransferases; *Smed-SETD8-1, Smed-SUV42-1, Smed-SETB1-1, Smed-CARM1-1, Smed-NSD2-2, Smed-SUV92-1, Smed-TRR-1, Smed-DOT1L-1, Smed-ANM8-1, Smed-ANMX-1, Smed-ANM1-1, Smed-ASH2-L-1, Smed-SETB1-2*, and *Smed-SUV92-2*. (D) Histone demethylases; *Smed-JARID-3, Smed-JmjC-1, Smed-JARID-2, Smed-JmjC-6, Smed-JmjC-4*, and *Smed-JmjC-7*. (E) Putative DNA demethylases; *Smed-RL7A, Smed-NHP2L1, Smed-RPS12*, and *Smed-RPL30*. (F) Endonucleases; Smed-XPG-1, Smed-XPG-2, and Smed-FEN1. (G) Histone phosphorylases; *Smed-TAF1-2 and Smed-TAF1-3*. Magenta square indicates stem-cell like expression pattern. Green square, central nervous system expression pattern.

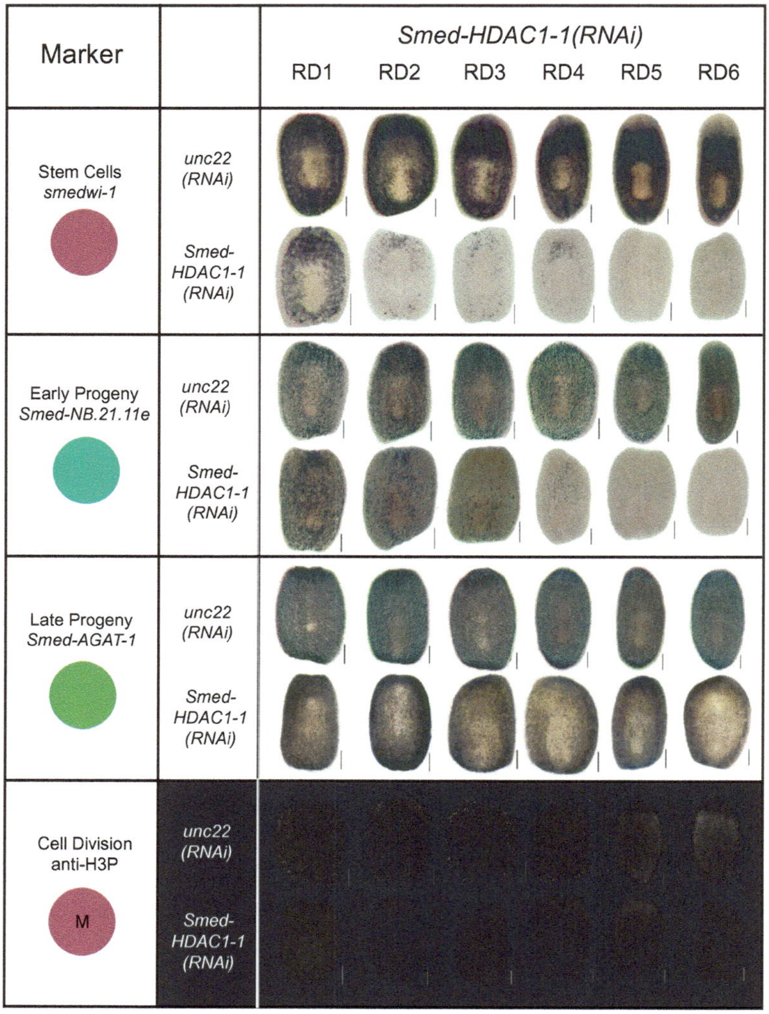

Sofia M.C. Robb and Alejandro Sánchez Alvarado, Figure 3.8 Stem cell lineage analysis of *Smed-HDAC1-1(RNAi)* intact and amputated animals. Intact *Smed-HDAC1-1(RNAi)* animals were fixed on D4, D7, and D10 and amputated. *Smed-HDAC1-1(RNAi)* animals (labeled "e" for experimental) were fixed on D11 through D16 (RD1 through RD6) and tested for the presence of the four markers *smedwi-1*, *Smed-NB.21.11e*, *Smed-AGAT-1*, and anti-H3P. *unc22(RNAi)* animals were used as a control for intact animals and for amputated animals (labeled "c" for control).

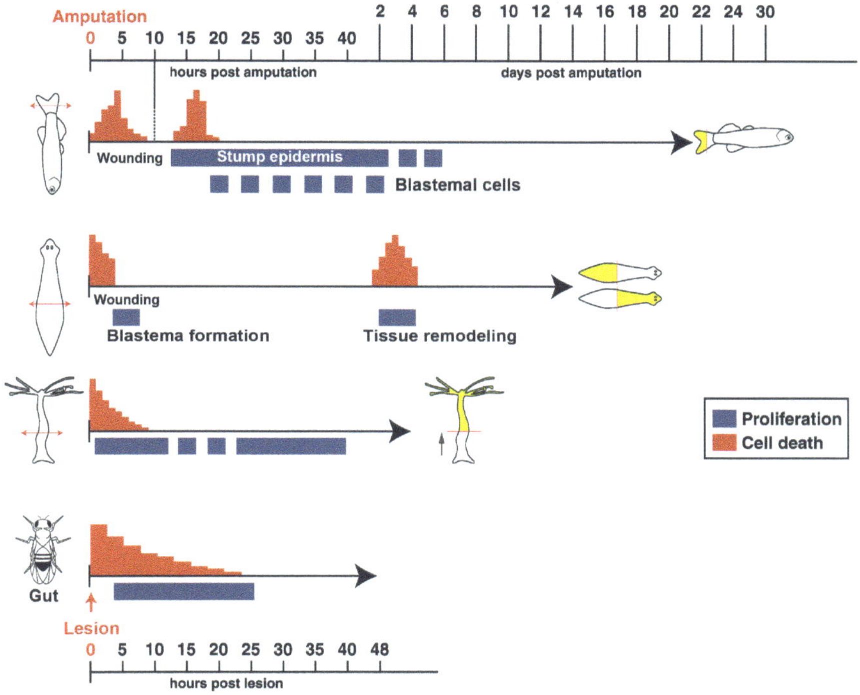

Sophie Vriz *et al.*, Figure 5.1 Injury-induced cell death (red) and cell proliferation (blue) events in *Danio rerio* regenerating its caudal fin, *S. mediterranean* regenerating its body, *Hydra vulgaris* regenerating its head after mid-gastric bisection, *Drosophila melanogaster* repairing its midgut. Fin regeneration in *Danio*, completed in 3 weeks, shows two peaks of apoptosis, the second one being required for the proliferation of the stump epidermis and the recruitment of blastema cells (Gauron et al., 2013). Body regeneration in *Schmidtea* (planarian), completed in 10–14 days, also exhibits two peaks of apoptosis, the second one associated with a round of proliferation, is involved in tissue remodeling (Pellettieri et al., 2010). Head regeneration in *Hydra*, completed in 3 days, displays one burst of apoptosis, directly promoting cell division (Chera et al., 2009). Infection with pathogenic bacteria leads to cell death of enterocytes in adult *Drosophila*, which stimulates the proliferation of intestine stem cells and the repair of the gut (Amcheslavsky et al., 2009; Buchon, Broderick, Poidevin, Pradervand, & Lemaitre, 2009). Time scales are indicated at the top and bottom; regenerated structures are colored yellow.

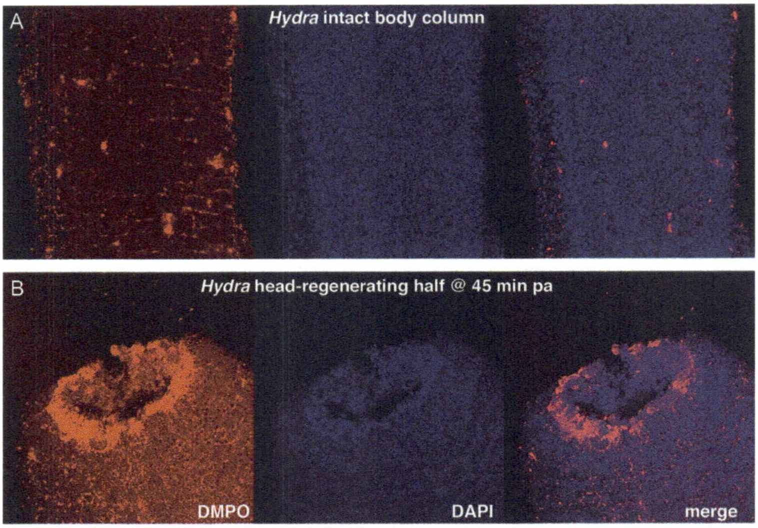

Sophie Vriz *et al.*, Figure 5.3 ROS signals in intact and bisected *Hydra*. ROS were detected by DMPO spin trap immunostaining in intact (A) or bisected (B) *Hydra* polyps (*sf-1* strain). In homeostatic conditions (A), few cells constitutively produce ROS along the body column. Upon bisection (B) ROS are produced locally at the wound edges, detected here after 45 min.

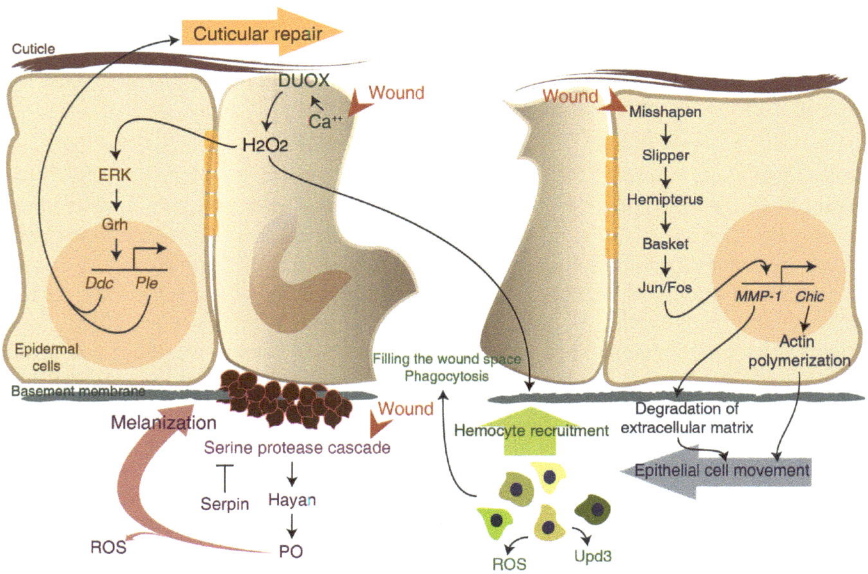

Won-Jae Lee and Masayuki Miura, Figure 6.2 Local wound responses following integumental wound in *Drosophila*. Upon integumental wound, complex biological processes, such as cuticular repair, ROS generation, epithelial cell movement, hemocyte recruitment, and melanization, take place immediately around the wound site. See text for more details.

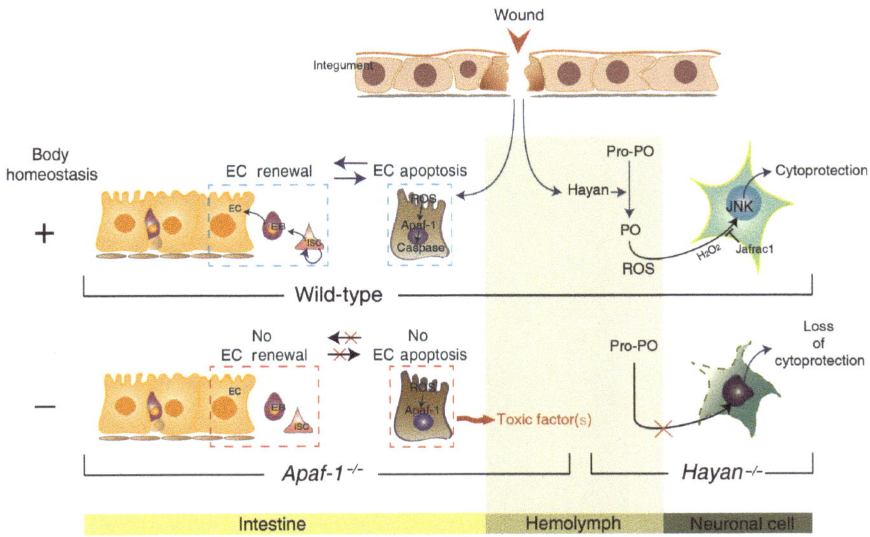

Won-Jae Lee and Masayuki Miura, Figure 6.4 Systemic wound responses following integumental wound in *Drosophila*. Following integumental wounding, two remote tissues distant from an initial wound site, that is, neuronal cells and intestinal epithelial cells, received local wound information. In the intestinal cells, caspase activation in ECs (EC apoptosis) and subsequent ISC proliferation/differentiation (EC renewal) are required for efficient host protection. Note the reciprocally activating interactions between "EC apoptosis" and "EC renewal." *Drosophila* mutant for the caspase activator, *Apaf-1*$^{-/-}$, was defective in gut-cell turnover (i.e., absence of EC apoptosis leading to inhibition of EC renewal) and showed increased wound-induced lethality. In this mutant animal, a lethal toxic factor was released into the hemolymph following integumental wounding. Following integumental wounding, Hayan-dependent conversion of Pro-PO to active PO is one of the earliest events to occur in hemolymph. Subsequently, active PO induces hemolymph ROS that directly induce JNK-dependent cytoprotections in neurons. H_2O_2-removing enzyme, jafrac1, seems to be involved in the redox balance in these cells. *Hayan* mutant animals, *Hayan*$^{-/-}$, are defective in hemolymph PO activation and neuronal JNK activation, which results in high susceptibility to integumental wound. See text for more details. ISC, intestinal stem cells; EB, enteroblast; EE, enteroendocrine; Pro-PO, pro-phenoloxidase; PO, phenoloxidase.

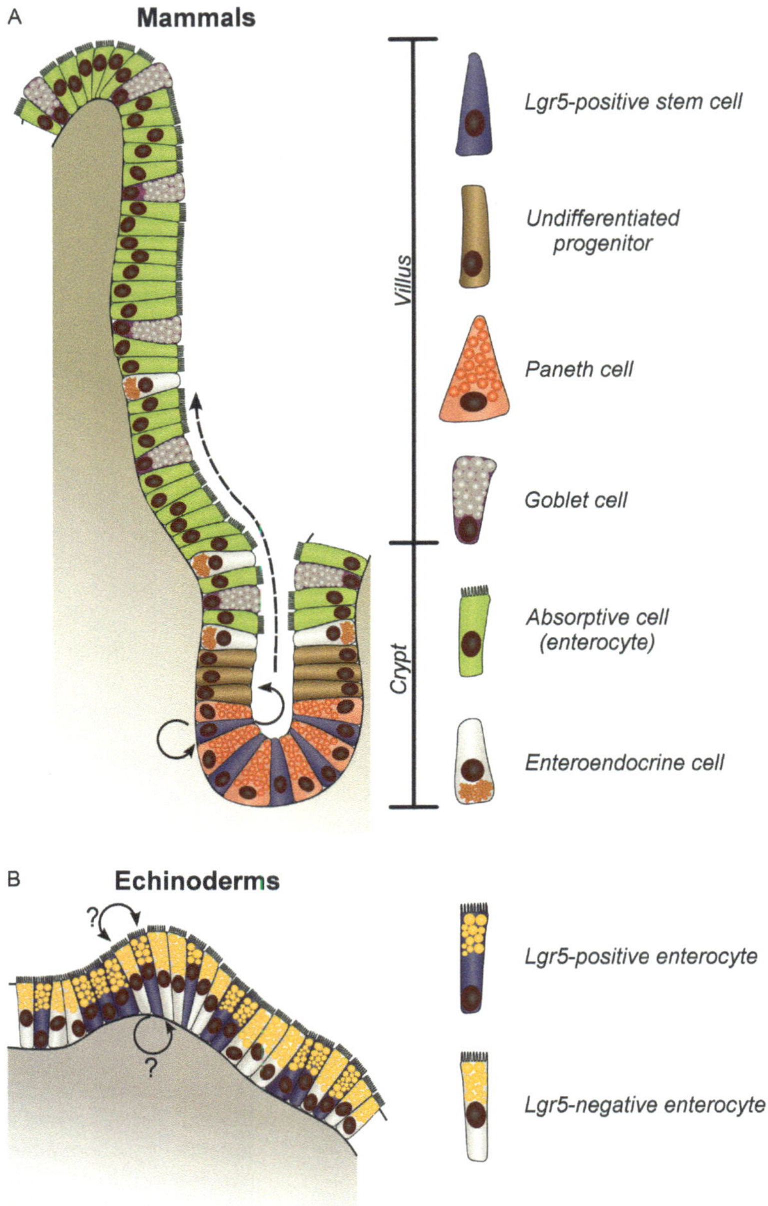

Vladimir S. Mashanov *et al.*, Figure 7.1 Diagram of organization of the mucosal epithelium in the mammalian small intestine (A) and in the digestive tube of a sea cucumber (B) highlighting the difference in cell renewal mechanisms. In mammals, Lgr5-positive intestinal stem cells are localized at the bottom of the crypt and both self-renew and produce rapidly dividing transit-amplifying progenitors (arrows), which gradually move apically (dashed arrow) and give rise to the specialized cells. Paneth cells are the only differentiated cell type that remains in the crypt and constitute a component of the stem cell niche. In the digestive epithelium of echinoderms, unlike in mammals, Lgr5-positive cells are not spatially segregated into distinct zones, but are instead interspersed among Lgr5-negative enterocytes. Lineage relationships still remain to be determined in this model (arrows with question marks).

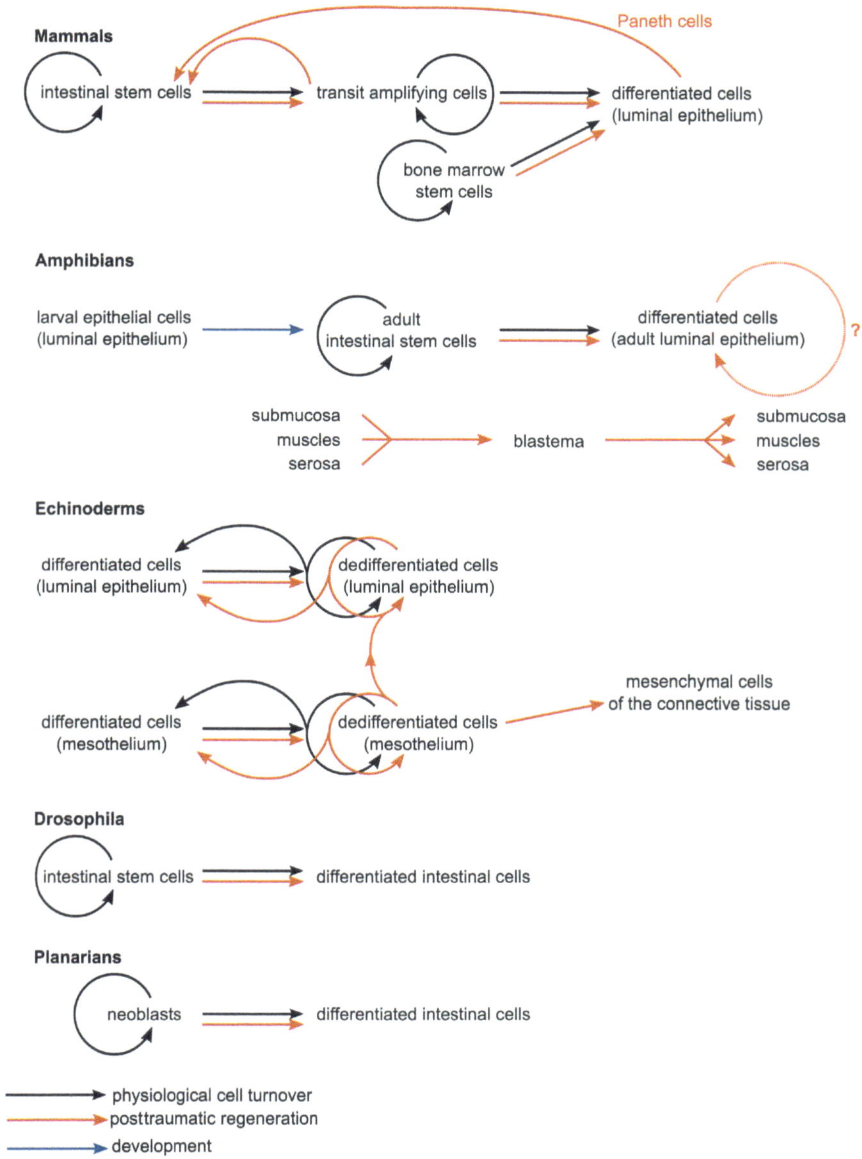

Vladimir S. Mashanov *et al.*, Figure 7.2 Diagram summarizing cell lineage relationships in normal (physiological) cell turnover and posttraumatic regeneration in the digestive tube of different bilaterian animal models. Straight arrows show cell lineage relationships in physiological cell turnover (black), posttraumatic regeneration (red), and development (blue). Circular arrows indicate self-renewal.

Vladimir S. Mashanov *et al.*, Figure 7.3 One of the best studied models of visceral regeneration: the sea cucumber *Holothuria glaberrima* Selenka. (A) Normal adult individual. (B) and (B′) An individual photographed early and late in the process of evisceration, respectively.

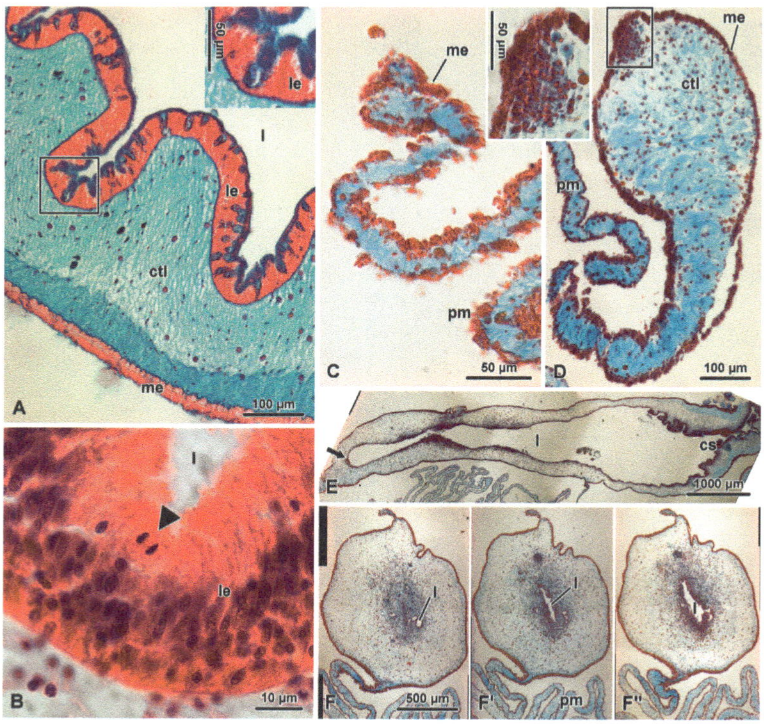

Vladimir S. Mashanov *et al.*, Figure 7.4 Histological organization of the digestive tube of the sea cucumber *H. glaberrima* in noneviscerated animals (A) and (B) and at different time points of regeneration (C–F″). (A) Cross-section of the intestine in a noneviscerated animal stained with alcian blue (pH 2.5) and nuclear fast red. The *inset* shows a higher magnification view of the luminal epithelium. (B) Dividing cell (anaphase, *arrowhead*) in the luminal epithelium of the normal intestine stained with hematoxylin and eosin. (C–F″) Regenerating digestive tube stained with Heindenhein's azan. (C and D) Cross-sections through the early gut rudiment on days 3 and 7 postevisceration, respectively. The *inset* in D shows a detailed view of the mesothelial cells ingressing into the underlying connective tissue. (E) Longitudinal section through the growing gut rudiment on day 14 postevisceration. The growing tip (*arrow*) is on the left, whereas the cloacal stump is on the right. (F–F″) Day 14 postevisceration; successive cross-sections of a posterior gut rudiment taken through the growing tip (F) and at distances of 40 μm (F′) and 110 μm (F″). *cs*, cloacal stump; *ctl*, connective tissue layer of the gut wall; *l*, gut lumen; *le*, luminal (mucosal) epithelium; *me*, mesothelium; *pm*, proximal mesentery.

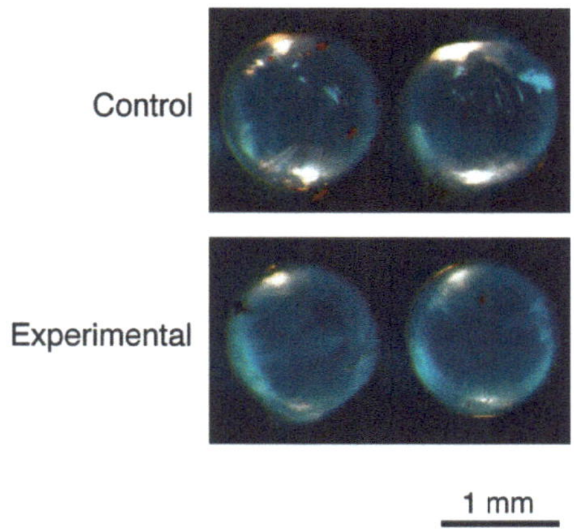

Konstantinos Sousounis *et al.*, Figure 8.1 Newt lens regeneration during aging. Control newt lenses taken from 14-year-old newts are similar in size, shape, and morphology to experimental newt lenses obtained from approximately 30-year=old newts that had regenerated the lens 17 times over a period of 16 years. Bar indicates size.

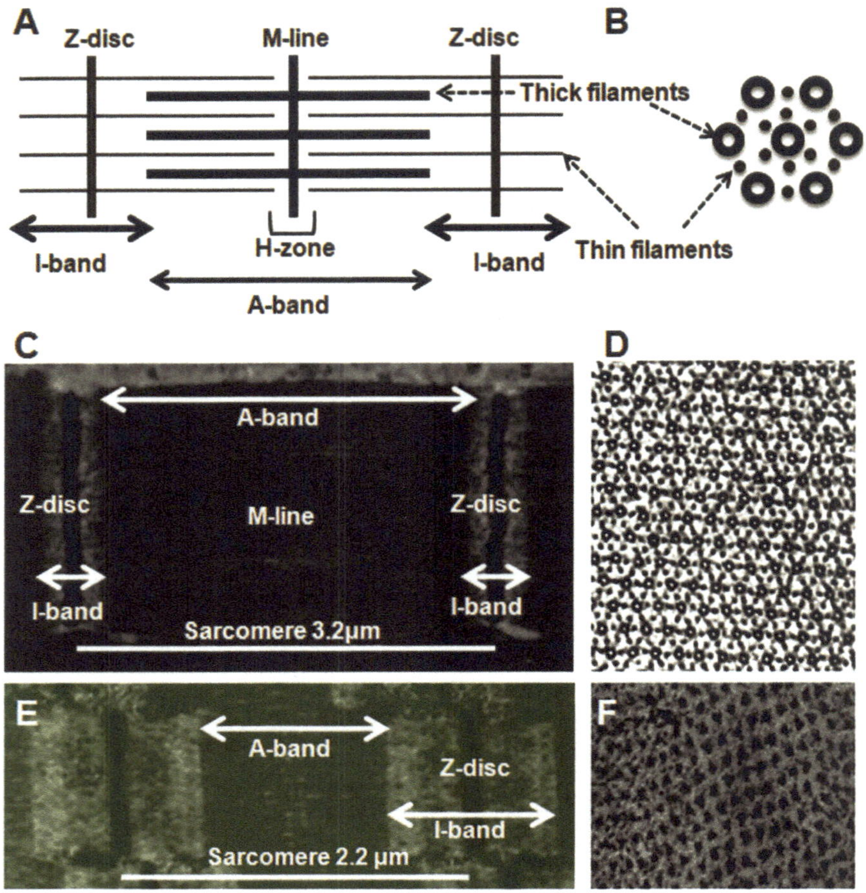

Mamta Rai *et al.*, Figure 9.1 Comparison of *Drosophila* IFM and human skeletal muscle sarcomere organization. (A) Diagram of a single sarcomere viewed in longitudinal and (B) in transverse section showing hexagonal arrangement. (C) Electron micrograph of *Drosophila* IFM sarcomere shown in longitudinal section and (D) in transverse section. (E) Electron micrograph of human skeletal muscles sarcomere shown in longitudinal and (F) in transverse section. *(Courtesy Prof. Gayathri N., NIMHANS, Bangalore, for E, and F).*

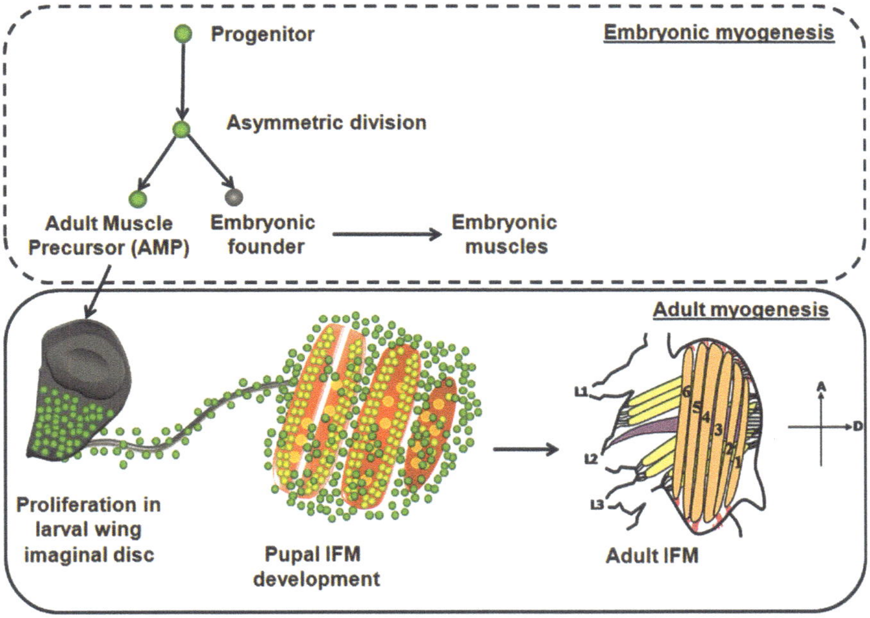

Mamta Rai *et al.*, Figure 9.2 Biphasic muscle development in *Drosophila*. In the embryo, muscle progenitors are specified to form both embryonic and adult muscles. The AMPs that will form the IFMs get segregated in the wing imaginal disc where they proliferate during larval stages and at pupal metamorphosis, they then migrate toward, and fuse with, the three larval template muscles to form the six adult DLMs, as part of the IFMs. Green balls: myoblasts; Yellow balls: fused myoblasts or nuclei; large yellow balls: larval nuclei; adult DLMs are labeled 1–6; DVMs are present beneath the DLM.

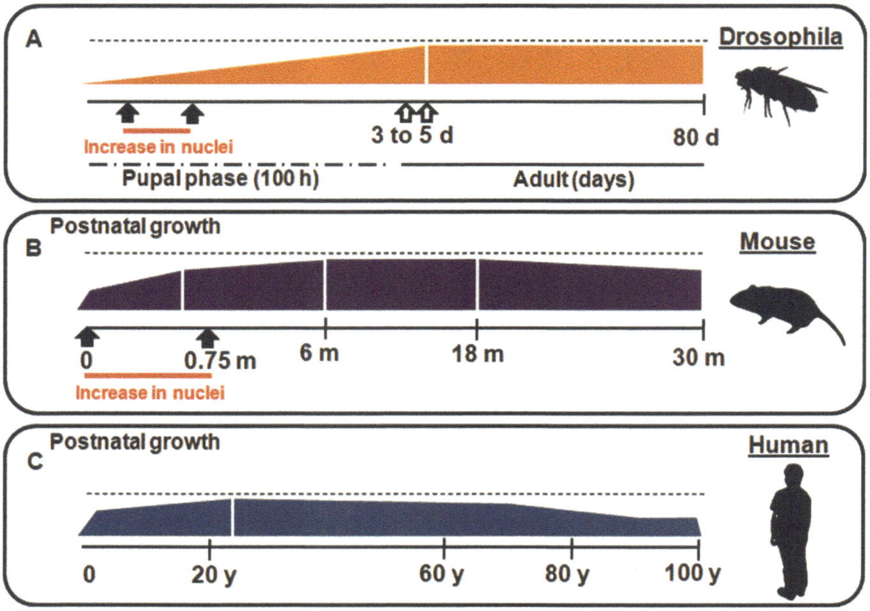

Mamta Rai *et al.*, Figure 9.3 Indication of growth pattern over time for increasing myofiber size, maintenance of adult muscles, and longevity in the three species: not to scale. (A) In *Drosophila*, IFM development starts at pupal metamorphosis, with myoblast fusion starting as early as 8 h (h) "After Puparium Formation" (APF) with increasing numbers of myonuclei continuing up to approximately 36 h APF; the duration of this hyperplasia phase is indicated by black arrows (Fernandes et al., 1991). The muscles then undergo hypertrophic growth until 3–5 days posteclosion; shown by white arrows, with 1d indicating the start of the adult muscle phase (Reedy & Beall, 1993), and myofibers retain their size throughout the life of the fly (Miller et al., 2008). (B) In mice, rapid postnatal growth of hindlimb muscles occurs for about 6 weeks, then growth (including elongation) slows and has almost ceased by 3 months. Initially, there is extensive fusion of myoblasts (satellite cells) with the growing myofibers (hyperplasia) but this stops around 3 weeks (0.75 month), with growth after this time being due almost entirely to hypertrophy (White et al., 2010). Age-related loss of muscle mass (sarcopenia) is evident by 24 months of age and is more pronounced by 29 months. (C) In humans, the muscles increase in size until after adolescence (around 18 years of age); after this time when elongation has ceased in adults, muscles may increase in cross-sectional area (especially in males) due to hypertrophy and retain their mass until about 50–60 years of age. After this age, sarcopenia can become apparent with muscle mass decreased by about 30% by 70 years (d, m, and y represents day(s), month(s), and year(s), respectively).

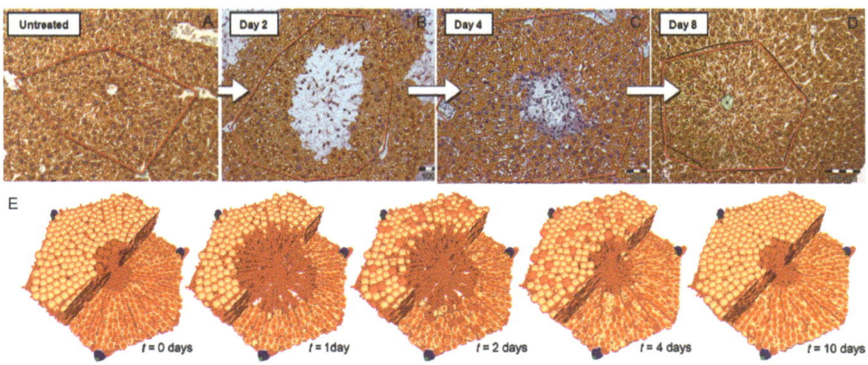

Osvaldo Chara *et al.*, Figure 10.4 *Liver regeneration model.* (A–D) Mouse liver lobules with CCl4 induced liver damage in the central region, visible as gray domain in (B, C). Light microscopy images for (A) Control, (B) 2 days, (C) 4 days, and (D) 8 days after administration of CCl4 illustrate how the liver cell mass regenerates. Lines in (A–D) denote the borders of a single liver lobule in cross section. (E) Simulation of liver regeneration due to liver cell proliferation within a digital liver lobule. *All images reprinted with permission from Höhme et al. (2010).*

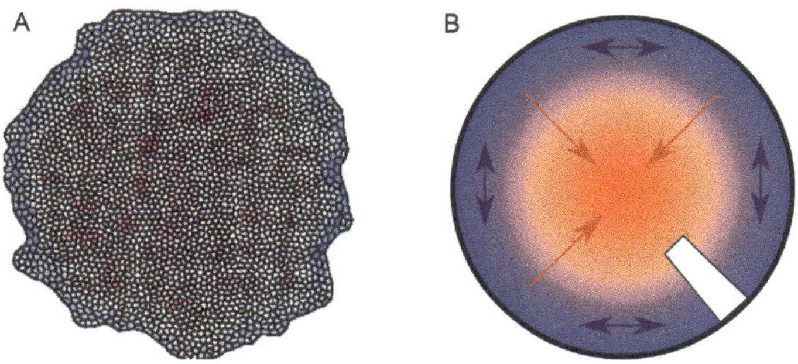

Osvaldo Chara *et al.*, Figure 10.5 *Mechanical feedback model.* (A–B) One of the proposed mechanisms for tissue size regulation combines a *central* growth factor source (not shown) which drives proliferation in the center of the wing disc with a *decentral* mechanical feedback mechanism through proliferation-inhibiting compression forces (radial/red arrows in B for direction and darkness/intensity of red color in the center of A and B for amplitude of forces) and proliferation-promoting stretch forces (tangential/blue arrows in B for direction and darkness/intensity of blue color in the periphery of A and B for amplitude of forces). (A) Snapshot of a cell-based simulation starting from a small and relaxed cell aggregate shows the buildup of the force pattern (Hufnagel et al., 2007). Ongoing growth in the center stretches outer cell layers tangentially which consequently compresses the inner cell mass such that the net proliferation rate decreases sharply at the target size. (B) Sketch of the force pattern and location of the hypothetical tissue loss that is discussed in a Gedanken experiment (see text). *Panel A was reprinted with permission from Hufnagel et al. (2007).*

FSC
www.fsc.org
MIX
Papier | Fördert gute Waldnutzung
FSC® C083411

Kolibri 360 GmbH
Ferdinand-Jühlke-Straße 7
99095 Erfurt, Deutschland
produktsicherheit@kolibri360.de